Multi-Carrier and Spread Spectrum Systems

Multi-Carrier and Spread Spectrum Systems

From OFDM and MC-CDMA to LTE and WiMAX

Second Edition

K. Fazel

Ericsson GmbH, Germany

and

S. Kaiser

DOCOMO Communications Laboratories Europe GmbH, Germany

A John Wiley and Sons, Ltd, Publication

This edition first published 2008
© 2008 John Wiley & Sons, Ltd

Registered office   John Wiley & Sons Ltd, The Atrium,
                       Southern Gate, Chichester,
                       West Sussex PO19 8SQ, United Kingdom

For details of our global editorial offices, for customer services and for information about how to apply for permission to reuse the copyright material in this book please see our website at www.wileyeurope.com or www.wiley.com.

Library of Congress Cataloging-in-Publication Data

Fazel, Khaled.
    Multi-carrier and spread spectrum systems: from OFDM and MC-CDMA to LTE and WiMAX / K. Fazel, S. Kaiser. – 2nd ed.
        p. cm.
    Includes bibliographical references and index.
    ISBN 978-0-470-99821-2
    1. Spread spectrum communications. 2. Multiplexing. I. Kaiser, Stefan, 1960–II. Title.
    TK5103.45.F39 2008
    621.382–dc22

                                                                            2008031728

A catalogue record for this book is available from the British Library.

ISBN 978-0-470-99821-2

Typeset by Laserwords Private Limited, Chennai, India
Printed and bound in Singapore by Markono Print Media Pte Ltd.

to

my parents, my wife Miriam,
my daughters Sarah, Sophia, and Susanna
(K.F.)

my wife Susanna,
my sons Lukas and Philipp and my daughter Anna
(S.K.)

Contents

Foreword

This book discusses multi-carrier modulation and spread spectrum techniques, recognized as the most promising candidate modulation methods for the 4th generation (4G) of mobile communications systems. The authors of this book were the first to propose MC-CDMA for the next generation of mobile communications, and are still continuing their contribution towards beyond 3G. Considering the requirements of 4G systems, multi-carrier and spread spectrum systems appear to be the most suitable as they provide higher flexibility, higher transmission rates, and frequency usage efficiency. This is the first book on these methods, providing the reader with the fundamentals of the technologies involved and the related applications.

The book deals with the principles through definitions of basic technologies and the multipath channel over which the signals are transmitted. It defines MC-CDMA as a frequency PN pattern and MC-DS-CDMA as a straight extension of DS-CDMA; and argues that these twin asymmetric technologies are most suitable for 4G since MC-CDMA is suitable for the downlink and MC-DS-CDMA is suitable for the uplink in the cellular systems. Although MC-CDMA performs better than MC-DS-CDMA, it needs chip synchronization between users, and is therefore difficult to deploy in the uplink. Thus, for this asymmetric structure it is very important to understand the multi-carrier spread spectrum methods. Hybrid multiple access schemes like Multi-Carrier FDMA, Multi-Carrier TDMA, and Ultra Wide Band systems are discussed as more extended systems. Implementation issues, including synchronization, channel estimation, and RF issues, are also discussed in depth. Wireless local area networks, broadcasting transmission, and cellular mobile radio are shown to realize seamless networking for 4G. Although cellular systems have not yet been combined with other wireless networks, different wireless systems should be seamlessly combined. The last part of this book discusses capacity and flexibility enhancement technologies like diversity techniques, space–time/frequency coding, and SDR (Software Defined Radio).

This book greatly assists not only theoretical researchers but also practicing engineers of the next generation of mobile communications systems.

Prof. Masao Nakagawa
Department of Information and Computer Science
Keio University, Japan

Preface (Second Edition)

The demand for high data rate wireless multi-media applications has increased significantly in the past few years. The wireless user's pressure towards faster communications, no matter whether mobile, nomadic, or fixed positioned, without extra cost is nowadays a reality. Finding an optimal solution for this dilemma is challenging, not only for manufacturers but also for network operators. The recent strategy followed within ETSI 3GPP LTE and the WiMAX Forum was an evolutionary concept, especially for mobile applications. Both have adopted a new PHY layer multi-carrier transmission with a MIMO scheme, a promising combination offering a high data rate at low cost.

Since the first edition of our book in 2003, the application field of multi-carrier transmission in mobile communications has been extended in two important areas: *3GPP Long Term Evolution* (LTE) and *WiMAX*. Parallel to that, the topic of MIMO in conjunction with OFDM has taken further steps.

New experiences gained during the above-mentioned standardization activities and the latest research results on MIMO-OFDM gave us sufficient background and material to extend this book towards its Second Edition, i.e. covering new application fields (LTE and WiMAX) and new trends on MIMO-OFDM technology. We hope that the Second Edition of this book will further contribute to a better understanding of the principles of multi-carrier transmission and of its variety of application fields, and, finally, it may motivate research toward new developments.

K. Fazel, S. Kaiser

Preface (First Edition)

Nowadays, multi-carrier transmission is considered to be an old concept. Its basic idea goes back to the mid-1960s. Nevertheless, behind any old technique there are always many simple and exciting ideas, the terrain for further developments of new efficient schemes.

Our first experience with the simple and exciting idea of OFDM started in early 1991 with *digital audio broadcasting (DAB)*. From 1992, our active participation in several research programmes on *digital terrestrial TV broadcasting (DVB-T)* gave us further opportunities to look at several aspects of the OFDM technique with its new advanced digital implementation possibilities. The experience gained from the joined specification of several OFDM-based demonstrators within the German HDTV-T and the EU-RACE dTTb research projects served as a basis for our commitment in 1995 to the final specifications of the DVB-T standard, relaying on the multi-carrier transmission technique.

Parallel to the HDTV-T and the dTTb projects, our further involvement from 1993 in the EU-RACE CODIT project, with the scope of building a first European 3G testbed, following the DS-CDMA scheme, inspired our interest in another old technique, *spread spectrum*, being as impressive as multi-carrier transmission. Although the final choice of the specification of the CODIT testbed was based on wideband CDMA, an alternative multiple-access scheme exploiting the new idea of combining OFDM with spread spectrum, i.e. *multi-carrier spread spectrum* (MC-SS), was considered as a potential candidate and discussed widely during the definition phase of the first testbed.

Our strong belief in the efficiency and flexibility of multi-carrier spread spectrum compared to W-CDMA for applications such as beyond 3G motivated us, from the introduction of this new multiple access scheme at PIMRC '93, to further contribute to it, and to investigate different corresponding system level aspects.

Due to the recognition of the merits of this combination by well-known international experts, since the PIMRC '93 conference, MC-SS has rapidly become one of the most widespread independent research topics in the field of mobile radio communications. The growing success of our organized series of international workshops on MC-SS since 1997, the large number of technical sessions devoted in international conferences to multi-carrier transmission, and the several special editions of the *European Transactions on Telecommunications (ETT)* on MC-SS highlight the importance of this combination for future wireless communications.

Several MC-CDMA demonstrators, e.g. one of the first built within DLR and its live demonstration during the 3rd international MC-SS workshop, a multitude of recent international research programmes like the research collaboration between DOCOMO Euro-Labs and DLR on the design of a future broadband air interface or the EU-IST MATRICE, 4MORE, and WINNER projects, and especially the NTT DOCOMO research initiative to build a demonstrator for beyond 3G systems based on the multi-carrier spread spectrum technique, emphasize the commitment of the international research community to this new topic.

Our experience gained during the above-mentioned research programmes, our current involvement in the ETSI-BRAN project, our yearly seminars organized within Carl Granz Gesellschaft (CCG) on digital TV broadcasting and on WLAN/WLL have given us sufficient background knowledge and material to take this initiative to collect in this book most of the important aspects on multi-carrier, spread spectrum, and multi-carrier spread spectrum systems.

We hope that this book will contribute to a better understanding of the principles of multi-carrier and spread spectrum and may motivate further investigation into and development of this new technology.

K. Fazel, S. Kaiser

Acknowledgements

The authors would like to express their sincere thanks to Prof. M. Nakagawa from Keio University, Japan, for writing the foreword. Many thanks go to Dr. H. Atarashi, Dr. N. Maeda, Dr. S. Abeta, and Dr. M. Sawahashi from NTT DOCOMO for providing us with material regarding their multi-carrier spread spectrum activities. Many thanks also for the support of Dr. E. Auer from Ericsson GmbH and for helpful technical discussions with members of the DOCOMO Euro-Labs and the Mobile Radio Transmission Group from DLR. Further thanks also go to Dr. I. Cosovic from DOCOMO Euro-Labs who provided us with results for the uplink, especially with pre-equalization.

K. Fazel, S. Kaiser

Introduction

The main feature of the next-generation wireless systems will be the convergence of mult-media services such as speech, audio, video, image, and data. This implies that a future wireless terminal, by guaranteeing high speed data, will be able to connect to different networks in order to support various services: switched traffic, IP data packets, and broadband streaming services such as video. The development of wireless terminals with generic protocols and multiple physical layers or software-defined radio interfaces is expected to allow users to seamlessly switch access between existing and future standards.

The rapid increase in the number of wireless mobile terminal subscribers, which currently exceeds 3 billion users, highlights the importance of wireless communications in this new millennium. This revolution in the information society has taken place, especially in Europe and Japan through a continuous evolution of emerging standards and products by keeping a seamless strategy for the choice of solutions and parameters. The continuous adaptation of wireless technologies to the user's rapidly changing demands has been one of the main drivers for the success of this evolution. Therefore, the worldwide wireless access system is and will continue to be characterized by a heterogeneous multitude of standards and systems. This plethora of wireless communication systems is not limited to cellular mobile telecommunication systems such as GSM, IS-95, D-AMPS, PDC, CDMA-2000, WCDMA/UMTS, HSDPA, HSUPA, or 3GPP LTE, but also includes wireless local area networks (WLANs), e.g. IEEE 802.11a/n and Bluetooth, broadband wireless access (BWA) such as WiMAX systems with their first goal to introduce wireless local loops (WLL) services based on ETSI HIPERMAN and IEEE 802.16x standards and in a future step to support full cellular mobility, as well as broadcast systems such as digital audio broadcasting (DAB) and digital video broadcasting (DVB-T and DVB-H).

These trends have accelerated since the beginning of the 1990s with the replacement of the first-generation analogue mobile networks by the current second-generation (2G) systems (GSM, IS-95, D-AMPS, and PDC), which opened the door for a fully digital network. This evolution is continuing with the deployment of the third-generation (3G) systems namely WCDMA/UMTS, HSDPA, HSUPA, and CDMA-2000, referred to as IMT 2000. The *3GPP Long Term Evolution* (LTE) standard with significantly higher data rates than in 3G systems can be considered as 3G evolution. In the meantime, the research community is focusing its activity towards the next-generation mobile systems beyond 3G (B3G), referred to as *IMT-Advanced* or fourth-generation (4G) systems, with even more

Multi-Carrier and Spread Spectrum Systems   Second Edition   K. Fazel and S. Kaiser
© 2008 John Wiley & Sons, Ltd

ambitious technological challenges. Note that especially within the *WINNER Project* [55], which is a European research project, partly funded by the European Commission, innovative solutions for IMT-Advanced are targeted. The WINNER concept covers data rates that are higher than with LTE. The WINNER concept is described in References [54] and [55].

The primary goal of next-generation wireless systems (IMT-Advanced) will not only be the introduction of new technologies to cover the need for higher data rates and new services but also the *integration* of existing technologies into a common platform. Hence, the selection of a *generic* air interface for future-generation wireless systems will be of great importance. Although the exact requirements for IMT-Advanced have not yet been commonly defined, its new air interface will fulfill at least the following requirements:

- *generic architecture*, enabling the integration of existing technologies;
- *high spectral efficiency*, offering higher data rates in a given scarce spectrum;
- *high scalability*, designing different cell configurations (hot spot, ad hoc) for improved coverage;
- *high adaptability and reconfigurability*, supporting different standards and technologies;
- *low latency*, allowing better service quality;
- *low cost*, enabling a rapid market introduction; and
- *future proof*, opening the door for new technologies.

From 2G to 3G and B3G Multiple Access Schemes

2G wireless systems are mainly characterized by the transition from analogue to a fully digital technology and comprise the GSM, IS-95, PDC, and D-AMPS standards.

Work on the pan-European digital cellular standard *Global System for Mobile Communications* (GSM) started in 1982 [18, 39], where now it accounts for about 85 % of the world mobile market. In 1989, the technical specifications of GSM were approved by the European Telecommunication Standard Institute (ETSI), where its commercial success began in 1993. Although GSM is optimized for circuit-switched services such as voice, it offers low rate data services up to 14.4 kbit/s. High speed data services with up to 171.2 kbit/s are possible with the enhancement of the GSM standard, namely the *General Packet Radio Service* (GPRS), by assigning multiple time slots to one link. GPRS uses the same modulation, frequency band, and frame structure as GSM. The *Enhanced Data Rate for Global Evolution* (EDGE) [6] system, which further improves the data rate up to 384 kbit/s, introduces a new spectrum efficient modulation scheme. Parallel to GSM, the American IS-95 standard [45] (recently renamed *cdmaOne*) was approved by the Telecommunication Industry Association (TIA) in 1993, where its first commercial application started in 1995. Like GSM, the first version of this standard (IS-95A) offers data services up to 14.4 kbit/s. In its second version (IS-95B) up to 64 kbit/s are possible.

Two further 2G mobile radio systems were introduced: *Digital Advanced Mobile Phone Services* (D-AMPS/IS-136), called TDMA in the USA, and the *Personal Digital Cellular* (PDC) in Japan [30]. The most convincing example of high speed mobile Internet services, called *i-mode*, was introduced 1999 in Japan in the PDC system. The high increase in customers and traffic in the PDC system urged the Japanese to start the first 3G WCDMA network in 2001 under the name *FOMA*.

Trends toward more capacity for mobile receivers, new multimedia services, new frequencies, and new technologies have motivated the idea of 3G systems. A unique international standard was targeted, referred to as *International Mobile Telecommunications 2000* (IMT-2000), realizing a new generation of mobile communications technology for a world in which personal communication services will dominate. The objectives of the third-generation standards, namely WCDMA/UMTS [1], HSDPA [2], HSUPA [3], and CDMA-2000 [46] went far beyond the second-generation systems, especially with respect to:

- the wide range of multimedia services (speech, audio, image, video, data) and bit rates (up to 14.4 Mbit/s for indoor and hot spot applications);
- the high quality of service requirements (better speech/image quality, lower bit error rate (BER), higher number of active users);
- operation in mixed cell scenarios (macro, micro, pico);
- operation in different environments (indoor/outdoor, business/domestic, cellular/cordless); and
- finally, flexibility in frequency (variable bandwidth), in data rate (variable), and in radio resource management (variable power/channel allocation).

The commonly used multiple access schemes for second- and third-generation wireless mobile communication systems are based on either *time division multiple access* (TDMA) or *code division multiple access* (CDMA), or the combined access schemes in conjunction with an additional *frequency division multiple access* (FDMA) component:

- The GSM standard, employed in the 900 MHz and 1800 MHz bands, first divides the allocated bandwidth into 200 kHz FDMA sub-channels. In each sub-channel, up to eight users share eight time slots in a TDMA manner [39].
- In the IS-95 standard up to 64 users share the 1.25 MHz channel by CDMA [45]. The system is used in the 850 MHz and 1900 MHz bands.
- The aim of D-AMPS (TDMA IS-136) is to coexist with the analogue AMPS, where the 30 kHz channel of AMPS is divided into three channels, allowing three users to share a single radio channel by allocating unique time slots to each user [29].
- The 3G standards (WCDMA/UMTS and CDMA-2000) adopted by ITU are both based on CDMA [1, 46]. For UMTS, the CDMA-FDD mode, which is known as wideband CDMA (WCDMA), employs separate 5 MHz channels for both the uplink and downlink directions. Within the 5 MHz bandwidth, each user is separated by a specific code, resulting in an end-user data rate of theoretically up to 2 Mbit/s per carrier. Further evolutions of WCDMA are the extensions towards *high speed downlink packet access* (HSDPA) with data rates of up to 14.4 Mbit/s and of *high speed uplink packet access* (HSUPA) with data rates of up to 5.74 Mbit/s. These improvements are obtained by introducing higher order modulation schemes, channel-dependent scheduling, and hybrid ARQ (HARQ) with soft combining together with multiple code allocation. The downlink additionally supports adaptive coding and modulation. The set of HSDPA and HSUPA is termed *high speed packet access* (HSPA) [2, 3].

Tables 1 and 2 summarize the key characteristics of 2G and 3G mobile communication systems.

Besides tremendous developments in mobile communication systems, in public and private environments, operators are offering wireless services using WLANs in selected spots such as hotels, train stations, airports, and conference rooms. As Table 3 shows, there is a similar objective to go higher in data rates with WLANs, where as multiple access schemes TDMA or CDMA are employed [32].

FDMA, TDMA, and CDMA are obtained if the transmission bandwidth, the transmission time, or the spreading code is related to the different users, respectively [5].

Table 1   Main parameters of 2G mobile radio systems

Parameter	2G systems		
	GSM (+ GPRS, EDGE)	IS-95/cdmaOne (+ CDMA-2000)	PDC
Carrier frequencies	900 MHz 1800 MHz	850 MHz 1900 MHz	850 MHz 1500 MHz
Peak data rate	64 kbit/s 171.2 kbit/s (GPRS) 384 kbit/s (EDGE)	64 kbit/s 144 kbit/s (CDMA-2000 1x)	28.8 kbit/s
Multiple access	TDMA	CDMA	TDMA
Services	Voice, low rate data	Voice, low rate data	Voice, low rate data

Table 2   Main parameters of 3G and B3G mobile radio systems

Parameter	3G and B3G systems		
	WCDMA/UMTS (+ HSDPA, HSUPA)	EV-DO Rev. 0 (+ Rev. A, Rev. B)	3GPP long term evolution (LTE)
Carrier frequencies	2 GHz	450 MHz, 800 MHz, 2 GHz	IMT bands
Peak data rate	384 kbit/s – 2 Mbit/s 14.4 Mbit/s (HSDPA) 5.74 Mbit/s (HSUPA)	2.4 Mbit/s (Rev. 0) 3.1 Mbit/s (Rev. A) 4.9–14.7 Mbit/s (Rev. B)	DL: 326.4 Mbit/s UL: 86.4 Mbit/s
Multiple access	CDMA	CDMA	DL: OFDMA UL: DFTS-OFDM
Services	Voice, data	Voice, date	High rate data, voice

Table 3 Main parameters of WLAN communication systems

Parameter	Bluetooth	IEEE 802.11b	IEEE 802.11a/g/h	IEEE 802.11n
Carrier frequency	2.4 GHz (ISM)	2.4 GHz (ISM)	2.4 GHz/5 GHz (ISM)	2.4 GHz/5 GHz (ISM)
Peak data rate	2 Mbit/s	5.5 Mbit/s	54 Mbit/s	248 Mbit/s
Multiple access	TDMA and FH-CDMA	DS-CDMA	OFDM-TDMA	OFDM-TDMA
Services	Data	Data	High rate data	High rate data

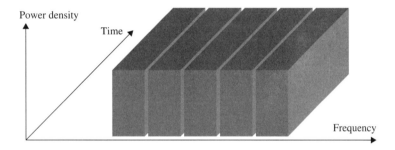

Figure 1 Principle of FDMA (with $N_c = 5$ sub-channels)

FDMA is a multiple access technology widely used in satellite, cable, and terrestrial radio networks. FDMA subdivides the total bandwidth into N_c narrowband sub-channels that are available during the whole transmission time (see Figure 1). This requires band-pass filters with sufficient stop band attenuation. Furthermore, a sufficient guard band is left between two adjacent spectra in order to cope with frequency deviations of local oscillators and to minimize interference from adjacent channels. The main advantages of FDMA are in its low required transmit power and in channel equalization that is either not needed or much simpler than with other multiple access techniques. However, its drawback in a cellular system might be the implementation of N_c modulators and demodulators at the base station (BS).

TDMA is a popular multiple access technique, which is used in several international standards. In a TDMA system all users employ the same band and are separated by allocating short and distinct time slots, one or several assigned to a user (see Figure 2). In TDMA, neglecting the overhead due to framing and burst formatting, the multiplexed signal bandwidth will be approximately N_c times higher than in an FDMA system, hence leading to quite complex equalization, especially for high data rate applications. The channel separation of TDMA and FDMA is based on the orthogonality of signals. Therefore, in a cellular system, the co-channel interference is only present from the re-use of frequency.

On the contrary, in CDMA systems all users transmit at the same time on the same carrier using a wider bandwidth than in a TDMA system (see Figure 3). The signals of users are distinguished by assigning different spreading codes with low cross-correlation

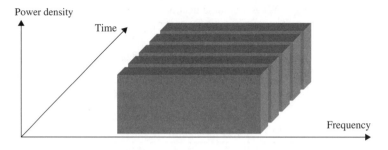

Figure 2   Principle of TDMA (with five time slots)

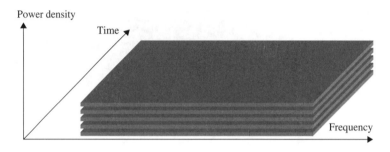

Figure 3   Principle of CDMA (with five spreading codes)

properties. Advantages of the spread spectrum technique are immunity against multi-path distortion, simple frequency planning, high flexibility, variable rate transmission, and resistance to interference.

In Table 4, the main advantages and drawbacks of FDMA, TDMA, and CDMA are summarized.

From 3G to 4G Multiple Access Schemes

Besides offering new services and applications, the success of the next generation of wireless systems (4G) will strongly depend on the choice of the concept and technology innovations in architecture, spectrum allocation, spectrum utilization, and exploitation [40, 41]. Therefore, new high performance physical layer and multiple access technologies are needed to provide high speed data rates with flexible bandwidth allocation. A low cost *generic radio interface*, being operational in various mixed-cell environments with scalable bandwidth and data rates, is needed to fulfill these requirements.

The technique of *spread spectrum* allows the above requirements to be at least partially fulfilled. As explained earlier, a multiple access scheme based on direct sequence code division multiple access (DS-CDMA) relies on spreading the data stream using an assigned spreading code for each user in the time domain [42, 47, 49, 50]. The capability of minimizing multiple access interference is given by the cross-correlation properties of the spreading codes. In the case of severe multi-path propagation in mobile communications, the capability of distinguishing one component from others in the composite received

Table 4 Advantages and drawbacks of different multiple access schemes

Multiple access scheme	Advantages	Drawbacks
FDMA	– Low transmit power – Robust to multi-path – Easy frequency planning – Low delay	– Low peak data rate – Loss due to guard bands – Sensitive to narrowband interference
TDMA	– High peak data rate – High multiplexing gain in the case of bursty traffic	– High transmit power – Sensitive to multi-path – Difficult frequency planning
CDMA	– Low transmit power – Robust to multi-path – Easy frequency planning – High scalability – Low delay	– Low peak data rate – Limited capacity per sector due to multiple access interference

signal is offered by the auto-correlation properties of the spreading codes [47]. The so-called rake receiver should contain multiple correlators, each matched to a different resolvable path in the received composite signal [42]. The performance of a DS-CDMA system will strongly depend on the number of active users, the channel characteristics, and the number of arms employed in the rake. The system capacity is limited by self-interference and multiple access interference, which results from the imperfect auto- and cross-correlation properties of spreading codes. Therefore, it will be difficult for a DS-CDMA receiver to make full use of the received signal energy scattered in the time domain and hence to handle full load conditions [42].

The technique of *multi-carrier transmission* has received wide interest, especially for high data rate broadcast applications. The history of orthogonal multi-carrier transmission dates back to the mid-1960s, when Chang published his papers on the synthesis of band-limited signals for multi-channel transmission [8, 9]. He introduced the basic principle of transmitting data simultaneously through a band-limited channel without interference between sub-channels (without *inter-channel interference*, ICI) and without interference between consecutive transmitted symbols (without *inter-symbol interference*, ISI) in the time domain. Later, Saltzberg performed further analyses [43]. A major contribution to multi-carrier transmission was presented in 1971 by Weinstein and Ebert [51] who used the Fourier transform for baseband processing instead of a bank of sub-carrier oscillators. To combat ICI and ISI, they introduced the well-known *guard time* between the transmitted symbols with raised cosine windowing.

The main advantages of multi-carrier transmission are its robustness in frequency selective fading channels and, in particular, the reduced signal processing complexity by equalization in the frequency domain.

The basic principle of multi-carrier modulation relies on the transmission of data by dividing a high-rate data stream into several low rate sub-streams. These sub-streams are modulated on different sub-carriers [4, 7, 12]. By using a large number of sub-carriers, a high immunity against multi-path dispersion can be provided since the useful symbol duration T_s on each sub-stream is much larger than the channel time dispersion. Hence, the effects of ISI are minimized. Since the amount of filters and oscillators necessary is considerable for a large number of sub-carriers, an efficient digital implementation of a special form of multi-carrier modulation, called orthogonal frequency division multiplexing (OFDM), with rectangular pulse-shaping and guard time was proposed in Reference [4]. OFDM can be easily realized by using the discrete Fourier transform (DFT). OFDM, having densely spaced sub-carriers with overlapping spectra of the modulated signals, abandons the use of steep band-pass filters to detect each sub-carrier as it is used in FDMA schemes. Therefore, it offers high spectral efficiency.

Today, progress in digital technology has enabled the realization of a DFT also for large numbers of sub-carriers (up to several thousand), through which OFDM has gained much importance. The breakthrough of OFDM came in the 1990s as it was the modulation chosen for ADSL in the USA [11] and was selected for the European DAB standard [14]. This success continued with the choice of OFDM for the European DVB-T standard [17] in 1995 and later for the WLAN standard IEEE 802.11a [32] and the interactive terrestrial return channel (DVB-RCT) [16] as well as the European DVB-H standard [15]. Further deployments of OFDM technology are in the cellular mobile radio standard 3GPP LTE and future broadband wireless access standards such as HIPERMAN and IEEE 802.16x/WiMAX [19, 33, 52, 53]. Table 5 summarizes the main characteristics of several broadcasting and WLAN standards employing OFDM.

The advantages of multi-carrier modulation on the one hand and the flexibility offered by the spread spectrum technique on the other hand have motivated many researchers to investigate the combination of both techniques, known as *Multi-Carrier Spread Spectrum* (MC-SS). This combination, published in 1993 by several authors independently [10, 13, 20, 27, 37, 48, 56], has introduced new multiple access schemes called MC-CDMA and MC-DS-CDMA. It allows one to benefit from several advantages of both multi-carrier modulation and spread spectrum systems by offering, for instance, high flexibility, high spectral efficiency, simple and robust detection techniques, and narrowband interference rejection capability.

Multi-carrier modulation and multi-carrier spread spectrum are today considered potential candidates to fulfill the requirements of next-generation (4G) high speed wireless multimedia communications systems, where *spectral efficiency* and *flexibility* are considered as the most important criteria for the choice of the air interface.

Table 5 Examples of wireless transmission systems employing OFDM

Parameter	DAB	DVB-T	IEEE 802.11a	IEEE 802.11n
Carrier frequency	VHF	VHF and UHF	2.4 and 5 GHz	2.4 and 5 GHz
Bandwidth	1.5 MHz	8 MHz (7 MHz)	20 MHz	40 MHz
Max. data rate	1.7 Mbit/s	31.7 Mbit/s	54 Mbit/s	248 Mbit/s
Number of sub-carriers (FFT size)	192 up to 1536 (256 up to 2048)	1705 and 6817 (2048 and 8196)	52 (64)	52 and 104 (64 and 128)

Multi-Carrier Spread Spectrum

Since 1993, various combinations of multi-carrier modulation with the spread spectrum technique have been introduced as multiple access schemes. It has been shown that multi-carrier spread spectrum (MC-SS) offers high spectral efficiency, robustness, and flexibility [31]. Two different philosophies exist, namely MC-CDMA (or OFDM-CDMA) and MC-DS-CDMA (see Figure 4 and Table 6).

MC-CDMA is based on a serial concatenation of direct sequence (DS) spreading with multi-carrier modulation [10, 20, 27, 56]. The high-rate DS spread data stream of

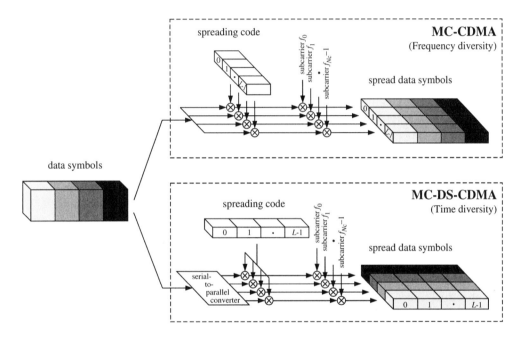

Figure 4   General principle of MC-CDMA and MC-DS-CDMA systems

Table 6   Main characteristics of different MC-SS concepts

Parameter	MC-CDMA	MC-DS-CDMA
Spreading	Frequency direction	Time direction
Sub-carrier spacing	$F_S = \dfrac{P_G}{N_c T_d}$	$F_S \geq \dfrac{P_G}{N_c T_d}$
Detection algorithm	MRC, EGC, ZF, MMSE equalization, IC, MLD	Correlation detector (coherent rake)
Specific characteristics	Very efficient for the synchronous downlink by using orthogonal codes	Designed especially for an asynchronous uplink
Applications	Synchronous uplink and downlink	Asynchronous uplink and downlink

processing gain P_G is multi-carrier modulated in the way that the chips of a spread data symbol are transmitted in parallel and the assigned data symbol is simultaneously transmitted on each sub-carrier (see Figure 4). As for DS-CDMA, a user may occupy the total bandwidth for the transmission of a single data symbol. Separation of the user signals is performed in the code domain. Each data symbol is copied on the sub-streams before multiplying it with a chip of the spreading code assigned to the specific user. This shows that an MC-CDMA system performs the spreading in the frequency direction and, thus, has an additional degree of freedom compared to a DS-CDMA system. Mapping of the chips in the frequency direction allows for simple methods of signal detection. This concept was proposed with OFDM for optimum use of the available bandwidth. The realization of this concept implies a guard time between adjacent OFDM symbols to prevent ISI or to assume that the symbol duration is significantly larger than the time dispersion of the channel. The number of sub-carriers N_c has to be chosen sufficiently large to guarantee frequency nonselective fading on each sub-channel. The application of orthogonal codes, such as Walsh–Hadamard codes for a synchronous system, e.g. the downlink of a cellular system, guarantees the absence of multiple access interference in an ideal channel and a minimum multiple access interference in a fading channel. For signal detection, single-user detection techniques such as maximum ratio combining (MRC), equal gain combining (EGC), zero forcing (ZF), or minimum mean square error (MMSE) equalization, as well as multi-user detection techniques like interference cancellation (IC) or maximum likelihood detection (MLD), can be applied.

As depicted in Figure 4, MC-DS-CDMA modulates sub-streams on sub-carriers with a carrier spacing proportional to the inverse of the chip rate. This will guarantee orthogonality between the spectra of the sub-streams [44]. If the spreading code length is smaller or equal to the number of sub-carriers N_c, a single data symbol is not spread in the frequency direction, instead it is spread in the time direction. Spread spectrum is obtained by modulating N_c time spread data symbols on parallel sub-carriers. By using high numbers of sub-carriers, this concept benefits from time diversity. However, due to the frequency nonselective fading per sub-channel, frequency diversity can only be exploited if channel coding with interleaving or sub-carrier hopping is employed, or if the same information is transmitted on several sub-carriers in parallel. Furthermore, higher frequency diversity could be achieved if the sub-carrier spacing is chosen larger than the chip rate. This concept was investigated for an asynchronous uplink scenario. For data detection, N_c coherent receivers can be used.

It can be noted that both schemes have a generic architecture. In the case where the number of sub-carriers $N_c = 1$, the classical DS-CDMA transmission scheme is obtained, whereas without spreading ($P_G = 1$) it results in a pure OFDM system.

By using a variable spreading factor in frequency and/or time and a variable sub-carrier allocation, the system can easily be adapted to different environments such as *multi-cell* and *single-cell* topologies, each with different coverage areas.

Today, in parallel with multi-carrier transmission, the field of multi-carrier spread spectrum communications is considered to be an independent and important research topic (see References [21] to [25], [28], and [38]). Several deep system analysis and comparisons of MC-CDMA and MC-DS-CDMA with DS-CDMA have been performed that show the superiority of MC-SS [26, 31, 34–36]. In addition, new application fields have been proposed such as high rate cellular mobile (4G), high rate wireless indoor, and fixed and broadband wireless access (F/BWA). In addition to system-level analysis, a multitude of research activities have been addressed to develop appropriate strategies for detection, interference cancellation, channel coding, modulation, synchronization (especially uplink), and low cost implementation design.

The Aim of This Book

The interest in multi-carrier transmission and in multi-carrier spread spectrum is still growing. Many researchers and system designers are involved in system aspects and the implementation of these new techniques. However, a comprehensive collection of their work is still missing.

The aim of this book is first to describe and analyze the basic concepts of multi-carrier transmission and its combination with spread spectrum, where the different architectures and the different detection strategies are detailed. Concrete examples of its applications for future cellular mobile communications systems are given. Then, we examine other derivatives of MC-SS (e.g. OFDMA, SS-MC-MA, and DFT-spread OFDM/interleaved FDMA) and other variants of the combination of OFDM with TDMA, which are today part of LTE, WiMAX, WLAN, and DVB-RCT systems. Basic OFDM implementation issues, valid for most of these combinations, such as channel coding, modulation, digital I/Q generation, synchronization, channel estimation, and effects of phase noise and nonlinearity are further analyzed.

Chapter 1 covers the fundamentals of today's wireless communications. First a detailed analysis of the radio channel (outdoor and indoor) and its modeling are presented. Then the principle of OFDM multi-carrier transmission is introduced. In addition, a general overview of the spread spectrum technique, especially of DS-CDMA, is given. Examples of applications of OFDM and DS-CDMA for broadcast, WLAN, and cellular systems (IS-95, CDMA-2000, WCDMA/UMTS, HSPA) are briefly presented.

Chapter 2 describes the combinations of multi-carrier transmission with the spread spectrum technique, namely MC-CDMA and MC-DS-CDMA. It includes a detailed description of the different detection strategies (single-user and multi-user) and presents their performance in terms of the bit error rate (BER), spectral efficiency, and complexity. Here a cellular system with a point to multi-point topology is considered. Both downlink and uplink architectures are examined.

Hybrid multiple access schemes based on OFDM, MC-SS, or spread spectrum are analyzed in Chapter 3. This chapter covers OFDMA, being a derivative of MC-CDMA, OFDM-TDMA, SS-MC-MA, DFT-spread OFDM/interleaved FDMA, and ultra wideband (UWB) schemes. All these multiple access schemes have recently received wide interest. Their concrete application fields are detailed in Chapter 5.

The issues of digital implementation of multi-carrier transmission systems, essential especially for system and hardware designers, are addressed in Chapter 4. Here, the different functions such as digital I/Q generation, analogue/digital conversion, digital multi-carrier modulation/demodulation, synchronization (time, frequency), channel estimation, coding/decoding, and other related RF issues such as nonlinearities, phase noise, and narrowband interference rejection are analyzed.

In Chapter 5, concrete application fields of OFDMA, OFDM-TDMA, and MC-SS for B3G cellular mobile (LTE and WiMAX), 4G (IMT-Advanced), wireless indoor (WLAN), and interactive multimedia communication (DVB-T return channel) are outlined, where for each of these systems the multi-carrier architecture and their main parameters are described. The capacity advantages of using adaptive channel coding and modulation, adaptive spreading, and scalable bandwidth allocation are discussed.

Finally, Chapter 6 covers further techniques that can be used to enhance system capacity or offer more flexibility for the implementation and deployment of the transmission systems examined in Chapter 5. Here, diversity techniques such as space–time/frequency coding and Tx/Rx antenna diversity in MIMO concepts and the software-defined radio (SDR) are introduced.

References

[1] 3GPP (TS 25.401), "UTRAN overall description," *Technical Specification*, Sophia Antipolis, France, 2002.

[2] 3GPP (TS 25.308), "High speed downlink packet access (HSDPA): overall description; Stage 2," *Technical Specification*, Sophia Antipolis, France, 2004.

[3] 3GPP (TS 25.309), "FDD enhanced uplink: overall description; Stage 2," *Technical Specification*, Sophia Antipolis, France, 2006.

[4] Alard M. and Lassalle R., "Principles of modulation and channel coding for digital broadcasting for mobile receivers," *European Broadcast Union Review*, no. 224, pp. 47–69, Aug. 1987.

[5] Bhargava V. K., Haccoun D., Matyas R., and Nuspl P. P., *Digital Communications by Satellite*, New York: John Wiley & Sons, Inc., 1981.

[6] Bi Q., Zysman G. I., and Menkes H, "Wireless mobile communications at the start of the 21st century," *IEEE Communications Magazine*, vol. 39, pp. 110–116, Jan. 2001.

[7] Bingham J. A. C., "Multicarrier modulation for data transmission: an idea whose time has come," *IEEE Communications Magazine*, vol. 28, pp. 5–14, May 1990.

[8] Chang R. W., "Synthesis of band-limited orthogonal signals for multi-channel data transmission," *Bell Labs Technical Journal*, no. 45, pp. 1775–1796, Dec. 1966.

[9] Chang R. W. and Gibby R. A., "A theoretical study of performance of an orthogonal multiplexing data transmission scheme," *IEEE Transactions on Communication Technology*, vol. 16, pp. 529–540, Aug. 1968.

[10] Chouly A., Brajal A., and Jourdan S., "Orthogonal multicarrier techniques applied to direct sequence spread spectrum CDMA systems," in *Proc. IEEE Global Telecommunications Conference (GLOBECOM '93)*, Houston, USA, pp. 1723–1728, Nov./Dec. 1993.

[11] Chow J. S., Tu J.-C., and Cioffi J. M., "A discrete multitone transceiver system for HDSL applications," *IEEE Journal on Selected Areas in Communications (JSAC)*, vol. 9, pp. 895–908, Aug. 1991.

[12] Cimini L. J., "Analysis and simulation of a digital mobile channel using orthogonal frequency division multiplexing," *IEEE Transactions on Communications*, vol. 33, pp. 665–675, July 1985.

[13] DaSilva V. and Sousa E. S., "Performance of orthogonal CDMA codes for quasi-synchronous communication systems," in *Proc. IEEE International Conference on Universal Personal Communications (ICUPC '93)*, Ottawa, Canada, pp. 995–999, Oct. 1993.

[14] ETSI DAB (EN 300 401), "Radio broadcasting systems; digital audio broadcasting (DAB) to mobile, portable and fixed receivers," Sophia Antipolis, France, April 2000.

[15] ETSI DVB-H (EN 302 304), "Digital video broadcasting (DVB); transmission system for handheld terminals (DVB-H)," Sophia Antipolis, France, Nov. 2004.

[16] ETSI DVB RCT (EN 301 958), "Interaction channel for digital terrestrial television (RCT) incorporating multiple access OFDM," Sophia Antipolis, France, March 2001.

[17] ETSI DVB-T (EN 300 744), "Digital video broadcasting (DVB); framing structure, channel coding and modulation for digital terrestrial television," Sophia Antipolis, France, July 1999.

[18] ETSI GSM Recommendations, 05 series, Sophia Antipolis, France, Sept. 1994.

[19] ETSI HIPERMAN (TR 101 856), "High performance metropolitan area network, requirements MAC and physical layer below 11 GHz band," Sophia Antipolis, France, 2004.

[20] Fazel K., "Performance of CDMA/OFDM for mobile communications system," in *Proc. IEEE International Conference on Universal Personal Communications (ICUPC '93)*, Ottawa, Canada, pp. 975–979, Oct. 93.

[21] Fazel K. and Fettweis G. (eds), *Multi-Carrier Spread Spectrum*, Boston: Kluwer Academic Publishers, 1997; also in *Proceedings of the 1st International Workshop on Multi-Carrier Spread Spectrum (MC-SS '97)*.

[22] Fazel K. and Kaiser S. (eds), *Multi-Carrier Spread Spectrum and Related Topics*, Boston: Kluwer Academic Publishers, 2000–2004; also in *Proceedings of the International Workshop on Multi-Carrier Spread Spectrum and Related Topics (MC-SS '1999–2003)*.

[23] Fazel K. and Kaiser S. (eds), *Multi-Carrier Spread Spectrum and Related Topics*, Dordrecht: Springer, 2006; also in *Proceedings of the International Workshop on Multi-Carrier Spread Spectrum and Related Topics (MC-SS 2005)*.

[24] Fazel K. and Kaiser S. (eds), *Special Issue on Multi-Carrier Spread Spectrum and Related Topics, European Transactions on Telecommunications (ETT)*, 2000–2006.

[25] Fazel K. and Kaiser S. (eds), *Special Issue on Multi-Carrier Spread Spectrum and Related Topics, European Transactions on Telecommunications (ETT)*, vol. 19, no. 5, 2008.

[26] Fazel K., Kaiser S., and Schnell M., "A flexible and high performance cellular mobile communications system based on orthogonal multi-carrier SSMA," *Wireless Personal Communications*, vol. 2, nos. 1 and 2, pp. 121–144, 1995.

[27] Fazel K. and Papke L., "On the performance of convolutionally-coded CDMA/OFDM for mobile communications system," in *Proc. IEEE International Symposium on Personal, Indoor and Mobile Radio Communications (PIMRC '93)*, Yokohama, Japan, pp. 468–472, Sept. 1993.

[28] Fazel K. and Prasad R. (eds), *Special Issue on Multi-Carrier Spread Spectrum, European Transactions on Telecommunications (ETT)*, vol. 10, no. 4, July/Aug. 1999.

[29] Goodman D. J., "Second generation wireless information network," *IEEE Transactions on Vehicular Technology*, vol. 40, no. 2, pp. 366–374, May 1991.

[30] Goodman D. J., "Trends in cellular and cordless communications," *IEEE Communications Magazine*, vol. 29, pp. 31–40, June 1991.

[31] Hara S. and Prasad R., "Overview of multicarrier CDMA," *IEEE Communications Magazine*, vol. 35, pp. 126–133, Dec. 1997.

[32] IEEE 802.11 (P802.11a/D6.0), "LAN/MAN specific requirements – Part 2: wireless MAC and PHY specifications – high speed physical layer in the 5 GHz band," IEEE 802.11, May 1999.

[33] IEEE 802.16d, "Air interface for fixed broadband wireless access systems," IEEE 802.16, May 2004.

[34] Kaiser S., "OFDM-CDMA versus DS-CDMA: performance evaluation for fading channels," in *Proc. IEEE International Conference on Communications (ICC '95)*, Seattle, USA, pp. 1722–1726, June 1995.

[35] Kaiser S., "On the performance of different detection techniques for OFDM-CDMA in fading channels," in *Proc. IEEE Global Telecommunications Conference (GLOBECOM '95)*, Singapore, pp. 2059–2063, Nov. 1995.

[36] Kaiser S., *Multi-Carrier CDMA Mobile Radio Systems – Analysis and Optimization of Detection, Decoding, and Channel Estimation*, Düsseldorf: VDI-Verlag, Fortschrittberichte VDI, series 10, no. 531, 1998, PhD thesis.

[37] Kondo S. and Milstein L. B., "On the use of multicarrier direct sequence spread spectrum systems," in *Proc. IEEE Military Communications Conference (MILCOM '93)*, Boston, USA, pp. 52–56, Oct. 1993.

[38] Linnartz J. P. and Hara S. (eds), *Special Issue on Multi-Carrier Communications, Wireless Personal Communications*, Kluwer Academic Publishers, vol. 2, nos. 1 and 2, 1995.

[39] Mouly M. and Paulet M.-B., *The GSM System for Mobile Communications*, Palaiseau, Published by authors, France, 1992.

[40] Pereira J. M., "Beyond third generation," in *Proc. International Symposium on Wireless Personal Multimedia Communications (WPMC '99)*, Amsterdam, The Netherlands, Sept. 1999.

[41] Pereira J. M., "Fourth generation: now it is personal!," in *Proc. IEEE International Symposium on Personal, Indoor and Mobile Radio Communications (PIMRC 2000)*, London, UK, pp. 1009–1016, Sept. 2000.

[42] Pickholtz R. L., Schilling D. L., and Milstein L. B., "Theory of spread spectrum communications – a tutorial," *IEEE Transactions on Communication Technology*, vol. 30, pp. 855–884, May 1982.

[43] Saltzberg, B. R., "Performance of an efficient parallel data transmission system," *IEEE Transactions on Communication Technology*, vol. 15, pp. 805–811, Dec. 1967.

[44] Sourour E. A. and Nakagawa M., "Performance of orthogonal multi-carrier CDMA in a multipath fading channel," *IEEE Transactions on Communications*, vol. 44, pp. 356–367, March 1996.

[45] TIA/EIA/IS-95, "Mobile station-base station compatibility standard for dual mode wideband spread spectrum cellular system," July 1993.

[46] TIA/EIA/IS-CDMA-2000, "Physical layer standard for CDMA-2000 spread spectrum systems," Aug. 1999.

[47] Turin G. L., "Introduction to spread spectrum anti-multi-path techniques and their application to urban digital radio," *Proceedings of the IEEE*, vol. 68, pp. 328–353, March 1980.

[48] Vandendorpe L., "Multitone direct sequence CDMA system in an indoor wireless environment," in *Proc. IEEE First Symposium of Communications and Vehicular Technology*, Delft, The Netherlands, pp. 4.1.1–4.1.8, Oct. 1993.

[49] Viterbi A. J., "Spread spectrum communications – myths and realities," *IEEE Communications Magazine*, vol. 17, pp. 11–18, May 1979.

[50] Viterbi A. J., *CDMA: Principles of Spread Spectrum Communication*, Reading: Addison-Wesley, 1995.

[51] Weinstein S. B. and Ebert P. M., "Data transmission by frequency-division multiplexing using the discrete Fourier transform," *IEEE Transactions on Communication Technology*, vol. 19, pp. 628–634, Oct. 1971.

[52] WiMAX Forum, "Mobile WiMAX – Part I: a technical overview and performance evaluation," *White Paper*, Aug. 2006.

[53] WiMAX Forum, www.wimaxforum.org.

[54] WINNER Project, "WINNER II system concept description," *Deliverable D6.13.14*, Nov. 2007.

[55] WINNER Project, www.ist-winner.org.

[56] Yee N., Linnartz J.-P., and Fettweis G., "Multi-carrier CDMA for indoor wireless radio networks," in *Proc. International Symposium on Personal, Indoor and Mobile Radio Communications (PIMRC '93)*, Yokohama, Japan, pp. 109–113, Sept. 1993.

1

Fundamentals

This chapter describes the fundamentals of today's wireless communications. First a detailed description of the radio channel and its modeling is presented, followed by the introduction of the principle of OFDM multi-carrier transmission. In addition, a general overview of the spread spectrum technique, especially DS-CDMA, is given and examples of potential applications for OFDM and DS-CDMA are analyzed. This introduction is essential for a better understanding of the idea behind the combination of OFDM with the spread spectrum technique, which is briefly introduced in the last part of this chapter.

1.1 Radio Channel Characteristics

Understanding the characteristics of the communications medium is crucial for the appropriate selection of transmission system architecture, dimensioning of its components, and optimizing system parameters. Especially mobile radio channels are considered to be the most difficult channels, since they suffer from many imperfections like multi-path fading, interference, Doppler shift, and shadowing. The choice of system components is totally different if, for instance, multi-path propagation with long echoes dominates the radio propagation.

Therefore, an accurate channel model describing the behavior of radio wave propagation in different environments such as mobile/fixed and indoor/outdoor is needed. This may allow one, through simulations, to estimate and validate the performance of a given transmission scheme in its several design phases.

1.1.1 Understanding Radio Channels

In mobile radio channels (see Figure 1-1), the transmitted signal suffers from different effects, which are characterized as follows.

Multi-path propagation occurs as a consequence of reflections, scattering, and diffraction of the transmitted electromagnetic wave at natural and man-made objects. Thus, at the receiver antenna, a multitude of waves arrives from many different directions with different delays, attenuations, and phases. The superposition of these waves results in amplitude and phase variations of the composite received signal.

Multi-Carrier and Spread Spectrum Systems Second Edition K. Fazel and S. Kaiser
© 2008 John Wiley & Sons, Ltd

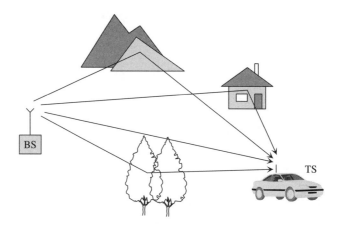

Figure 1-1   Time-variant multi-path propagation

Doppler spread is caused by moving objects in the mobile radio channel. Changes in the phases and amplitudes of the arriving waves lead to time-variant multi-path propagation. Even small movements on the order of the wavelength may result in a totally different wave superposition. The varying signal strength due to time-variant multi-path propagation is referred to as fast fading.

Shadowing is caused by obstruction of the transmitted waves by, for example, hills, buildings, walls, and trees, which results in more or less strong attenuation of the signal strength. Compared to fast fading, longer distances have to be covered to change the shadowing constellation significantly. The varying signal strength due to shadowing is called slow fading and can be described by a log-normal distribution [41].

Path loss indicates how the mean signal power decays with distance between the transmitter and receiver. In free space, the mean signal power decreases with the square of the distance between the base station (BS) and terminal station (TS). In a mobile radio channel, where often no line of sight (NLOS) exists, signal power decreases with a power higher than two and is typically in the order of three to five.

Variations of the received power due to shadowing and path loss can be efficiently counteracted by power control. In the following, the mobile radio channel is described with respect to its fast fading characteristic.

1.1.2   Channel Modeling

The mobile radio channel can be characterized by the time-variant channel impulse response $h(\tau, t)$ or by the time-variant channel transfer function $H(f, t)$, which is the Fourier transform of $h(\tau, t)$. The channel impulse response represents the response of the channel at time t due to an impulse applied at time $t - \tau$. The mobile radio channel is assumed to be a wide-sense stationary random process, i.e. the channel has a fading statistic that remains constant over short periods of time or small spatial distances. In environments with multi-path propagation, the channel impulse response is composed of a large number of scattered impulses received over N_p different paths:

$$h(\tau, t) = \sum_{p=0}^{N_p-1} a_p e^{j(2\pi f_{D,p}t + \varphi_p)} \delta(\tau - \tau_p), \tag{1.1}$$

where

$$\delta(\tau - \tau_p) = \begin{cases} 1 & \text{if } \tau = \tau_p \\ 0 & \text{otherwise} \end{cases} \tag{1.2}$$

and a_p, $f_{D,p}$, φ_p, and τ_p are the amplitude, Doppler frequency, phase, and propagation delay, respectively associated with path p, $p = 0, \ldots, N_p - 1$. The assigned channel transfer function is

$$H(f, t) = \sum_{p=0}^{N_p-1} a_p e^{j(2\pi(f_{D,p}t - f\tau_p) + \varphi_p)}. \tag{1.3}$$

The delays are measured relative to the first detectable path at the receiver. The Doppler frequency

$$f_{D,p} = \frac{v f_c \cos(\alpha_p)}{c} \tag{1.4}$$

depends on the velocity v of the terminal station, the speed of light c, the carrier frequency f_c, and the angle of incidence α_p of a wave assigned to path p. A channel impulse response with a corresponding channel transfer function is illustrated in Figure 1-2.

The delay power density spectrum $\rho(\tau)$ that characterizes the frequency selectivity of the mobile radio channel gives the average power of the channel output as a function of the delay τ. The mean delay $\bar{\tau}$, the root mean square (RMS) delay spread τ_{RMS}, and the maximum delay τ_{max} are characteristic parameters of the delay power density spectrum. The mean delay is

$$\bar{\tau} = \frac{\displaystyle\sum_{p=0}^{N_p-1} \tau_p \Omega_p}{\displaystyle\sum_{p=0}^{N_p-1} \Omega_p}, \tag{1.5}$$

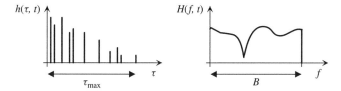

Figure 1-2   Time-variant channel impulse response and channel transfer function with frequency-selective fading

where

$$\Omega_p = |a_p|^2 \tag{1.6}$$

is the power of path p. The RMS delay spread is defined as

$$\tau_{RMS} = \sqrt{\frac{\displaystyle\sum_{p=0}^{N_p-1} \tau_p^2 \Omega_p}{\displaystyle\sum_{p=0}^{N_p-1} \Omega_p} - \overline{\tau}^2}. \tag{1.7}$$

Similarly, the Doppler power density spectrum $S(f_D)$ characterizes the time variance of the mobile radio channel and gives the average power of the channel output as a function of the Doppler frequency f_D. The frequency dispersive properties of multi-path channels are most commonly quantified by the maximum occurring Doppler frequency f_{Dmax} and the Doppler spread $f_{Dspread}$. The Doppler spread is the bandwidth of the Doppler power density spectrum and can take on values up to two times $|f_{Dmax}|$, i.e.

$$f_{Dspread} \leq 2|f_{D\,max}|. \tag{1.8}$$

1.1.3 Channel Fade Statistics

The statistics of the fading process characterize the channel and are of importance for channel model parameter specifications. A simple and often used approach is obtained from the assumption that there is a large number of scatterers in the channel that contribute to the signal at the receiver side. The application of the central limit theorem leads to a complex-valued Gaussian process for the channel impulse response. In the absence of line of sight (LOS) or a dominant component, the process is zero-mean. The magnitude of the corresponding channel transfer function

$$a = a(f, t) = |H(f, t)| \tag{1.9}$$

is a random variable, for brevity denoted by a, with a Rayleigh distribution given by

$$p(a) = \frac{2a}{\Omega} e^{-a^2/\Omega}, \tag{1.10}$$

where

$$\Omega = E\{a^2\} \tag{1.11}$$

is the average power. The phase is uniformly distributed in the interval $[0, 2\pi]$.

In the case where the multi-path channel contains a LOS or dominant component in addition to the randomly moving scatterers, the channel impulse response can no longer be modeled as zero-mean. Under the assumption of a complex-valued Gaussian process for the channel impulse response, the magnitude a of the channel transfer function has a Rice distribution given by

$$p(a) = \frac{2a}{\Omega} e^{-(a^2/\Omega + K_{Rice})} I_0 \left(2a \sqrt{\frac{K_{Rice}}{\Omega}} \right). \tag{1.12}$$

The Rice factor K_{Rice} is determined by the ratio of the power of the dominant path to the power of the scattered paths. I_0 is the zero-order modified Bessel function of first kind. The phase is uniformly distributed in the interval $[0, 2\pi]$.

1.1.4 Inter-Symbol (ISI) and Inter-Channel Interference (ICI)

The delay spread can cause inter-symbol interference (ISI) when adjacent data symbols overlap and interfere with each other due to different delays on different propagation paths. The number of interfering symbols in a single-carrier modulated system is given by

$$N_{\text{ISI, single carrier}} = \left\lceil \frac{\tau_{\max}}{T_d} \right\rceil. \tag{1.13}$$

For high data rate applications with very short symbol duration $T_d < \tau_{\max}$, the effect of ISI and, with that, the receiver complexity can increase significantly. The effect of ISI can be counteracted by different measures such as time or frequency domain equalization. In spread spectrum systems, rake receivers with several arms are used to reduce the effect of ISI by exploiting the multi-path diversity such that individual arms are adapted to different propagation paths.

If the duration of the transmitted symbol is significantly larger than the maximum delay $T_d \gg \tau_{\max}$, the channel produces a negligible amount of ISI. This effect is exploited with multi-carrier transmission where the duration per transmitted symbol increases with the number of sub-carriers N_c and, hence, the amount of ISI decreases. The number of interfering symbols in a multi-carrier modulated system is given by

$$N_{\text{ISI, multi carrier}} = \left\lceil \frac{\tau_{\max}}{N_c T_d} \right\rceil. \tag{1.14}$$

Residual ISI can be eliminated by the use of a guard interval (see Section 1.2).

The maximum Doppler spread in mobile radio applications using single-carrier modulation is typically much less than the distance between adjacent channels, such that the effect of interference on adjacent channels due to Doppler spread is not a problem for single-carrier modulated systems. For multi-carrier modulated systems, the sub-channel spacing F_s can become quite small, such that Doppler effects can cause significant ICI. As long as all sub-carriers are affected by a common Doppler shift f_D, this Doppler shift can be compensated for in the receiver and ICI can be avoided. However, if Doppler spread in the order of several percent of the sub-carrier spacing occurs, ICI may degrade the system performance significantly. To avoid performance degradations due to ICI or more complex receivers with ICI equalization, the sub-carrier spacing F_s should be chosen as

$$F_s \gg f_{D\max}, \tag{1.15}$$

such that the effects due to Doppler spread can be neglected (see Chapter 4). This approach corresponds with the philosophy of OFDM described in Section 1.2 and is followed in current OFDM-based wireless standards.

Nevertheless, if a multi-carrier system design is chosen such that the Doppler spread is in the order of the sub-carrier spacing or higher, a rake receiver in the frequency domain can be used [25]. With the frequency domain rake receiver each branch of the rake resolves a different Doppler frequency.

1.1.5 Examples of Discrete Multi-Path Channel Models

Various discrete multi-path channel models for indoor and outdoor cellular systems with different cell sizes have been specified. These channel models define the statistics of the discrete propagation paths. An overview of widely used discrete multi-path channel models is given in the following.

COST 207 [9]

The COST 207 channel models specify four outdoor macro cell propagation scenarios by continuous, exponentially decreasing delay power density spectra. Implementations of these power density spectra by discrete taps are given by using up to 12 taps. Examples for settings with 6 taps are listed in Table 1-1. In this table for several propagation environments the corresponding path delay and power profiles are given. Hilly terrain causes the longest echoes.

The classical Doppler spectrum with uniformly distributed angles of arrival of the paths can be used for all taps for simplicity. Optionally, different Doppler spectra are defined for the individual taps in Reference [9]. The COST 207 channel models are based on channel measurements with a bandwidth of $8-10$ MHz in the 900 MHz band used for 2 G systems such as GSM.

Table 1-1   Settings for the COST 207 channel models with 6 taps [9]

Path #	Rural Area (RA)		Typical Urban (TU)		Bad Urban (BU)		Hilly Terrain (HT)	
	Delay	Power	Delay	Power	Delay	Power	Delay	Power
	(μs)	(dB)	(μs)	(dB)	(μs)	(dB)	(μs)	(dB)
1	0	0	0	−3	0	−2.5	0	0
2	0.1	−4	0.2	0	0.3	0	0.1	−1.5
3	0.2	−8	0.5	−2	1.0	−3	0.3	−4.5
4	0.3	−12	1.6	−6	1.6	−5	0.5	−7.5
5	0.4	−16	2.3	−8	5.0	−2	15.0	−8.0
6	0.5	−20	5.0	−10	6.6	−4	17.2	−17.7

COST 231 [10] and COST 259 [11]

These COST actions which are the continuation of COST 207 extend the channel characterization to DCS 1800, DECT, HIPERLAN, and WCDMA/UMTS channels, taking into account macro, micro, and pico cell scenarios. Channel models with spatial resolution have been defined in COST 259. The spatial component is introduced by the definition of several clusters with local scatterers, which are located in a circle around the base station. Three types of channel models are defined. The macro cell type has cell sizes from 500 m up to 5000 m and a carrier frequency of 900 MHz or 1.8 GHz. The micro cell type is defined for cell sizes of about 300 m and a carrier frequency of 1.2 GHz or 5 GHz. The pico cell type represents an indoor channel model with cell sizes smaller than 100 m in industrial buildings and in the order of 10 m in an office. The carrier frequency is 2.5 GHz or 24 GHz.

COST 273

The COST 273 action additionally takes multi-antenna channel models into account, which are not covered by the previous COST actions.

CODIT [8]

These channel models define typical outdoor and indoor propagation scenarios for macro, micro, and pico cells. The fading characteristics of the various propagation environments are specified by the parameters of the Nakagami m distribution. Every environment is defined in terms of a number of scatterers, which can take on values up to 20. Some channel models consider also the angular distribution of the scatterers. They have been developed for the investigation of 3 G system proposals. Macro cell channel type models have been developed for carrier frequencies around 900 MHz with 7 MHz bandwidth. The micro and pico cell channel type models have been developed for carrier frequencies between 1.8 GHz and 2 GHz. The bandwidths of the measurements are in the range of 10–100 MHz for macro cells and around 100 MHz for pico cells.

JTC [32]

The JTC channel models define indoor and outdoor scenarios by specifying 3 to 10 discrete taps per scenario. The channel models are designed to be applicable for wideband digital mobile radio systems anticipated as candidates for the PCS (personal communications systems) common air interface at carrier frequencies of about 2 GHz.

UMTS/UTRA [21, 49]

Test propagation scenarios have been defined for UMTS and UTRA system proposals, which are developed for frequencies around 2 GHz. The modeling of the multi-path propagation corresponds to that used by the COST 207 channel models.

LTE [1]

3GPP LTE has defined delay profiles for low, medium and high delay spread environments. The delay profiles are summarized in Table 1-2. The three models are defined on a 10 ns sampling grid. The detailed parameters of the LTE channel models are given in Table 1-3.

Table 1-2 Delay power profiles of the LTE channel models [1]

Model	Number of paths	RMS delay spread	Maximum delay
Extended Pedestrian A (EPA)	7	45 ns	410 ns
Extended Vehicular A (EVA)	9	357 ns	2.51 μs
Extended Typical Urban (ETU)	9	991 ns	5 μs

Table 1-3 LTE channel models for pedestrian, vehicular, and typical urban propagation scenarios [1]

Path number	Extended Pedestrian A (EPA)		Extended Vehicular A (EVA)		Extended Typical Urban (ETU)	
	Delay (ns)	Power (dB)	Delay (ns)	Power (dB)	Delay (ns)	Power (dB)
1	0	0	0	0	0	−1
2	30	−1	30	−1.5	50	−1
3	70	−2	150	−1.4	120	−1
4	90	−3	310	−3.6	200	0
5	110	−8	370	−0.6	230	0
6	190	−17.2	710	−9.1	500	0
7	410	−20.8	1090	−7	1600	−3
8			1730	−12	2300	−5
9			2510	−16.9	5000	−7

Table 1-4 Doppler frequencies defined for the LTE channel models [1]

	Low Doppler frequency	Medium Doppler frequency	High Doppler frequency
Frequency	5 Hz	70 Hz	300 Hz
Velocity	2.7 km/h at 2 GHz 6.4 km/h at 850 MHz	40.8 km/h at 2 GHz 88.9 km/h at 850 MHz	162 km/h at 2 GHz 381.2 km/h at 850 MHz

The classical Doppler spectrum with uniformly distributed angles of arrival of the paths is applied in the LTE channel models. The classical Doppler spectrum is also referred to as Clark's spectrum or Jake's spectrum. The classical Doppler spectrum is characterized by the maximum Doppler frequency. Three typical maximum Doppler frequencies are specified for the LTE channel models, as shown in Table 1-4. LTE baseline combinations of channel models and Doppler frequencies are:

- **[EPA 5 Hz]** Extended Pedestrian A with 5 Hz Doppler frequency
- **[EVA 5 Hz]** Extended Vehicular A with 5 Hz Doppler frequency
- **[EVA 70 Hz]** Extended Vehicular A with 70 Hz Doppler frequency
- **[ETU 70 Hz]** Extended Typical Urban with 70 Hz Doppler frequency
- **[ETU 300 Hz]** Extended Typical Urban with 300 Hz Doppler frequency

Multi-antenna channel models for LTE are defined by correlation matrices applied to the channel models described in Table 1-3. MIMO channel models are defined for high, medium, and low correlations between antennas. The correlation matrix for the base station with two antennas is

$$R_{BS} = \begin{pmatrix} 1 & \alpha \\ \alpha^* & 1 \end{pmatrix}. \tag{1.16}$$

In the case of one antenna at the base station the correlation matrix R_{BS} is equal to 1. At the mobile terminal station the correlation matrix is given by

$$R_{TS} = \begin{pmatrix} 1 & \beta \\ \beta^* & 1 \end{pmatrix}. \tag{1.17}$$

The parameters α and β are defined in Table 1-5.

For the 2×2 case the spatial channel correlation matrix R_{spat} is defined as

$$R_{spat} = R_{BS} \otimes R_{TS} = \begin{pmatrix} 1 & \alpha \\ \alpha^* & 1 \end{pmatrix} \otimes \begin{pmatrix} 1 & \beta \\ \beta^* & 1 \end{pmatrix} = \begin{pmatrix} 1 & \beta & \alpha & \alpha\beta \\ \beta^* & 1 & \alpha\beta^* & \alpha \\ \alpha^* & \alpha^*\beta & 1 & \beta \\ \alpha^*\beta^* & \alpha^* & \beta^* & 1 \end{pmatrix}. \tag{1.18}$$

where $\otimes$ is the Kronecker product. For the 1×2 case the spatial channel correlation matrix R_{spat} results in

$$R_{spat} = \begin{pmatrix} 1 & \beta \\ \beta^* & 1 \end{pmatrix}. \tag{1.19}$$

Table 1-5 Correlation parameters for the LTE MIMO channel models [1]

Low correlation		Medium correlation		High correlation	
α	β	α	β	α	β
0	0	0.3	0.9	0.9	0.9

WiMAX [16, 17, 31]

IEEE 802.16x uses two categories of channel models, (i) one describing the fixed positioned TS channel with NLOS and (ii) the second one defining the behavior of the mobile TS.

(i) Fixed Positioned TS

The channel model adopted in IEEE 802.16 for fixed BWA is described in References [31] and [17]. Two main parameters are characterizing this model: (a) mean path loss versus area type and (b) multi-path propagation profiles. For a suburban area the mean path loss in dB is given by ($d > 100$ m) [16]

$$\text{Path loss}_{NLOS} \approx 12.5 + 20 \log_{10}\left(\frac{f_c}{MHz}\right) + 10\eta \log_{10}\left(\frac{10d}{km}\right), \qquad (1.20)$$

where f_c is the carrier frequency, d (in m) the distance between the TS and the BS, h_{BS} describes the BS antenna height, and η is a factor that depends on the geographical terrain properties:

$$\eta = a - b \times h_{BS} + \frac{c}{h_{BS}}. \qquad (1.21)$$

The parameters of η are given in Table 1-6 for different terrain types [16].

Regarding multi-path propagation, IEEE 802.16 has adopted the so-called Stanford University Interim (SUI) channel models given in Reference [17]. The model for each area type has defined two multi-path profiles, each containing three paths. The parameters of this model are given in Table 1-7. Note that the variations of the multi-path behavior is modeled for each path by a Rayleigh or Rice fading distribution, as described earlier.

Table 1-6   Mean power attenuation parameters

Parameter	Strong hilly	Weak hilly	Flat
a	4.6	4.0	3.6
b	0.0075	0.0065	0.005
c	12.6	17.1	20

Table 1-7   Multi-path profiles for different terrains for fixed positioned TS

Ch. #	Area type	Path 1		Path 2		Path 3	
		Delay (μs)	Mean attenuation (dB)	Delay (μs)	Mean attenuation (dB)	Delay (μs)	Mean attenuation (dB)
SUI1	Flat	0	0	0.4	−15	0.9	−20
SUI2	Flat	0	0	0.4	−12	1.1	−15
SUI3	Weak hilly	0	0	0.4	−5	0.9	−10
SUI4	Weak hilly	0	0	1.5	−4	4	−8
SUI5	Strong hilly	0	0	4	−5	10	−10
SUI6	Strong hilly	0	0	14	−10	20	−14

Table 1-8 Multi-path profiles for different terrains for mobile TS

Path number	1	2	3	4	5	6
Delay (μs)	0	0.3	8.9	12.9	17.1	20
Mean attenuation (dB)	−2.5	0	−12.8	−10	−25.2	−16
Doppler	Jake's model					

(ii) Mobile TS
The adopted model in IEEE 802.16e for mobile BWA is based on the UMTS (ITU/IMT 2000 [31, 37]) channel model. The maximum vehicle speed considered is up to 125 km/h, where its mean path loss can be approximated by

$$\text{Path loss}_{Mobile} \approx 59 + 21 \log_{10}\left(\frac{f_c}{MHz}\right) + 38 \log_{10}\left(\frac{d}{km}\right), \tag{1.22}$$

where f_c represents the carrier frequency and d the distance between the TS and the BS. The multi-path propagation model parameters [31] are given in Table 1-8.

HIPERLAN/2 [38]
Five typical indoor propagation scenarios for wireless LANs in the 5 GHz frequency band have been defined. Each scenario is described by 18 discrete taps of the delay power density spectrum. The time variance of the channel (Doppler spread) is modeled by a classical Jake's spectrum with a maximum terminal speed of 3 km/h.

Further channel models exist which are, for instance, given in Reference [19] for DVB-T.

1.1.6 Multi-Carrier Channel Modeling

Multi-carrier systems can either be simulated in the time domain or, more computationally efficient, in the frequency domain. Preconditions for the frequency domain implementation are the absence of ISI and ICI, the frequency nonselective fading per sub-carrier, and the time-invariance during one OFDM symbol. A proper system design approximately fulfills these preconditions. The discrete channel transfer function adapted to multi-carrier signals results in

$$H_{n,i} = H(nF_s, iT_s')$$

$$= \sum_{p=0}^{N_p-1} a_p e^{j(2\pi(f_{D,p}iT_s' - nF_s\tau_p)+\varphi_p)}$$

$$= a_{n,i} e^{j\varphi_{n,i}}, \tag{1.23}$$

where the continuous channel transfer function $H(f, t)$ is sampled in time at the OFDM symbol rate $1/T_s'$ and in frequency at the sub-carrier spacing F_s. The duration T_s' is the total OFDM symbol duration including the guard interval. Finally, a symbol transmitted on sub-channel n of the OFDM symbol i is multiplied by the resulting fading amplitude $a_{n,i}$ and rotated by a random phase $\varphi_{n,i}$.

The advantage of the frequency domain channel model is that the IFFT and FFT operation for OFDM and inverse OFDM can be avoided and the fading operation results in one complex-valued multiplication per sub-carrier. The discrete multi-path channel models introduced in Section 1.1.5 can directly be applied to Equation (1.23). A further simplification of the channel modeling for multi-carrier systems is given by using the so-called uncorrelated fading channel models.

1.1.6.1 Uncorrelated Fading Channel Models for Multi-Carrier Systems

These channel models are based on the assumption that the fading on adjacent data symbols after inverse OFDM and de-interleaving can be considered as uncorrelated [33]. This assumption holds when, for example, a frequency and time interleaver with sufficient interleaving depth is applied. The fading amplitude $a_{n,i}$ is chosen from a distribution $p(a)$ according to the considered cell type and the random phase $\varphi_{n,i}$ is uniformly distributed in the interval $[0,2\pi]$. The resulting complex-valued channel fading coefficient is thus generated independently for each sub-carrier and OFDM symbol. For a propagation scenario in a macro cell without LOS, the fading amplitude $a_{n,i}$ is generated by a Rayleigh distribution and the channel model is referred to as uncorrelated Rayleigh fading channel. For smaller cells where often a dominant propagation component occurs, the fading amplitude is chosen from a Rice distribution. The advantages of the uncorrelated fading channel models for multi-carrier systems are their simple implementation in the frequency domain and the simple reproducibility of the simulation results.

1.1.7 Diversity

The coherence bandwidth $(\Delta f)_c$ of a mobile radio channel is the bandwidth over which the signal propagation characteristics are correlated and can be approximated by

$$(\Delta f)_c \approx \frac{1}{\tau_{max}}. \tag{1.24}$$

The channel is frequency-selective if the signal bandwidth B is larger than the coherence bandwidth $(\Delta f)_c$. On the other hand, if B is smaller than $(\Delta f)_c$, the channel is frequency nonselective or flat. The coherence bandwidth of the channel is of importance for evaluating the performance of spreading and frequency interleaving techniques that try to exploit the inherent frequency diversity D_f of the mobile radio channel. In the case of multi-carrier transmission, frequency diversity is exploited if the separation of sub-carriers transmitting the same information exceeds the coherence bandwidth. The maximum achievable frequency diversity D_f is given by the ratio between the signal bandwidth B and the coherence bandwidth,

$$D_f = \frac{B}{(\Delta f)_c}. \tag{1.25}$$

The coherence time of the channel $(\Delta t)_c$ is the duration over which the channel charac-teristics can be considered as time-invariant and can be approximated by

$$(\Delta t)_c \approx \frac{1}{2 f_{D\,\text{max}}}. \tag{1.26}$$

If the duration of the transmitted symbol is larger than the coherence time, the channel is time-selective. On the other hand, if the symbol duration is smaller than $(\Delta t)_c$, the channel is time-nonselective during one symbol duration. The coherence time of the channel is of importance for evaluating the performance of coding and interleaving techniques that try to exploit the inherent time diversity D_t of the mobile radio channel. Time diversity can be exploited if the separation between time slots carrying the same information exceeds the coherence time. A number of N_s successive time slots create a time frame of duration T_{fr}. The maximum time diversity D_t achievable in one time frame is given by the ratio between the duration of a time frame and the coherence time,

$$D_t = \frac{T_{fr}}{(\Delta t)_c}. \tag{1.27}$$

A system exploiting frequency and time diversity can achieve the overall diversity

$$D_O = D_f D_t. \tag{1.28}$$

The system design should allow one to optimally exploit the available diversity D_O. For instance, in systems with multi-carrier transmission the same information should be transmitted on different sub-carriers and in different time slots, achieving uncorrelated faded replicas of the information in both dimensions.

Uncoded multi-carrier systems with flat fading per sub-channel and time-invariance during one symbol cannot exploit diversity and have a poor performance in time- and frequency-selective fading channels. Additional methods have to be applied to exploit diversity. One approach is the use of data spreading where each data symbol is spread by a spreading code of length L. This, in combination with interleaving, can achieve the performance results that are given for $D_O \geq L$ by the closed-form solution for the BER for diversity reception in Rayleigh fading channels according to [45]

$$P_b = \left[\frac{1-\gamma}{2}\right]^L \sum_{l=0}^{L-1} \binom{L-1+l}{l} \left[\frac{1+\gamma}{2}\right]^l, \tag{1.29}$$

where $\binom{n}{k}$ represents the combinatory function,

$$\gamma = \sqrt{\frac{1}{1+\sigma^2}}, \tag{1.30}$$

and σ^2 is the variance of the noise. As soon as the interleaving is not perfect or the diversity offered by the channel is smaller than the spreading code length L, or

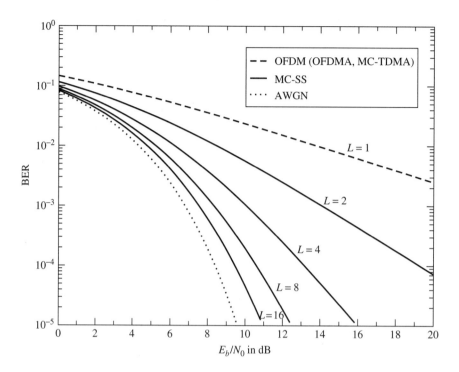

Figure 1-3   Diversity in OFDM and MC-SS systems in a Rayleigh fading channel

MC-CDMA with multiple access interference is applied, Equation (1.29) is a lower bound. For $L = 1$, the performance of an OFDM system without forward error correction (FEC) is obtained, which cannot exploit any diversity. The BER according to Equation (1.29) of an OFDM (OFDMA, MC-TDMA) system and a multi-carrier spread spectrum (MC-SS) system with different spreading code lengths L is shown in Figure 1-3. No other diversity techniques are applied. QPSK modulation is used for symbol mapping. The mobile radio channel is modeled as uncorrelated Rayleigh fading channel (see Section 1.1.6). As these curves show, for large values of L, the performance of MC-SS systems approaches that of an AWGN channel.

Another form of achieving diversity in OFDM systems is channel coding by FEC, where the information of each data bit is spread over several code bits. Additional to the diversity gain in fading channels, a coding gain can be obtained due to the selection of appropriate coding and decoding algorithms.

1.2   Multi-Carrier Transmission

The principle of multi-carrier transmission is to convert a serial high rate data stream on to multiple parallel low rate sub-streams. Each sub-stream is modulated on another sub-carrier. Since the symbol rate on each sub-carrier is much less than the initial serial data symbol rate, the effects of delay spread, i.e. ISI, significantly decrease, reducing the

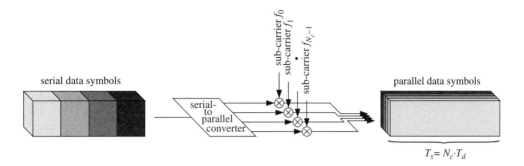

Figure 1-4   Multi-carrier modulation with $N_c = 4$ sub-channels

complexity of the equalizer. OFDM is a low complex technique used to modulate multiple sub-carriers efficiently by using digital signal processing [6, 15, 29, 51, 54].

An example of multi-carrier modulation with four sub-channels $N_c = 4$ is depicted in Figure 1-4. Note that the three-dimensional time/frequency/power density representation is used to illustrate the principle of various multi-carrier and multi-carrier spread spectrum systems. A cuboid indicates the three-dimensional time/frequency/power density range of the signal, in which most of the signal energy is located and does not make any statement about the pulse or spectrum shaping.

An important design goal for a multi-carrier transmission scheme based on OFDM in a mobile radio channel is that the channel can be considered as time-invariant during one OFDM symbol and that fading per sub-channel can be considered as flat. Thus, the OFDM symbol duration should be smaller than the coherence time $(\Delta t)_c$ of the channel and the sub-carrier bandwidth should be smaller than the coherence bandwidth $(\Delta f)_c$ of the channel. By fulfilling these conditions, the realization of low complex receivers is possible.

1.2.1 Orthogonal Frequency Division Multiplexing (OFDM)

A communication system with multi-carrier modulation transmits N_c complex-valued source symbols[1] S_n, $n = 0, \ldots, N_c - 1$, in parallel on to N_c sub-carriers. The source symbols may, for instance, be obtained after source and channel coding, interleaving, and symbol mapping. The source symbol duration T_d of the serial data symbols results after serial-to-parallel conversion in the OFDM symbol duration

$$T_s = N_c T_d. \tag{1.31}$$

The principle of OFDM is to modulate the N_c sub-streams on sub-carriers with a spacing of

$$F_s = \frac{1}{T_s} \tag{1.32}$$

---

[1] Variables that can be interpreted as values in the frequency domain like the source symbols S_n, each modulating another sub-carrier frequency, are written with capital letters.

in order to achieve orthogonality between the signals on the N_c sub-carriers, presuming a rectangular pulse shaping. The N_c parallel modulated source symbols S_n, $n = 0, \ldots, N_c - 1$, are referred to as an OFDM symbol. The complex envelope of an OFDM symbol with rectangular pulse shaping has the form

$$x(t) = \frac{1}{N_c} \sum_{n=0}^{N_c - 1} S_n e^{j 2\pi f_n t}, \quad 0 \leq t < T_s. \tag{1.33}$$

The N_c sub-carrier frequencies are located at

$$f_n = \frac{n}{T_s}, \quad n = 0, \ldots, N_c - 1. \tag{1.34}$$

The normalized power density spectrum of an OFDM symbol with 16 sub-carriers versus the normalized frequency $f T_d$ is depicted as solid curve in Figure 1-5. Note that in this figure the power density spectrum is shifted to the center frequency. The symbols S_n, $n = 0, \ldots, N_c - 1$, are transmitted with equal power. The dotted curve illustrates the power density spectrum of the first modulated sub-carrier and indicates the construction of the overall power density spectrum as the sum of N_c individual power density spectra, each shifted by F_s. For large values of N_c, the power density spectrum becomes flatter in the normalized frequency range of $-0.5 \leq f T_d \leq 0.5$ containing the N_c sub-channels.

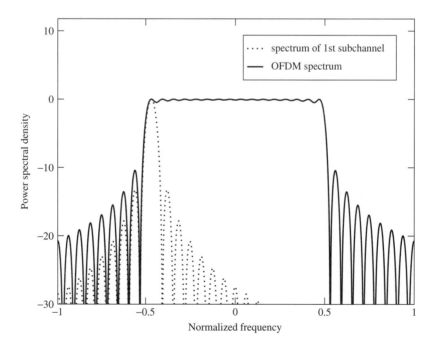

Figure 1-5   OFDM spectrum with 16 sub-carriers

Only sub-channels near the band edges contribute to the out-of-band power emission. Therefore, as N_c becomes large, the power density spectrum approaches that of single-carrier modulation with ideal Nyquist filtering.

A key advantage of using OFDM is that multi-carrier modulation can be implemented in the discrete domain by using an IDFT, or a more computationally efficient IFFT. When sampling the complex envelope $x(t)$ of an OFDM symbol with rate $1/T_d$ the samples are

$$x_v = \frac{1}{N_c} \sum_{n=0}^{N_c-1} S_n e^{j2\pi nv/N_c}, \quad v = 0, \ldots, N_c - 1. \tag{1.35}$$

The sampled sequence x_v, $v = 0, \ldots, N_c - 1$, is the IDFT of the source symbol sequence S_n, $n = 0, \ldots, N_c - 1$. The block diagram of a multi-carrier modulator employing OFDM based on an IDFT and a multi-carrier demodulator employing inverse OFDM based on a DFT is illustrated in Figure 1-6.

When the number of sub-carriers increases, the OFDM symbol duration T_s becomes large compared to the duration of the impulse response $\tau_{\max}$ of the channel, and the amount of ISI reduces. However, to completely avoid the effects of ISI and thus, to maintain the orthogonality between the signals on the sub-carriers, i.e. to also avoid ICI, a guard interval of duration

$$T_g \geq \tau_{\max} \tag{1.36}$$

has to be inserted between adjacent OFDM symbols. The guard interval is a cyclic extension of each OFDM symbol, which is obtained by extending the duration of an OFDM symbol to

$$T_s' = T_g + T_s. \tag{1.37}$$

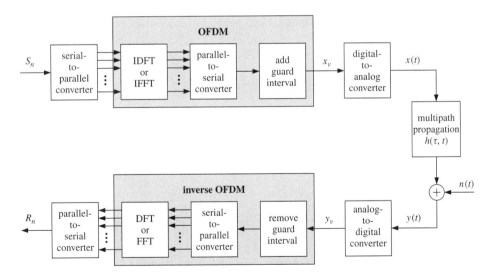

Figure 1-6   Digital multi-carrier transmission system applying OFDM

The discrete length of the guard interval has to be

$$L_g \geq \left\lceil \frac{\tau_{max} N_c}{T_s} \right\rceil \tag{1.38}$$

samples in order to prevent ISI. The sampled sequence with cyclic extended guard interval results in

$$x_v = \frac{1}{N_c} \sum_{n=0}^{N_c-1} S_n e^{j2\pi n v / N_c}, \quad v = -L_g, \ldots, N_c - 1. \tag{1.39}$$

This sequence is passed through a digital-to-analogue converter whose output ideally would be the signal waveform $x(t)$ with increased duration T'_s. The signal is upconverted and the RF signal is transmitted to the channel (see Chapter 4 regarding RF up/down conversion).

The output of the channel, after RF down-conversion, is the received signal waveform $y(t)$ obtained from convolution of $x(t)$ with the channel impulse response $h(\tau, t)$ and addition of a noise signal $n(t)$, i.e.

$$y(t) = \int_{-\infty}^{\infty} x(t - \tau) h(\tau, t) \, d\tau + n(t). \tag{1.40}$$

The received signal $y(t)$ is passed through an analogue-to-digital converter, whose output sequence y_v, $v = -L_g, \ldots, N_c - 1$, is the received signal $y(t)$ sampled at rate $1/T_d$. Since ISI is only present in the first L_g samples of the received sequence, these L_g samples are removed before multi-carrier demodulation. The ISI-free part $v = 0, \ldots, N_c - 1$, of y_v is multi-carrier demodulated by inverse OFDM exploiting a DFT. The output of the DFT is the multi-carrier demodulated sequence R_n, $n = 0, \ldots, N_c - 1$, consisting of N_c complex-valued symbols

$$R_n = \sum_{v=0}^{N_c-1} y_v e^{-j2\pi n v / N_c}, \quad n = 0, \ldots, N_c - 1. \tag{1.41}$$

Since ICI can be avoided due to the guard interval, each sub-channel can be considered separately. Furthermore, when assuming that the fading on each sub-channel is flat and ISI is removed, a received symbol R_n is obtained from the frequency domain representation according to

$$R_n = H_n S_n + N_n, \quad n = 0, \ldots, N_c - 1, \tag{1.42}$$

where H_n is the flat fading factor and N_n represents the noise of the nth sub-channel. The flat fading factor H_n is the sample of the channel transfer function $H_{n,i}$ according to Equation (1.23) where the time index i is omitted for simplicity. The variance of the noise is given by

$$\sigma^2 = E\{|N_n|^2\}. \tag{1.43}$$

When ISI and ICI can be neglected, the multi-carrier transmission system shown in Figure 1-6 can be viewed as a discrete time and frequency transmission system with a set of N_c parallel Gaussian channels with different complex-valued attenuations H_n (see Figure 1-7).

A time/frequency representation of an OFDM symbol is shown in Figure 1-8(a). A block of subsequent OFDM symbols, where the information transmitted within these OFDM symbols belongs together, e.g. due to coding and/or spreading in the time and frequency directions, is referred to as an OFDM frame. An OFDM frame consisting of N_s OFDM symbols with frame duration

$$T_{fr} = N_s T_s'$$ (1.44)

is illustrated in Figure 1-8(b).

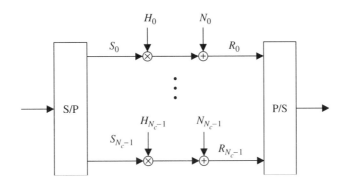

Figure 1-7   Simplified multi-carrier transmission system using OFDM

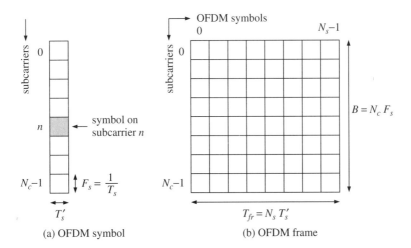

(a) OFDM symbol

(b) OFDM frame

Figure 1-8   Time/frequency representation of an OFDM symbol and an OFDM frame

The following matrix–vector notation is introduced to describe multi-carrier systems concisely. Vectors are represented by boldface small letters and matrices by boldface capital letters. The symbol $(\cdot)^T$ denotes the transposition of a vector or a matrix. The complex-valued source symbols S_n, $n = 0, \ldots, N_c - 1$, transmitted in parallel in one OFDM symbol, are represented by the vector

$$\mathbf{s} = (S_0, S_1, \ldots, S_{N_c-1})^T. \tag{1.45}$$

The $N_c \times N_c$ channel matrix

$$\mathbf{H} = \begin{pmatrix} H_{0,0} & 0 & \cdots & 0 \\ 0 & H_{1,1} & & 0 \\ \vdots & & \ddots & \vdots \\ 0 & 0 & \cdots & H_{N_c-1,N_c-1} \end{pmatrix} \tag{1.46}$$

is of the diagonal type in the absence of ISI and ICI. The diagonal components of $\mathbf{H}$ are the complex-valued flat fading coefficients assigned to the N_c sub-channels. The vector

$$\mathbf{n} = (N_0, N_1, \ldots, N_{N_c-1})^T \tag{1.47}$$

represents the additive noise. The received symbols obtained after inverse OFDM are given by the vector

$$\mathbf{r} = (R_0, R_1, \ldots, R_{N_c-1})^T \tag{1.48}$$

and are obtained by

$$\mathbf{r} = \mathbf{Hs} + \mathbf{n}. \tag{1.49}$$

1.2.2 Advantages and Drawbacks of OFDM

This section summarizes the strengths and weaknesses of multi-carrier modulation based on OFDM.

Advantages:

- High spectral efficiency due to nearly rectangular frequency spectrum for high numbers of sub-carriers.
- Simple digital realization by using the FFT operation.
- Low complex receivers due to the avoidance of ISI and ICI with a sufficiently long guard interval.
- Flexible spectrum adaptation can be realized, e.g. notch filtering.
- Different modulation schemes can be used on individual sub-carriers which are adapted to the transmission conditions on each sub-carrier, e.g. water filling.

Disadvantages:

- Multi-carrier signals with high peak-to-average power ratio (PAPR) require high linear amplifiers. Otherwise, performance degradations occur and the out-of-band power will be enhanced.

- Loss in spectral efficiency due to the guard interval.
- More sensitive to Doppler spreads than single-carrier modulated systems.
- Phase noise caused by the imperfections of the transmitter and receiver oscillators influences the system performance.
- Accurate frequency and time synchronization is required.

1.2.3 Applications and Standards

The key parameters of various multi-carrier-based communications standards for broadcasting (DAB and DVB), WLAN, and WLL are summarized in Tables 1-9 to 1-11.

Table 1-9   Broadcasting standards DAB and DVB-T

Parameter	DAB			DVB-T	
Bandwidth	1.5 MHz			8 MHz	
Number of sub-carriers N_c	192 (256 FFT)	384 (512 FFT)	1536 (2 k FFT)	1705 (2 k FFT)	6817 (8 k FFT)
Symbol duration T_s	125 μs	250 μs	1 ms	224 μs	896 μs
Carrier spacing F_s	8 kHz	4 kHz	1 kHz	4.464 kHz	1.116 kHz
Guard time T_g	31 μs	62 μs	246 μs	$T_s/32$, $T_s/16$, $T_s/8$, $T_s/4$	
Modulation	D-QPSK			QPSK, 16-QAM, 64-QAM	
FEC coding	Convolutional with code rate 1/3 up to 3/4			Reed Solomon + convolutional with code rate 1/2 up to 7/8	
Max. data rate	1.7 Mbit/s			31.7 Mbit/s	

Table 1-10   Wireless local area network (WLAN) standard

Parameter	IEEE 802.11a
Bandwidth	20 MHz
Number of sub-carriers N_c	52 (48 data + 4 pilots) (64 FFT)
Symbol duration T_s	4 μs
Carrier spacing F_s	312.5 kHz
Guard time T_g	0.8 μs
Modulation	BPSK, QPSK, 16-QAM, and 64-QAM
FEC coding	Convolutional with code rate 1/2 up to 3/4
Max. data rate	54 Mbit/s

Table 1-11 Wireless local loop (WLL) standards

Parameter	IEEE 802.16d, ETSI HIPERMAN	
Bandwidth	From 1.5 to 28 MHz	
Number of sub-carriers N_c	256 (OFDM mode)	2048 (OFDMA mode)
Symbol duration T_s	From 8 to 125 µs (Depending on bandwidth)	From 64 to 1024 µs (Depending on bandwidth)
Guard time T_g	From 1/32 up to 1/4 of T_s	
Modulation	QPSK, 16-QAM, and 64-QAM	
FEC coding	Reed Solomon + convolutional with code rate 1/2 up to 5/6	
Max. data rate (in a 7 MHz channel)	Up to 26 Mbit/s	

1.3 Spread Spectrum Techniques

Spread spectrum systems have been developed since the mid-1950s. The initial applications have been military anti-jamming tactical communications, guidance systems, and experimental anti-multi-path systems [44, 48].

Literally, a spread spectrum system is a system in which the transmitted signal is spread over a wide frequency band, much wider than the minimum bandwidth required to transmit the information being sent (see Figure 1-9). Band spreading is accomplished by means of a code which is independent of the data. A reception synchronized to the code is used to de-spread and recover the data at the receiver [52, 53].

There are many application fields for spreading the spectrum [14]:

– Anti-jamming
– Interference rejection
– Low probability of intercept

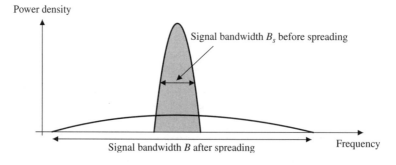

Figure 1-9 Power spectral density after direct sequence spreading

– Multiple access
– Multi-path reception
– Diversity reception
– High resolution ranging
– Accurate universal timing

There are two primary spread spectrum concepts for multiple access: direct sequence code division multiple access (DS-CDMA) and frequency hopping code division multiple access (FH-CDMA).

The general principle behind DS-CDMA is that the information signal with bandwidth B_s is spread over a bandwidth B, where $B \gg B_s$. The processing gain is specified as

$$P_G = \frac{B}{B_s}. \tag{1.50}$$

The higher the processing gain, the lower the power density needed to transmit the information. If the bandwidth is very large, the signal can be transmitted in such a way that it appears like noise. Here, for instance, ultra wideband (UWB) systems (see Chapter 3) can be mentioned as an example [42]. One basic design problem with DS-CDMA is that, when multiple users access the same spectrum, it is possible that a single user could mask all other users at the receiver side if its power level is too high. Hence, accurate power control is an inherent part of any DS-CDMA system [44].

For signal spreading, pseudo-random noise (PN) codes with good cross- and auto-correlation properties are used [43]. A PN code is made up of a number of *chips* for mixing the data with the code (see Figure 1-10). In order to recover the received signal, the code with which the signal was spread in the transmitter is reproduced in the receiver and mixed with the spread signal. If the incoming signal and the locally generated PN code are synchronized, the original signal after correlation can be recovered. In a multi-user environment, the user signals are distinguished by different PN codes and the receiver needs only knowledge of the user's PN code and has to synchronize with it. This principle of user separation is referred to as DS-CDMA. The longer the PN code, the more noise-like signals appear. The drawback is that synchronization becomes more difficult unless synchronization information such as pilot signals is sent to aid acquisition.

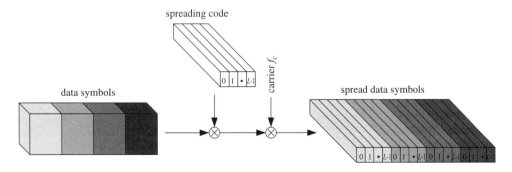

Figure 1-10   Principle of DS-CDMA

Frequency hopping (FH) is similar to direct sequence spreading where a code is used to spread the signal over a much larger bandwidth than that required to transmit the signal. However, instead of spreading the signal over a continuous bandwidth by mixing the signal with a code, the signal bandwidth is unchanged and is hopped over a number of channels, each having the same bandwidth as the transmitted signal. Although at any instant the transmit power level in any narrowband region is higher than with DS-CDMA, the signal is present in a particular channel for a very small time period.

For detection, the receiver must know the hopping pattern in advance, otherwise it will be very difficult to detect the signal. It is the function of the PN code to ensure that all frequencies in the total available bandwidth are optimally used.

There are two kinds of frequency hopping [14]: slow frequency hopping (SFH) and fast frequency hopping (FFH). With SFH many symbols are transmitted per hop. FFH means that there are many hops per symbol. FFH is more resistant to jamming but it is more complex to implement since fast frequency synthesizers are required.

In order to reduce complexity, a hybrid DS/FH scheme can be considered. Here, the signal is first spread over a bandwidth as in DS-CDMA and then hopped over a number of channels, each with a bandwidth equal to the bandwidth of the DS spread signal. This allows one to use a much larger bandwidth than with conventional DS spreading by using low cost available components. For instance, if we have a 1 GHz spectrum available, a PN code generator producing 10^9 chips/s or hopping achieving 10^9 hops/s might not be practicable. Alternatively, we could use two code generators: one for spreading the signal and the other for producing the hopping pattern. Both codes could be generated using low cost components.

1.3.1 Direct Sequence Code Division Multiple Access

The principle of DS-CDMA is to spread a data symbol with a spreading sequence $c^{(k)}(t)$ of length L,

$$c^{(k)}(t) = \sum_{l=0}^{L-1} c_l^{(k)} p_{T_c}(t - lT_c), \tag{1.51}$$

assigned to user k, $k = 0, \ldots, K - 1$, where K is the total number of active users. The rectangular pulse $p_{T_c}(t)$ is equal to 1 for $0 \leq t < T_c$ and zero otherwise. T_c is the chip duration and $c_l^{(k)}$ are the chips of the user specific spreading sequence $c^{(k)}(t)$. After spreading, the signal $x^{(k)}(t)$ of user k is given by

$$x^{(k)}(t) = d^{(k)} \sum_{l=0}^{L-1} c_l^{(k)} p_{T_c}(t - lT_c), \quad 0 \leq t < T_d, \tag{1.52}$$

for one data symbol duration $T_d = LT_c$, where $d^{(k)}$ is the transmitted data symbol of user k. The multiplication of the information sequence with the spreading sequence is done bit-synchronously and the overall transmitted signal $x(t)$ of all K synchronous users (case downlink of a cellular system) results in

$$x(t) = \sum_{k=0}^{K-1} x^{(k)}(t). \tag{1.53}$$

The proper choice of spreading sequences is a crucial problem in DS-CDMA, since the multiple access interference strongly depends on the cross-correlation function (CCF) of the used spreading sequences. To minimize the multiple access interference, the CCF values should be as small as possible [46]. In order to guarantee equal interference among all transmitting users, the cross-correlation properties between different pairs of spreading sequences should be similar. Moreover, the autocorrelation function (ACF) of the spreading sequences should have low out-of-phase peak magnitudes in order to achieve a reliable synchronization.

The received signal $y(t)$ obtained at the output of the radio channel with impulse response $h(t)$ can be expressed as

$$y(t) = x(t) \otimes h(t) + n(t) = r(t) + n(t)$$

$$= \sum_{k=0}^{K-1} r^{(k)}(t) + n(t) \qquad (1.54)$$

where $r^{(k)}(t) = x^{(k)}(t) \otimes h(t)$ is the noise-free received signal of user k, $n(t)$ is the additive white Gaussian noise (AWGN), and $\otimes$ denotes the convolution operation. The impulse response of the matched filter (MF) $h_{MF}^{(k)}(t)$ in the receiver of user k is adapted to both the transmitted waveform including the spreading sequence $c^{(k)}(t)$ and to the channel impulse response $h(t)$,

$$h_{MF}^{(k)}(t) = c^{(k)*}(-t) \otimes h^*(-t). \qquad (1.55)$$

The notation x^* denotes the conjugate of the complex value x. The signal $z^{(k)}(t)$ after the matched filter of user k can be written as

$$z^{(k)}(t) = y(t) \otimes h_{MF}^{(k)}(t)$$

$$= r^{(k)}(t) \otimes h_{MF}^{(k)}(t) + \sum_{\substack{g \neq k \\ g=0}}^{K-1} r^{(g)}(t) \otimes h_{MF}^{(k)}(t) + n(t) \otimes h_{MF}^{(k)}(t). \qquad (1.56)$$

After sampling at the time-instant $t = 0$, the decision variable $\rho^{(k)}$ for user k results in

$$\rho^{(k)} = z^{(k)}(0)$$

$$= \int_0^{T_d+\tau_{max}} r^{(k)}(\tau) h_{MF}^{(k)}(\tau) d\tau + \sum_{\substack{g \neq k \\ g=0}}^{K-1} \int_0^{T_d+\tau_{max}} r^{(g)}(\tau) h_{MF}^{(k)}(\tau) d\tau$$

$$+ \int_0^{T_d+\tau_{max}} n(\tau) h_{MF}^{(k)}(\tau) \, d\tau, \qquad (1.57)$$

where τ_{max} is the maximum delay spread of the radio channel.

Finally, a threshold detection on $\rho^{(k)}$ is performed to obtain the estimated information symbol $\hat{d}^{(k)}$. The first term in the above equation is the desired signal part of user k, whereas the second term corresponds to the multiple access interference and the third term is the additive noise. It should be noted that due to the multiple access interference the estimate of the information bit might be wrong with a certain probability even at high SNRs, leading to the well-known error-floor in the BER curves of DS-CDMA systems.

Ideally, the matched filter receiver resolves all multi-path propagation in the channel. In practice a good approximation of a matched filter receiver is a rake receiver [45, 48] (see Section 1.3.1.2). A rake receiver has D arms to resolve D echoes, where D might be limited by the implementation complexity. In each arm d, $d = 0, \ldots, D - 1$, the received signal $y(t)$ is delayed and de-spread with the code $c^{(k)}(t)$ assigned to user k and weighted with the conjugate instantaneous value h_d^*, $d = 0, \ldots, D - 1$, of the time-varying complex channel attenuation of the assigned echo. Finally, the rake receiver combines the results obtained from each arm and makes a final decision.

1.3.1.1 DS-CDMA Transmitter

Figure 1-11 shows a direct sequence spread spectrum transmitter [45]. It consists of a forward error correction (FEC) encoder, mapping, spreader, pulse shaper, and analogue front-end (IF/RF part). Channel coding is required to protect the transmitted data against channel errors. The encoded and mapped data are spread with the code $c^{(k)}(t)$ over a much wider bandwidth than the bandwidth of the information signal. As the power of the output signal is distributed over a wide bandwidth, the power density of the output signal is much lower than that of the input signal. Note that the multiplication process is done with a spreading sequence with no DC component.

The chip rate directly influences the bandwidth and with that the processing gain; i.e. the wider the bandwidth, the better is the resolution in multi-path detection. Since the total transmission bandwidth is limited, a pulse shaping filtering is employed (e.g. a root Nyquist filter) so that the frequency spectrum is used efficiently.

1.3.1.2 DS-CDMA Receiver

In Figure 1-12, the receiver block diagram of a DS-CDMA signal is plotted [45]. The received signal is first filtered and then digitally converted with a sampling rate of $1/T_c$. It is followed by a rake receiver. The rake receiver is necessary to combat multi-path, i.e. to combine the power of each received echo path. The echo paths are detected with a

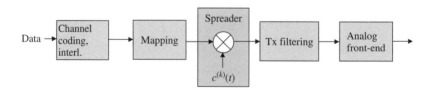

Figure 1-11   DS spread spectrum transmitter block diagram

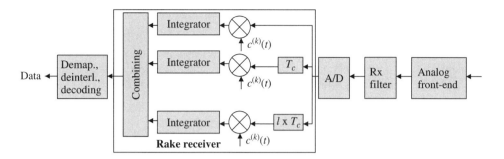

Figure 1-12 DS-CDMA rake receiver block diagram

resolution of T_c. Therefore, each received signal of each path is delayed by lT_c and correlated with the assigned code sequence. The total number of resolution paths depends on the processing gain. Typically 3 or 4 arms are used in practice. After correlation, the power of all detected paths is combined and, finally, the de-mapping and FEC de-coding are performed to assure the data integrity.

1.3.2 Advantages and Drawbacks of DS-CDMA

Conventional DS-CDMA systems offer several advantages in cellular environments including easy frequency planning, high immunity against interference if a high processing gain is used, and flexible data rate adaptation.

Besides these advantages, DS-CDMA suffers from several problems in multi-user wireless communications systems with limited available bandwidth [28]:

- *Multiple access interference (MAI).* As the number of simultaneously active users increases, the performance of the DS-CDMA system decreases rapidly, since the capacity of a DS-CDMA system with moderate processing gain (limited spread bandwidth) is limited by MAI.
- *Complexity.* In order to exploit all multi-path diversity it is necessary to apply a matched filter receiver approximated by a rake receiver with a sufficient number of arms, where the required number of arms is $D = \tau_{\max}/T_c + 1$ [45]. In addition, the receiver has to be matched to the time-variant channel impulse response. Thus, proper channel estimation is necessary. This leads to additional receiver complexity with adaptive receiver filters and a considerable signaling overhead.
- *Single-/multi-tone interference.* In the case of single-tone or multi-tone interference the conventional DS-CDMA receiver spreads the interference signal over the whole transmission bandwidth B whereas the desired signal part is de-spread. If this interference suppression is not sufficient, additional operations have to be done at the receiver, such as notch filtering in the time domain (based on the least mean square algorithm) or in the frequency domain (based on the fast Fourier transform) to partly decrease the amount of interference [34, 39]. Hence, this extra processing leads to additional receiver complexity.

1.3.3 Applications of Spread Spectrum

To illustrate the importance of the spread spectrum technique in today's wireless communications we will briefly introduce two examples of its deployment in cellular mobile communications systems. Here we will describe the main features of the IS-95 standard and the third-generation CDMA standards (CDMA-2000, W-CDMA).

1.3.3.1 IS-95, CDMA-2000, EV-DO

The first commercial cellular mobile radio communication system based on spread spectrum was the IS-95 standard [47], also referred to as cdmaOne. This standard was developed in the USA after the introduction of GSM in Europe. IS-95 is based on frequency division duplex (FDD). The available bandwidth is divided into channels with 1.25 MHz (nominal 1.23 MHz) bandwidth.

As shown in Figure 1-13, in the downlink, binary PN codes are used to distinguish signals received at the terminal station from different base stations. All CDMA signals share a quadrature pair of PN codes. Signals from different cells and sectors are distinguished by the time offset from the basic code. The PN codes used are generated by linear shift registers that produce a code with a period of 32 768 chips. Two codes are generated, one for each quadrature carrier (I and Q) of the QPSK type of modulation.

As mentioned earlier, signals (traffic or control) transmitted from a single antenna (e.g. a base station sector antenna) in a particular CDMA radio channel share a common PN code phase. The traffic and control signals are distinguished at the terminal station receiver by using a binary Walsh–Hadamard (WH) orthogonal code with a spreading factor of 64.

The transmitted downlink information (e.g. voice of rate 9.6 kbit/s) is first convolutionally encoded with rate 1/2 and memory 9 (see Figure 1-13). To provide communication privacy, each user's signal is scrambled with a user-addressed long PN code sequence. Each data symbol is spread using orthogonal WH codes of length 64. After superposition

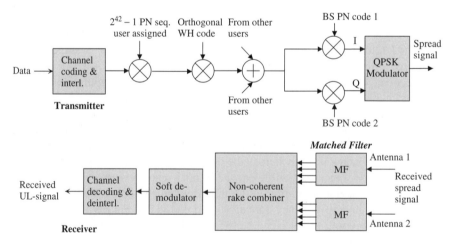

Figure 1-13　Simplified block diagram of the IS-95 base station transceiver

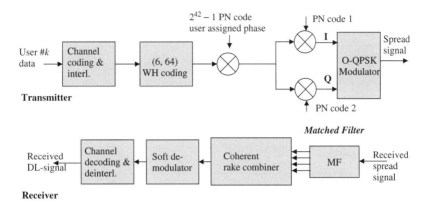

Figure 1-14   Simplified block diagram of the IS-95 terminal station transceiver

of the spread data of all active users, the resulting signal is transmitted to the in-phase and to the quadrature components, i.e. QPSK modulated by a pair of PN codes with an assigned offset. Furthermore, in the downlink a pilot signal is transmitted by each cell site and is used as a coherent carrier reference for demodulation by all mobile receivers. The pilot channel signal is the *zero* WH code sequence.

The transmitted uplink information is concatenated encoded (see Figure 1-14). The outer code is a convolutional code of rate 1/3 and memory 9. The encoded information is grouped into six symbol groups which are used to select one of the different WH inner code words of length 64 (rate 6/64). The signal from each terminal station is distinguished by the use of a very long ($2^{42} - 1$) PN code (privacy code) with a user address-determined time offset. Finally, the same information is transmitted in the in-phase (I) and quadrature (Q) component of an offset QPSK type modulator, where the I and Q components are multiplied by different long codes.

In Table 1-12 important parameters of the IS-95 standard are summarized. Note that in IS-95 the WH code in the uplink is used for FEC, which together with convolutional coding results in a very low code rate, hence guaranteeing very good protection. This is different from the downlink, where the WH code is used for signal spreading. Furthermore, the use of WH codes in the uplink allows one to perform noncoherent detection at the base station. It saves the transmission of pilot symbols from terminal stations.

The CDMA-2000, developed within 3GPP2 (a parallel Working Group of 3GPP) is a simple migration of the IS-95 standard towards 3G regarding not only the networking but also with respect to the radio interface. The same bandwidth and the same carrier frequency can be re-used. To provide a higher data rate, the CDMA-2000 can allocate several parallel carrier frequencies (called also Multi-Carrier CDMA) for the same user; for instance with 3 carriers (each with 1.25 MHz, chip rate 1.2288 Mchip/s, maximum data rate of 307 kbps) a total chip rate of 3.6864 Mchip/s can be achieved, resulting in a data rate of 3×307 kbps.

EV-DO is a further extension of the CDMA-2000 standard and is optimized for higher rate data services. This is also specified within the 3GPP2 Working Group. It employs high order modulation for both downlink and uplink. In its first version, Revision 0, a

Table 1-12   Radio link parameters of IS-95

Parameter	IS-95 (cdmaOne)
Bandwidth	1.25 MHz
Chip rate	1.2288 Mchip/s
Duplex scheme	Frequency division duplex (FDD)
Spreading code short/long	Walsh–Hadamard orthogonal code/PN code
Modulation	Coherent QPSK for the downlink, Noncoherent offset QPSK for the uplink
Channel coding	DL: convolutional $R = 1/2$, memory 9; UL: convolutional $R = 1/3$, memory 9 with WH (6, 64)
Processing gain	19.3 dB
Maximum data rate	14.4 kbit/s for data and 9.6 kbit/s for voice
Diversity	Rake + antenna
Power control	Fast power control based on signal-to-interference ratio (SIR) measurement

data rate up to 2.5 Mbps in a 1.25 MHz bandwidth in the downlink was offered. With its Revision A, a higher capacity for both uplink (up to 1.8 Mbps) and downlink (up to 3.1 Mbps) was offered, while with Revision B a much higher data rate is expected. In Table 1-13 the important parameters of ED-VO are summarized.

1.3.3.2   WCDMA/UMTS, HSPA

The major services of the second-generation mobile communication systems are limited to voice, facsimile, and low rate data transmission. With a variety of new high speed

Table 1-13   Radio link parameters of EV-DO

Parameter	EV-DO, Revision A
Bandwidth	1.25 MHz
Duplex scheme	FDD
Multiplexing	CDMA with TDM
Modulation	Downlink: QPSK, 8-PSK, 16-QAM
	Uplink: BPSK, QPSK, 8-PSK
Channel coding	As in IS-95
Data rates	Downlink: 3.1 Mbps
	Uplink: 1.8 Mbps

multi-media services such as high speed internet and video/high quality image transmission the need for higher data rates increases. The research activity on UMTS started in Europe at the beginning of the 1990s. Several EU-RACE projects such as CODIT and A-TDMA were dealing deeply with the study of the third-generation mobile communications systems. Within the CODIT project a wideband CDMA testbed was built, showing the feasibility of a flexible CDMA system [4]. Further detailed parameters for the 3 G system were specified within the EU-ACTS FRAMES project [5]. In 1998, ETSI decided to adopt wideband CDMA (WCDMA) as the technology for UMTS in the frequency division duplex bands [21]. Later on, ARIB approved WCDMA as standard in Japan as well, where both ETSI and ARIB use the same WCDMA concept.

The third-generation mobile communication systems, called International Mobile Telecommunications-2000 (IMT-2000) or Universal Mobile Telecommunications System (UMTS) in Europe, are designed to support wideband services with data rates up to 2 Mbit/s. The carrier frequency allocated for UMTS in Europe is about 2 GHz. In the case of FDD, the allocated total bandwidth is 2×60 MHz: the uplink carrier frequency is 1920–1980 MHz and the downlink carrier frequency is 2110–2170 MHz.

In Table 1-14 key parameters of WCDMA/UMTS are outlined. In Figures 1-15 and 1-16, simplified block diagrams of a base station and a terminal station are illustrated. In contrast to IS-95, the WCDMA/UMTS standard applies variable length orthogonal spreading codes and coherent QPSK detection for both uplink and downlink directions [2]. The generation of the orthogonal variable spreading code [13] is illustrated in Figure 1-17. Note that for scrambling and spreading, complex codes are employed.

The increasing demand in data rates resulted in the UMTS extensions referred to as *high speed downlink packet access* (HSDPA) and *high speed uplink packet access* (HSUPA). Both extensions are based on the Release 99 (R99) single-carrier WCDMA with an increase in data rate up to 14.4 Mbit/s with HSDPA (Release 5, R5) and 5.7 Mbit/s with HSUPA (Release 6, R6). These improvements are obtained by introducing higher order modulation, channel dependent scheduling and *hybrid ARQ* (HARQ) with soft combining together with multiple code allocation. The downlink additionally supports adaptive cod-

Table 1-14 Radio link parameters of UMTS

Parameter	WCDMA/UMTS
Bandwidth	5 MHz
Duplex scheme	FDD and TDD
Spreading code short/long	Tree-structured orthogonal variable spreading factor (VSF)/PN codes
Modulation	Coherent QPSK (downlink and uplink)
Channel coding	Voice: convolutional $R = 1/3$, memory 9
	Data: concatenated Reed Solomon (RS) + convolutional
	High rate high quality services: convolutional Turbo codes
Diversity	Rake + antenna
Power control	Fast power control based on SIR measurement

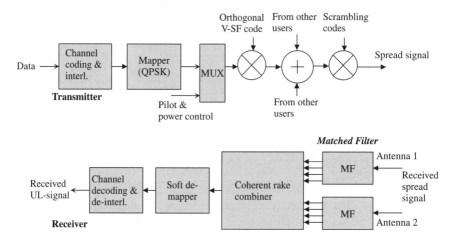

Figure 1-15   Simplified block diagram of a WCDMA/UMTS base station transceiver

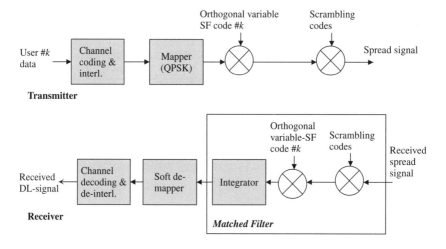

Figure 1-16   Simplified block diagram of a WCDMA/UMTS terminal station transceiver

ing and modulation. The set of HSDPA and HSUPA is termed *high speed packed access* (HSPA). In Table 1-15 the important parameters of HSPA are summarized.

1.4   Multi-Carrier Spread Spectrum

The success of the spread spectrum techniques for second-generation mobile radio and OFDM for digital broadcasting and wireless LANs motivated many researchers to investigate the combination of both techniques. The combination of DS-CDMA and multi-carrier modulation was proposed in 1993 [7, 12, 22, 24, 35, 50, 55]. Two different realizations of multiple access exploiting multi-carrier spread spectrums are detailed in this section.

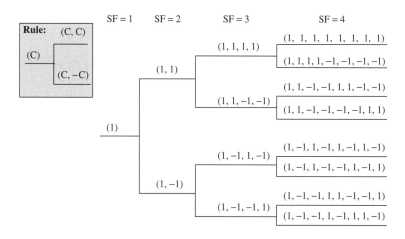

Figure 1-17 Variable length orthogonal spreading code generation

Table 1-15 Radio link parameters of HSPA

Parameter	HSPA (HSDPA, HSUPA)
Bandwidth	5 MHz
Duplex scheme	FDD
Multiplexing	CDMA with TDM
Modulation	Downlink: QPSK, 16QAM
	Uplink: BPSK, QPSK
Channel coding	Convolutional Turbo code
Data rates	Downlink: 14.4 Mbps
	Uplink: 5.7 Mbps

1.4.1 Principle of Various Schemes

The first realization is referred to as MC-CDMA, also known as OFDM-CDMA. The second realization is termed as MC-DS-CDMA. In both schemes, the different users share the same bandwidth at the same time and separate the data by applying different user-specific spreading codes, i.e. the separation of the user signals is carried out in the code domain. Moreover, both schemes apply multi-carrier modulation to reduce the symbol rate and, thus, the amount of ISI per sub-channel. This ISI reduction is significant in spread spectrum systems where high chip rates occur.

The difference between MC-CDMA and MC-DS-CMDA is the allocation of the chips to the sub-channels and OFDM symbols. This difference is illustrated in Figures 1-18 and 1-19. The principle of MC-CDMA is to map the chips of a spread data symbol in frequency direction over several parallel sub-channels while MC-DS-CDMA maps the chips of a spread data symbol in the time direction over several multi-carrier symbols.

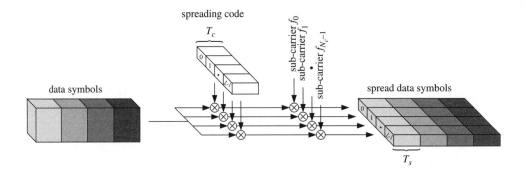

Figure 1-18   MC-CDMA signal generation for one user

MC-CDMA transmits a data symbol of a user simultaneously on several narrowband sub-channels. These sub-channels are multiplied by the chips of the user-specific spreading code, as illustrated in Figure 1-18. Multi-carrier modulation is realized by using the low complex OFDM operation. Since the fading on the narrowband sub-channels can be considered flat, simple equalization with one complex-valued multiplication per sub-channel can be realized. MC-CDMA offers a flexible system design, since the spreading code length does not have to be chosen equal to the number of sub-carriers, allowing adjustable receiver complexities. This flexibility is described in detail in Chapter 2.

MC-DS-CDMA serial-to-parallel converts the high rate data symbols into parallel low rate sub-streams before spreading the data symbols on each sub-channel with a user-specific spreading code in the time direction, which corresponds to direct sequence spreading on each sub-channel. The same spreading codes can be applied on the different sub-channels. The principle of MC-DS-CDMA is illustrated in Figure 1-19.

MC-DS-CDMA systems have been proposed with different multi-carrier modulation schemes, also without OFDM, such that within the description of MC-DS-CDMA the general term multi-carrier symbol instead of OFDM symbol is used. The MC-DS-CDMA schemes can be subdivided into schemes with broadband sub-channels and schemes with

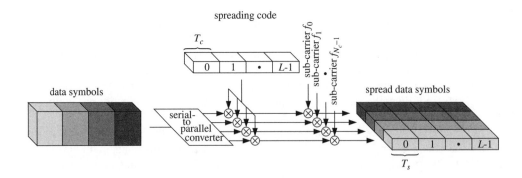

Figure 1-19   MC-DS-CDMA signal generation for one user

narrowband sub-channels. Systems with broadband sub-channels typically apply only few numbers of sub-channels, where each sub-channel can be considered as a classical DS-CDMA system with reduced data rate and ISI, depending on the number of parallel DS-CDMA systems. MC-DS-CDMA systems with narrowband sub-channels typically use high numbers of sub-carriers and can be efficiently realized by using the OFDM operation. Since each sub-channel is narrowband and spreading is performed in the time direction, these schemes can only achieve a time diversity gain if no additional measures such as coding or interleaving are applied.

Both multi-carrier spread spectrum concepts are described in detail in Chapter 2.

1.4.2 Advantages and Drawbacks

In Table 1-16, the main advantages and drawbacks of MC-CDMA and MC-DS-CDMA are summarized. A first conclusion from this table can be derived:

- The high spectral efficiency and the low receiver complexity of MC-CDMA makes it a good candidate for the downlink of a cellular system.
- The low peak to average power ratio (PAPR) property of MC-DS-CDMA with low numbers of subcarriers makes it more appropriate for the uplink of a multi-user system.

1.4.3 Examples of Future Application Areas

Multi-carrier spread spectrum concepts have been developed for a wide variety of applications.

Table 1-16   Advantages and drawbacks of MC-CDMA and MC-DS-CDMA

MC-CDMA		MC-DS-CDMA	
Advantages	Disadvantages	Advantages	Disadvantages
– Simple implementation with Hadamard Transform and FFT – Low complex receivers – High spectral efficiency – High frequency diversity gain due to spreading in the frequency direction	– High PAPR, especially in the uplink – Synchronous transmission	– Low PAPR in the uplink – High time diversity gain due to spreading in the time direction	– ISI and/or ICI can occur, resulting in more complex receivers – Less spectral efficient if other multi-carrier modulation schemes than OFDM are used

Cellular Mobile Radio
Due to the high spectral efficiency of MC-CDMA, it is a an interesting concept for a high rate downlink of future mobile radio systems [3]. In the uplink, MC-DS-CDMA seems to be a promising candidate since it has a lower PAPR compared to MC-CDMA, thus increasing the power efficiency of the mobile terminal. In Reference [23] a further concept of an MC-CDMA system for a mobile cellular system has been proposed.

DVB-T Return Link
The DVB-T interactive point to multi-point (PMP) network is intended to offer a variety of services requiring different data rates [18]. Therefore, the multiple access scheme needs to be flexible in terms of data rate assignment to each subscriber. As in the downlink terrestrial channel, its return channels suffer especially from high multi-path propagation delays. A derivative of MC-CDMA, namely OFDMA, is already adopted in the standard. Several orthogonal sub-carriers are assigned to each terminal station. However, the assignment of these sub-carriers over the time is hopped following a given spreading code.

MMDS/LMDS (FWA)
The aim of microwave/local multi-point distribution systems (MMDS/LMDS) or fixed broadband wireless access (FWA) systems is to provide wireless high speed services with, for example, IP/ATM to fixed positioned terminal stations with a coverage area from 2 km up to 20 km. In order to maintain reasonably low RF costs and good penetration of the radio signals for residential applications, the FWA systems typically use below 10 GHz carrier frequencies, e.g. the MMDS band (2.5–2.7 GHz) or around 5 GHz. As in the DVB-T return channel, OFDMA with frequency hopping for FWA below 10 GHz is proposed [20, 30]. However, for microwave frequencies above 10 GHz, e.g. LMDS, the main channel impairment will be the high amount of co-channel interference (CCI) due to the dense frequency re-use in a cellular environment. In Reference [36] a system architecture based on MC-CDMA for FWA/LMDS applications is proposed. The suggested system provides a high capacity, is robust against multi-path effects, and can offer service coverage not only to subscribers with LOS but also to subscribers without LOS.

Aeronautical Communications
An increase in air traffic will lead to bottlenecks in air traffic handling en route and on ground. Airports have been identified as one of the most capacity-restricted factors in the future if no countermeasures are taken. New digital standards should replace current analogue air traffic control systems. Different concepts for future air traffic control based on multi-carrier spread spectrum systems have been proposed [26, 27].

More potential application fields for multi-carrier spread spectrum systems are in wireless indoor communications [55] and broadband underwater acoustic communications [40].

References

[1] 3GPP (TR 36.803), "Evolved universal terrestrial radio access (E-UTRA); user equipment (UE) radio transmission and reception (Release 8)," *Technical Specification*, Sophia Antipolis, France, 2007.

[2] Adachi F., Sawahashi M., and Suda H., "Wideband CDMA for next generation mobile communications systems," *IEEE Communications Magazine*, vol. 26, pp. 56–69, June 1988.

[3] Atarashi H., Maeda N., Abeta S., and Sawahashi M., "Broadband packet wireless access based on VSF-OFCDM and MC/DS-CDMA," in *Proc. IEEE International Symposium on Personal, Indoor and Mobile Radio Communications (PIMRC 2002)*, Lisbon, Portugal, pp. 992–997, Sept. 2002.

[4] Baier A., Fiebig U. -C., Granzow W., Koch W., Teder P., and Thielecke J., "Design study for a CDMA-based third-generation mobile radio system, "*IEEE Journal on Selected Areas in Communications*, vol. 12, pp. 733–734, May 1994.

[5] Berruto E., Gudmundson M., Menolascino R., Mohr W., and Pizarroso M., "Research activities on UMTS radio interface, network architectures, and planning," *IEEE Communications Magazine*, vol. 36, pp. 82–95, Feb. 1998.

[6] Bingham J. A. C., "Multicarrier modulation for data transmission: an idea whose time has come," *IEEE Communications Magazine*, vol. 28, pp. 5–14, May 1990.

[7] Chouly A., Brajal A., and Jourdan S., "Orthogonal multicarrier techniques applied to direct sequence spread spectrum CDMA systems," in *Proc. IEEE Global Telecommunications Conference (GLOBECOM '93)*, Houston, USA, pp. 1723–1728, Nov./Dec. 1993.

[8] CODIT, "Final propagation model," *Report R2020/TDE/PS/DS/P/040/b1*, 1994.

[9] COST 207, "Digital land mobile radio communications," *Final Report*, 1989.

[10] COST 231, "Digital mobile radio towards future generation systems," *Final Report*, 1996.

[11] COST 259, "Wireless flexible personalized communications," *Final Report*, L. M. Correira (ed.), John Wiley & Sons, Ltd, 2001.

[12] DaSilva V. and Sousa E. S., "Performance of orthogonal CDMA codes for quasi-synchronous communication systems," in *Proc. IEEE International Conference on Universal Personal Communications (ICUPC '93)*, Ottawa, Canada, pp. 995–999, Oct. 1993.

[13] Dinan E. H. and Jabbari B., "Spreading codes for direct sequence CDMA and wideband CDMA cellular networks," *IEEE Communications Magazine*, vol. 26, pp. 48–54, June 1988.

[14] Dixon R. C., *Spread Spectrum Systems*, New York: John Wiley & Sons, Inc., 1976.

[15] Engels M. (ed.), *Wireless OFDM Systems: How to Make Them Work*, Boston: Kluwer Academic Publishers, 2002.

[16] Erceg V., *et al.*, "An empirically based path loss model for wireless channels in suburban environments," *IEEE Journal on Selected Areas in Communications*, vol. 17, July 1999.

[17] Erceg V., *et al.*, "Channel models for fixed wireless applications," IEEE 802.16 BWA Working Group, July 2001.

[18] ETSI DVB-RCT (EN 301 958), "Interaction channel for digital terrestrial television (RCT) incorporating multiple access OFDM," Sophia Antipolis, France, March 2001.

[19] ETSI DVB-T (EN 300 744), "Digital video broadcasting (DVB); framing structure, channel coding and modulation for digital terrestrial television," Sophia Antipolis, France, July 1999.

[20] ETSI HIPERMAN (TS 102 177), "High performance metropolitan local area networks, Part 1: Physical layer," Sophia Antipolis, France, 2004.

[21] ETSI UMTS (TR 101 112), "Universal mobile telecommunications system (UMTS)," Sophia Antipolis, France, 1998.

[22] Fazel K., "Performance of CDMA/OFDM for mobile communication system," in *Proc. IEEE International Conference on Universal Personal Communications (ICUPC '93)*, Ottawa, Canada, pp. 975–979, Oct. 1993.

[23] Fazel K., Kaiser S., and Schnell M., "A flexible and high performance cellular mobile communications system based on multi-carrier SSMA," *Wireless Personal Communications*, vol. 2, nos. 1 & 2, pp. 121–144, 1995.

[24] Fazel K. and Papke L., "On the performance of convolutionally-coded CDMA/OFDM for mobile communication system," in *Proc. IEEE International Symposium on Personal, Indoor and Mobile Radio Communications (PIMRC '93)*, Yokohama, Japan, pp. 468–472, Sept. 1993.

[25] Fettweis G., Bahai A. S., and Anvari K., "On multi-carrier code division multiple access (MC-CDMA) modem design," in *Proc. IEEE Vehicular Technology Conference (VTC '94)*, Stockholm, Sweden, pp. 1670–1674, June 1994.

[26] Haas E., Lang H., and Schnell M., "Development and implementation of an advanced airport data link based on multi-carrier communications," *European Transactions on Telecommunications (ETT)*, vol. 13, no. 5, pp. 447–454, Sept./Oct. 2002.

[27] Haindl B., "Multi-carrier CDMA for air traffic control air/ground communication," in *Proc. International Workshop on Multi-Carrier Spread Spectrum and Related Topics (MC-SS 2001)*, Oberpfaffenhofen, Germany, pp. 77–84, Sept. 2001.

[28] Hara H. and Prasad R., "Overview of multicarrier CDMA," *IEEE Communications Magazine*, vol. 35, pp. 126–133, Dec. 1997.

[29] Heiskala J. and Terry J., *OFDM Wireless LANs: A Theoretical and Practical Guide*, Indianapolis: SAMS, 2002.

[30] IEEE 802.16d, "Air interface for fixed broadband wireless access systems," IEEE 802.16, May 2004.

[31] ITU-R Recommendation M.1225, "Guidelines for evaluation of radio transmission technologies for IMT-2000," ITU-R, 1997.

[32] Joint Technical Committee (JTC) on Wireless Access, *Final Report on RF Channel Characterization*, JTC(AIR)/93.09.23-238R2, Sept. 1993.

[33] Kaiser S., *Multi-Carrier CDMA Mobile Radio Systems–Analysis and Optimization of Detection, Decoding, and Channel Estimation*, Düsseldorf: VDI-Verlag, Fortschritt-Berichte VDI, series 10, no. 531, 1998, PhD thesis.

[34] Ketchum J. W. and Proakis J. G., "Adaptive algorithms for estimating and suppressing narrow band interference in PN spread spectrum systems," *IEEE Transactions on Communications*, vol. 30, pp. 913–924, May 1982.

[35] Kondo S. and Milstein L. B., "On the use of multicarrier direct sequence spread spectrum systems," in *Proc. IEEE Military Communications Conference (MILCOM '93)*, Boston, USA, pp. 52–56, Oct. 1993.

[36] Li J. and Kaverhard M., "Multicarrier orthogonal-CDMA for fixed wireless access applications," *International Journal of Wireless Information Network*, vol. 8, no. 4, pp. 189–201, Oct. 2001.

[37] Maucher J. and Furrer J., *WiMAX, Der IEEE 802.16 Standard: Technik, Anwendung, Potential*, Heise Publisher, Hannover, 2007.

[38] Medbo J. and Schramm P., "Channel models for HIPERLAN/2 in different indoor scenarios," *Technical Report ETSI EP BRAN*, 3ERI085B, March 1998.

[39] Milstein L. B., "Interference rejection techniques in spread spectrum communications," *Proceedings of the IEEE*, vol. 76, pp. 657–671, June 1988.

[40] Ormondroyd R. F., Lam W. K. and Davies J., "A multi-carrier spread spectrum approach to broadband underwater acoustic communications," in *Proc. International Workshop on Multi-Carrier Spread Spectrum and Related Topics (MC-SS '99)*, Oberpfaffenhofen, Germany, pp. 63–70, Sept. 1999.

[41] Parsons D., *The Mobile Radio Propagation Channel*, New York: John Wiley & Sons, Inc., 1992.

[42] Petroff A. and Withington P., "Time modulated ultra-wideband (TM-UWB) overview," in *Proc. Wireless Symposium 2000*, San Jose, USA, Feb. 2000.

[43] Pickholtz R. L., Milstein L. B., and Schilling D. L., "Spread spectrum for mobile communications," *IEEE Transactions on Vehicular Technology*, vol. 40, no. 2, pp. 313–322, May 1991.

[44] Pickholtz R. L., Schilling D. L. and Milstein L. B., "Theory of spread spectrum communications – a tutorial," *IEEE Transactions on Communications*, vol. 30, pp. 855–884, May 1982.

[45] Proakis J. G., *Digital Communications*, New York: McGraw-Hill, 1995.

[46] Sarwate D. V. and Pursley M. B., "Crosscorrelation properties of pseudo-random and related sequences," *Proceedings of the IEEE*, vol. 88, pp. 593–619, May 1998.

[47] TIA/EIA/IS-95, "Mobile station-base station compatibility standard for dual mode wideband spread spectrum cellular system," July 1993.

[48] Turin G. L., "Introduction to spread spectrum anti-multipath techniques and their application to urban digital radio," *Proceedings of the IEEE*, vol. 68, pp. 328–353, March 1980.

[49] UTRA, *Submission of Proposed Radio Transmission Technologies*, SMG2, 1998.

[50] Vandendorpe L., "Multitone direct sequence CDMA system in an indoor wireless environment," in *Proc. IEEE First Symposium of Communications and Vehicular Technology*, Delft, The Netherlands, pp. 4.1.1–4.1.8, Oct. 1993.

[51] van Nee R. and Prasad R., *OFDM for Wireless Multimedia Communications*, Boston: Artech House Publishers, 2000.

[52] Viterbi A. J., "Spread spectrum communications – myths and realities," *IEEE Communications Magazine*, pp. 11–18, May 1979.

[53] Viterbi A. J., *CDMA: Principles of Spread Spectrum Communication*, Reading: Addison-Wesley, 1995.

[54] Weinstein S. B. and Ebert P. M., "Data transmission by frequency-division multiplexing using the discrete Fourier transform," *IEEE Transactions on Communication Technology*, vol. 19, pp. 628–634, Oct. 1971.

[55] Yee N., Linnartz J. P., and Fettweis G., "Multi-carrier CDMA in indoor wireless radio networks," in *Proc. IEEE International Symposium on Personal, Indoor and Mobile Radio Communications (PIMRC '93)*, Yokohama, Japan, pp. 109–113, Sept. 1993.

2

MC-CDMA and MC-DS-CDMA

In this chapter, the different concepts of the combination of multi-carrier transmission with spread spectrum, namely MC-CDMA and MC-DS-CDMA, are detailed and analyzed. Several single-user and multi-user detection strategies and their performance in terms of BER and spectral efficiency in a mobile communications system are examined.

2.1 MC-CDMA

2.1.1 Signal Structure

The basic MC-CDMA signal is generated by a serial concatenation of classical DS-CDMA and OFDM. Each chip of the direct sequence spread data symbol is mapped on to a different sub-carrier. Thus, with MC-CDMA the chips of a spread data symbol are transmitted in parallel on different sub-carriers, in contrast to a serial transmission with DS-CDMA. Let's assume K be the number of simultaneously active users[1] in an MC-CDMA mobile radio system.

Figure 2-1 shows multi-carrier spectrum spreading of one complex-valued data symbol $d^{(k)}$ assigned to user k. The rate of the serial data symbols is $1/T_d$. For brevity, but without loss of generality, the MC-CDMA signal generation is described for a single data symbol per user as far as possible, such that the data symbol index can be omitted. In the transmitter, the complex-valued data symbol $d^{(k)}$ is multiplied with the user specific spreading code

$$\mathbf{c}^{(k)} = (c_0^{(k)}, c_1^{(k)}, \ldots, c_{L-1}^{(k)})^T \tag{2.1}$$

of length $L = P_G$, where P_G is the processing gain. The chip rate of the serial spreading code $\mathbf{c}^{(k)}$ before serial-to-parallel conversion is

$$\frac{1}{T_c} = \frac{L}{T_d} \tag{2.2}$$

and is L times higher than the data symbol rate $1/T_d$. The complex-valued sequence obtained after spreading is given in vector notations by

$$\mathbf{s}^{(k)} = d^{(k)}\mathbf{c}^{(k)} = (S_0^{(k)}, S_1^{(k)}, \ldots, S_{L-1}^{(k)})^T. \tag{2.3}$$

[1] Values and functions related to user k are marked by the index $^{(k)}$, where k may take on the values $0, \ldots, K-1$.

Multi-Carrier and Spread Spectrum Systems Second Edition   K. Fazel and S. Kaiser
© 2008 John Wiley & Sons, Ltd

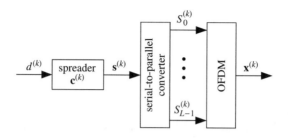

Figure 2-1   Multi-carrier spread spectrum signal generation

A multi-carrier spread spectrum signal is obtained after modulating the components $S_l^{(k)}, l = 0, \ldots, L - 1$, in parallel on to L sub-carriers. With multi-carrier spread spectrum systems, each data symbol is spread over L sub-carriers. In cases where the number of sub-carriers N_c of one OFDM symbol is equal to the spreading code length L, the OFDM symbol duration with a multi-carrier spread spectrum including a guard interval results in

$$T_s' = T_g + LT_c. \tag{2.4}$$

In this case one data symbol per user is transmitted in one OFDM symbol.

2.1.2   Downlink Signal

In the synchronous downlink, it is computationally efficient to add the spread signals of the K users before the OFDM operation as depicted in Figure 2-2. The superposition of the K sequences $\mathbf{s}^{(k)}$ results in the sequence

$$\mathbf{s} = \sum_{k=0}^{K-1} \mathbf{s}^{(k)} = (S_0, S_1, \ldots, S_{L-1})^T. \tag{2.5}$$

An equivalent representation for $\mathbf{s}$ in the downlink is

$$\mathbf{s} = \mathbf{Cd}, \tag{2.6}$$

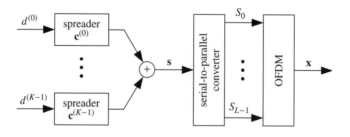

Figure 2-2   MC-CDMA downlink transmitter

where

$$\mathbf{d} = (d^{(0)}, d^{(1)}, \ldots, d^{(K-1)})^T \tag{2.7}$$

is the vector with the transmitted data symbols of the K active users and $\mathbf{C}$ is the spreading matrix given by

$$\mathbf{C} = (\mathbf{c}^{(0)}, \mathbf{c}^{(1)}, \ldots, \mathbf{c}^{(K-1)}). \tag{2.8}$$

The MC-CDMA downlink signal is obtained after processing the sequence $\mathbf{s}$ in the OFDM block according to Equation (1.33). By assuming that the guard time is long enough to absorb all echoes, the received vector of the transmitted sequence $\mathbf{s}$ after inverse OFDM and frequency de-interleaving is given by

$$\mathbf{r} = \mathbf{H}\mathbf{s} + \mathbf{n} = (R_0, R_1, \ldots, R_{L-1})^T, \tag{2.9}$$

where $\mathbf{H}$ is the $L \times L$ channel matrix and $\mathbf{n}$ is the noise vector of length L. The vector $\mathbf{r}$ is fed to the data detector in order to get a hard or soft estimate of the transmitted data. For the description of the multi-user detection techniques, an equivalent notation for the received vector $\mathbf{r}$ is introduced,

$$\mathbf{r} = \mathbf{A}\mathbf{d} + \mathbf{n} = (R_0, R_1, \ldots, R_{L-1})^T. \tag{2.10}$$

The system matrix $\mathbf{A}$ for the downlink is defined as

$$\mathbf{A} = \mathbf{H}\mathbf{C}. \tag{2.11}$$

2.1.3 Uplink Signal

In the uplink, the MC-CDMA transmit signal is obtained directly after processing the sequence $\mathbf{s}^{(k)}$ of user k in the OFDM block according to Equation (1.33). After inverse OFDM and frequency de-interleaving on the receiver side, the received vector is given by

$$\mathbf{r} = \sum_{k=0}^{K-1} \mathbf{H}^{(k)} \mathbf{s}^{(k)} + \mathbf{n} = (R_0, R_1, \ldots, R_{L-1})^T, \tag{2.12}$$

where $\mathbf{H}^{(k)}$ contains the coefficients of the sub-channels assigned to user k. The uplink is assumed to be synchronous in order to achieve the high spectral efficiency of OFDM. The vector $\mathbf{r}$ is fed to the data detector in order to get a hard or soft estimate of the transmitted data. The system matrix

$$\mathbf{A} = (\mathbf{a}^{(0)}, \mathbf{a}^{(1)}, \ldots, \mathbf{a}^{(K-1)}) \tag{2.13}$$

comprises K user-specific vectors

$$\mathbf{a}^{(k)} = \mathbf{H}^{(k)} \mathbf{c}^{(k)} = (H_{0,0}^{(k)} c_0^{(k)}, H_{1,1}^{(k)} c_1^{(k)}, \ldots, H_{L-1,L-1}^{(k)} c_{L-1}^{(k)})^T. \tag{2.14}$$

2.1.4 Spreading Techniques

The spreading techniques in MC-CDMA schemes differ in the selection of the spreading code and the type of spreading. Different strategies exist to map the spreading codes in time and frequency directions with MC-CDMA. Moreover, the constellation points of the transmitted signal can be improved by modifying the phase of the symbols to be distinguished by the spreading codes.

2.1.4.1 Spreading Codes

Various spreading codes exist which can be distinguished with respect to orthogonality, correlation properties, implementation complexity, and peak-to-average power ratio (PAPR). The selection of the spreading code depends on the given scenario. In the synchronous downlink, orthogonal spreading codes are of advantage, since they reduce the multiple access interference compared to nonorthogonal sequences. However, in the uplink, the orthogonality between the spreading codes gets lost due to different distortions of the individual codes. Thus, simple PN sequences can be chosen for spreading in the uplink. If the transmission is asynchronous, Gold codes have good cross-correlation properties. In cases where pre-equalization is applied in the uplink, orthogonality can be achieved at the receiver antenna, such that in the uplink orthogonal spreading codes can also be of advantage.

Moreover, the selection of the spreading code has influence on the PAPR of the transmitted signal (see Chapter 4). Especially in the uplink, the PAPR can be reduced by selecting, for example, Golay or Zadoff–Chu codes [8, 39, 40, 43, 56]. Spreading codes applicable in MC-CDMA systems are summarized in the following.

Walsh–Hadamard Codes

Orthogonal Walsh–Hadamard codes are simple to generate recursively by using the following Hadamard matrix generation,

$$C_L = \begin{bmatrix} C_{L/2} & C_{L/2} \\ C_{L/2} & -C_{L/2} \end{bmatrix}, \quad \forall L = 2^m, \quad m \geq 1, \quad C_1 = 1. \tag{2.15}$$

The maximum number of available orthogonal spreading codes is L, which determines the maximum number of active users K.

The Hadamard matrix generation described in Equation (2.15) can also be used to perform an L-ary Walsh–Hadamard modulation, which in combination with PN spreading can be applied in the uplink of an MC-CDMA system [14, 15].

Fourier Codes

The columns of an FFT matrix can also be considered as spreading codes, which are orthogonal to each other. The chips are defined as

$$c_l^{(k)} = e^{-j2\pi lk/L}. \tag{2.16}$$

Thus, if Fourier spreading is applied in MC-CDMA systems, the FFT for spreading and the IFFT for the OFDM operation cancels out if the FFT and IFFT have the same size;

i.e. the spreading is performed over all sub-carriers [7]. Thus, the resulting scheme is a single-carrier system with cyclic extension and frequency domain equalizer. This scheme has a dynamic range of single-carrier systems. The computational efficient implementation of the more general case where the FFT spreading is performed over groups of sub-carriers that are interleaved equidistantly is described in Reference [8]. A comparison of the amplitude distributions between Hadamard codes and Fourier codes shows that Fourier codes result in an equal or lower peak-to-average power ratio [9]. Fourier codes are applied for spreading in DFT-spread OFDM introduced in Chapter 3 and applied in the uplink of LTE.

Pseudo Noise (PN) Spreading Codes

The property of a PN sequence is that the sequence appears to be noise-like if the construction is not known at the receiver. They are typically generated by using shift registers. Often used PN sequences are maximum-length shift register sequences, known as m-sequences. A sequence has a length of

$$n = 2^m - 1 \tag{2.17}$$

bits and is generated by a shift register of length m with linear feedback [44]. The sequence has a period length of n and each period contains 2^{m-1} ones and $2^{m-1} - 1$ zeros; i.e. it is a balanced sequence.

Gold Codes

PN sequences with better cross-correlation properties than m-sequences are part of the so-called Gold sequences [44]. A set of n Gold sequences is derived from a preferred pair of m-sequences of length $L = 2^n - 1$ by taking the modulo-2 sum of the first preferred m-sequence with the n cyclically shifted versions of the second preferred m-sequence. By including the two preferred m-sequences, a family of $n + 2$ Gold codes is obtained. Gold codes have a three-valued cross-correlation function with values $\{-1, -t(m), t(m) - 2\}$, where

$$t(m) = \begin{cases} 2^{(m+1)/2} + 1 & \text{for } m \text{ odd} \\ 2^{(m+2)/2} + 1 & \text{for } m \text{ even} \end{cases} . \tag{2.18}$$

Golay Codes

Orthogonal Golay complementary codes can recursively be obtained by

$$\mathbf{C}_L = \begin{bmatrix} \mathbf{C}_{L/2} & \overline{\mathbf{C}}_{L/2} \\ \mathbf{C}_{L/2} & -\overline{\mathbf{C}}_{L/2} \end{bmatrix}, \quad \forall L = 2^m, \quad m \geq 1, \quad \mathbf{C}_1 = 1, \tag{2.19}$$

where the complementary matrix $\overline{\mathbf{C}}_L$ is defined by reverting the original matrix $\mathbf{C}_L$. If

$$\mathbf{C}_L = [\mathbf{A}_L \ \mathbf{B}_L], \tag{2.20}$$

and $\mathbf{A}_L$ and $\mathbf{B}_L$ are $L \times L/2$ matrices, then

$$\overline{\mathbf{C}}_L = [\mathbf{A}_L \ -\mathbf{B}_L]. \tag{2.21}$$

Zadoff-Chu Codes

The Zadoff–Chu codes have optimum correlation properties and are a special case of generalized chirp-like sequences. They are defined as

$$c_l^{(k)} = \begin{cases} e^{j2\pi k(ql+l^2/2)/L} & \text{for } L \text{ even} \\ e^{j2\pi k(ql+l(l+1)/2)/L} & \text{for } L \text{ odd} \end{cases}, \tag{2.22}$$

where q is any integer and k is an integer, prime with L. If L is a prime number, a set of Zadoff–Chu codes is composed of $L - 1$ sequences. Zadoff–Chu codes have an optimum periodic auto-correlation function and a low constant magnitude periodic cross-correlation function.

Low-Rate Convolutional Codes

Low-rate convolutional codes can be applied in CDMA systems as spreading codes with inherent coding gain [54]. These codes have been applied as an alternative to the use of a spreading code followed by a convolutional code. In MC-CDMA systems, low-rate convolutional codes can achieve good performance results for moderate numbers of users in the uplink [33, 35, 50]. The application of low-rate convolutional codes is limited to very moderate numbers of users since, especially in the downlink, signals are not orthogonal between the users, resulting in possibly severe multiple access interference. Therefore, they cannot reach the high spectral efficiency of MC-CDMA systems with separate coding and spreading.

2.1.4.2 Peak-to-Average Power Ratio (PAPR)

The variation of the envelope of a multi-carrier signal can be defined by the peak-to-average power ratio (PAPR), which is given by

$$PAPR = \frac{\max |x_v|^2}{\dfrac{1}{N_c} \displaystyle\sum_{v=0}^{N_c-1} |x_v|^2}. \tag{2.23}$$

The values x_v, $v = 0, \ldots, N_c - 1$, are the time samples of an OFDM symbol. An additional measure to determine the envelope variation is the crest factor (CF), which is

$$CF = \sqrt{PAPR}. \tag{2.24}$$

By appropriately selecting the spreading code, it is possible to reduce the PAPR of the multi-carrier signal [4, 40, 43]. This PAPR reduction can be of advantage in the uplink, where low power consumption is required in the terminal station.

Uplink PAPR

The uplink signal assigned to user k results in

$$x_v = x_v^{(k)}. \tag{2.25}$$

Table 2-1 PAPR bounds of MC-CDMA uplink
signals; $N_c = L$

Spreading code	PAPR
Walsh–Hadamard	$\leq 2L$
Golay	≤ 4
Zadoff–Chu	2
Gold	$\leq 2\left(t(m) - 1 - \dfrac{t(m) + 2}{L}\right)$

The PAPR for different spreading codes can be upper-bounded for the uplink by [39]

$$PAPR \leq \frac{2 \max\left\{\left|\sum_{l=0}^{L-1} c_l^{(k)} e^{j2\pi lt/T_s}\right|^2\right\}}{L}, \tag{2.26}$$

assuming that $N_c = L$. Table 2-1 summarizes the PAPR bounds for MC-CDMA uplink
signals with different spreading codes.

The PAPR bound for Golay codes and Zadoff–Chu codes is independent of the spreading code length. When N_c is a multiple of L, the PAPR of the Walsh–Hadamard code is
upper-bounded by $2N_c$.

Downlink PAPR

The time samples of a downlink multi-carrier symbol assuming synchronous transmission
are given as

$$x_v = \sum_{k=0}^{K-1} x_v^{(k)}. \tag{2.27}$$

The PAPR of an MC-CDMA downlink signal with K users and $N_c = L$ can be upper-
bounded by [39]

$$PAPR \leq \frac{2 \max\left\{\sum_{k=0}^{K-1}\left|\sum_{l=0}^{L-1} c_l^{(k)} e^{j2\pi lt/T_s}\right|^2\right\}}{L}. \tag{2.28}$$

2.1.4.3 One- and Two-Dimensional Spreading

Spreading in MC-CDMA systems can be carried out in the frequency direction, the time
direction or two-dimensional in the time and frequency direction. An MC-CDMA system
with spreading only in the time direction is equal to an MC-DS-CDMA system. Spreading
in two dimensions exploits time and frequency diversity and is a flexible and powerful
alternative to the conventional approach with spreading in the frequency or time direction
only. A two-dimensional spreading code is a spreading code of length L where the chips
are distributed in the time and frequency directions. Two-dimensional spreading can be

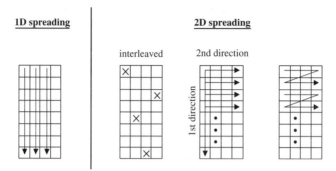

Figure 2-3   One- and two-dimensional spreading schemes

performed by a two-dimensional spreading code or by two cascaded one-dimensional spreading codes. An efficient realization of two-dimensional spreading is to use a one-dimensional spreading code followed by a two-dimensional interleaver, as illustrated in Figure 2-3 [26]. With two cascaded one-dimensional spreading codes, spreading is first carried out in one dimension with the first spreading code of length L_1. In the next step, the data-modulated chips of the first spreading code are again spread with the second spreading code in the second dimension. The length of the second spreading code is L_2. The total spreading length with two cascaded one-dimensional spreading codes results in

$$L = L_1 L_2. \tag{2.29}$$

If the two cascaded one-dimensional spreading codes are Walsh–Hadamard codes, the resulting two-dimensional code is again a Walsh–Hadamard code with total length L. For large L, two-dimensional spreading can outperform one-dimensional spreading in an uncoded MC-CDMA system [16, 46].

Two-dimensional spreading for maximum diversity gain is efficiently realized by using a sufficiently long spreading code with $L \geq D_O$, where D_O is the maximum achievable two-dimensional diversity (see Section 1.1.7). The spread sequence of length L has to be appropriately interleaved in time and frequency, such that all chips of this sequence are faded independently as far as possible.

Another approach with two-dimensional spreading is to locate the chips of the two-dimensional spreading code as close together as possible in order to get all chips similarly faded and, thus, preserve orthogonality of the spreading codes at the receiver as far as possible [3, 42]. Due to reduced multiple access interference, low complex receivers can be applied. However, the diversity gain due to spreading is reduced such that powerful channel coding is required. If the fading over all chips of a spreading code is flat, the performance of conventional OFDM without spreading is the lower bound for this spreading approach; i.e. the BER performance of an MC-CDMA system with two-dimensional spreading and Rayleigh fading which is flat over the whole spreading sequence results in the performance of OFDM with $L = 1$, shown in Figure 1-3. One- or two-dimensional spreading concepts with interleaving of the chips in time and/or frequency are lower-bounded by the diversity performance curves in Figure 1-3, which correspond to the chosen spreading code length L.

2.1.4.4 Rotated Constellations

With spreading codes like Walsh–Hadamard codes, the achievable diversity gain degrades if the signal constellation points of the resulting spread sequence **s** in the downlink concentrate their energy in less than L sub-channels, which in the worst case is only in one sub-channel while the signal on all other sub-channels is zero. Here we consider a full loaded scenario with $K = L$. The idea of rotated constellations [8] is to guarantee the existence of M^L distinct constellation points at each sub-carrier for a transmitted alphabet size of M and a spreading code length of L and to note that all points are nonzero. Thus, if all except one sub-channel are faded out, detection of all data symbols is still possible.

With rotated constellations, the L data symbols are rotated before spreading such that the data symbol constellations are different for each of the L data symbols of the transmit symbol vector **s**. This can be achieved by rotating the phase of the transmit symbol alphabet of each of the L spread data symbols by a fraction proportional to $1/L$. The rotation factor for user k is

$$r^{(k)} = e^{j2\pi k/(M_{rot} L)}, \tag{2.30}$$

where M_{rot} is a constant whose choice depends on the symbol alphabet. For example, $M_{rot} = 2$ for BPSK and $M_{rot} = 4$ for QPSK. For M-PSK modulation, the constant $M_{rot} = M$. The constellation points of the Walsh–Hadamard spread sequence **s** with BPSK modulation with and without rotation is illustrated in Figure 2-4 for a spreading code length of $L = 4$.

Spreading with rotated constellations can achieve better performance than the use of nonrotated spreading sequences. The performance improvements strongly depend on the chosen symbol mapping scheme. Large symbol alphabets reduce the degree of freedom for placing the points in a rotated signal constellation and decrease the gains. Moreover, the performance improvements with rotated constellations strongly depend on the chosen detection techniques. For higher order symbol mapping schemes, relevant performance improvements require the application of powerful multi-user detection techniques. The achievable performance improvements in SNR with rotated constellations can be in the order of several dB at a BER of 10^{-3} for an uncoded MC-CDMA system with QPSK in fading channels.

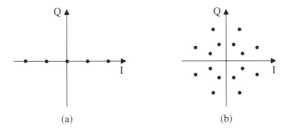

(a)   (b)

Figure 2-4 Constellation points after Hadamard spreading: (a) nonrotated and (b) rotated, both for BPSK and $L = 4$

2.1.5 Detection Techniques

Data detection techniques can be classified as either single-user detection or multi-user detection. The approach using single-user detection detects the user signal of interest by not taking into account any information about multiple access interference. In MC-CDMA mobile radio systems, single-user detection is realized by one tap equalization to compensate for the distortion due to flat fading on each sub-channel, followed by user-specific de-spreading. As in OFDM, the one tap equalizer is simply one complex-valued multiplication per sub-carrier. If the spreading code structure of the interfering signals is known, single-user detection is sub-optimum since the multiple access interference should not be considered in advance as noise-like. The sub-optimality of single-user detection can be overcome with multi-user detection where the *a priori* knowledge about the spreading codes of the interfering users is exploited in the detection process.

The performance improvements with multi-user detection compared to single-user detection are achieved at the expense of higher receiver complexity. The methods of multi-user detection can be divided into interference cancellation (IC) and joint detection. The principle of IC is to detect the information of the interfering users with single-user detection and to reconstruct the interfering contribution in the received signal before subtracting the interfering contribution from the received signal and detecting the information of the user of interest. The optimal detector applies joint detection with maximum likelihood detection. Since the complexity of maximum likelihood detection grows exponentially with the number of users, its use is limited in practice to applications with a small number of users. Simpler joint detection techniques can be realized by using block linear equalizers.

An MC-CDMA receiver in the terminal station of user k is depicted in Figure 2-5.

2.1.5.1 Single-User Detection

The principle of single-user detection is to detect the user signal of interest by not taking into account any information about the multiple access interference. A receiver with single-user detection of the data symbols of user k is shown in Figure 2-6.

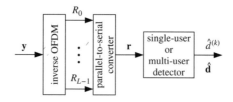

Figure 2-5   MC-CDMA receiver in the terminal station

Figure 2-6   MC-CDMA single-user detection

After inverse OFDM the received sequence **r** is equalized by employing a bank of adaptive one tap equalizers to combat the phase and amplitude distortions caused by the mobile radio channel on the sub-channels. The one tap equalizer is simply realized by one complex-valued multiplication per sub-carrier. The received sequence at the output of the equalizer has the form

$$\mathbf{u} = \mathbf{Gr} = (U_0, U_1, \ldots, U_{L-1})^T. \tag{2.31}$$

The diagonal equalizer matrix

$$\mathbf{G} = \begin{pmatrix} G_{0,0} & 0 & \cdots & 0 \\ 0 & G_{1,1} & & 0 \\ \vdots & & \ddots & \vdots \\ 0 & 0 & \cdots & G_{L-1,L-1} \end{pmatrix} \tag{2.32}$$

of dimension $L \times L$ represents the L complex-valued equalizer coefficients of the sub-carriers assigned to **s**. The complex-valued output **u** of the equalizer is de-spread by correlating it with the conjugate complex user-specific spreading code $\mathbf{c}^{(k)*}$. The complex-valued soft decided value at the output of the despreader is

$$v^{(k)} = \mathbf{c}^{(k)*} \mathbf{u}^T. \tag{2.33}$$

The hard decided value of a detected data symbol is given by

$$\hat{d}^{(k)} = Q\{v^{(k)}\}, \tag{2.34}$$

where $Q\{\cdot\}$ is the quantization operation according to the chosen data symbol alphabet.

The term equalizer is generalized in the following, since the processing of the received vector **r** according to typical diversity combining techniques is also investigated using the single-user detection scheme shown in Figure 2-6.

In the uplink **G** and **H** are user-specific.

Maximum Ratio Combining (MRC)

MRC weights each sub-channel with its respective conjugate complex channel coefficient, leading to

$$G_{l,l} = H_{l,l}^*, \tag{2.35}$$

where $H_{l,l}, l = 0, \ldots, L - 1$, are the diagonal components of **H**. The drawback of MRC in MC-CDMA systems in the downlink is that it destroys the orthogonality between spreading codes and, thus, additionally enhances the multiple access interference. In the uplink, MRC is the most promising single-user detection technique since the spreading codes do not superpose in an orthogonal fashion at the receiver and maximization of the signal-to-interference ratio is optimized.

Equal Gain Combining (EGC)

EGC compensates only for the phase rotation caused by the channel by choosing the equalization coefficients as

$$G_{l,l} = \frac{H_{l,l}^*}{|H_{l,l}|}. \tag{2.36}$$

EGC is the simplest single-user detection technique, since it only needs information about the phase of the channel.

Zero Forcing (ZF)

ZF applies channel inversion and can eliminate multiple access interference by restoring the orthogonality between the spread data in the downlink with an equalization coefficient chosen as

$$G_{l,l} = \frac{H_{l,l}^*}{|H_{l,l}|^2}. \tag{2.37}$$

The drawback of ZF is that the equalizer enhances noise, especially for small amplitudes of $H_{l,l}$.

Minimum Mean Square Error (MMSE) Equalization

Equalization according to the MMSE criterion minimizes the mean square value of the error

$$\varepsilon_l = S_l - G_{l,l} R_l \tag{2.38}$$

between the transmitted signal and the output of the equalizer. The mean square error

$$J_l = E\{|\varepsilon_l|^2\} \tag{2.39}$$

can be minimized by applying the orthogonality principle, stating that the mean square error J_l is minimum if the equalizer coefficient $G_{l,l}$ is chosen such that the error ε_l is orthogonal to the received signal R_l^*, i.e.

$$E\{\varepsilon_l R_l^*\} = 0. \tag{2.40}$$

The equalization coefficient based on the MMSE criterion for MC-CDMA systems results in

$$G_{l,l} = \frac{H_{l,l}^*}{|H_{l,l}|^2 + \sigma^2}. \tag{2.41}$$

The computation of the MMSE equalization coefficients requires knowledge about the actual variance of the noise σ^2. For SNR $\to \infty$, the MMSE equalizer becomes identical to the ZF equalizer. To overcome the additional complexity for the estimation of σ^2, a low complex sub-optimum MMSE equalization can be realized [24].

With sub-optimum MMSE equalization, the equalization coefficients are designed such that they perform optimally only in the most critical cases for which successful transmission should be guaranteed. The variance σ^2 is set equal to a threshold λ at which the optimal MMSE equalization guarantees the maximum acceptable BER. The equalization coefficient with sub-optimal MMSE equalization results in

$$G_{l,l} = \frac{H_{l,l}^*}{|H_{l,l}|^2 + \lambda} \tag{2.42}$$

and requires only information about $H_{l,l}$. The value λ has to be determined during the system design.

A controlled equalization can be applied in the receiver, which performs slightly worse than sub-optimum MMSE equalization [26]. Controlled equalization applies zero forcing on sub-carriers where the amplitude of the channel coefficients exceeds a predefined threshold a_{th}. All other sub-carriers apply equal gain combining in order to avoid noise amplification.

2.1.5.2 Multi-User Detection

Maximum Likelihood Detection

The optimum multi-user detection technique exploits the maximum *a posteriori* (MAP) criterion or the maximum likelihood criterion respectively. In this section, two optimum maximum likelihood detection algorithms are shown, namely the maximum likelihood sequence estimation (MLSE), which optimally estimates the transmitted data sequence $\mathbf{d} = (d^{(0)}, d^{(1)}, \ldots, d^{(K-1)})^T$, and the maximum likelihood symbol-by-symbol estimation (MLSSE), which optimally estimates the transmitted data symbol $d^{(k)}$. It is straightforward that both algorithms can be extended to a MAP sequence estimator and to a MAP symbol-by-symbol estimator by taking into account the *a priori* probability of the transmitted sequence and symbol respectively. When all possible transmitted sequences and symbols respectively are equally probable *a priori*, the estimator based on the MAP criterion and the one based on the maximum likelihood criterion are identical. The possible transmitted data symbol vectors are $\mathbf{d}_\mu, \mu = 0, \ldots, M^K - 1$, where M^K is the number of possible transmitted data symbol vectors and M is the number of possible realizations of $d^{(k)}$.

Maximum Likelihood Sequence Estimation (MLSE)
MLSE minimizes the sequence error probability, i.e. the data symbol vector error probability, which is equivalent to maximizing the conditional probability $P\{\mathbf{d}_\mu | \mathbf{r}\}$ that $\mathbf{d}_\mu$ was transmitted given the received vector $\mathbf{r}$. The estimate of $\mathbf{d}$ obtained with MLSE is

$$\hat{\mathbf{d}} = \arg \max_{\mathbf{d}_\mu} P\{\mathbf{d}_\mu | \mathbf{r}\}, \tag{2.43}$$

with arg denoting the argument of the function. If the noise N_l is additive white Gaussian, Equation (2.43) is equivalent to finding the data symbol vector $\mathbf{d}_\mu$ that minimizes the squared Euclidean distance

$$\Delta^2(\mathbf{d}_\mu, \mathbf{r}) = ||\mathbf{r} - \mathbf{A}\mathbf{d}_\mu||^2 \tag{2.44}$$

between the received and all possible transmitted sequences. The most likely transmitted data vector is

$$\hat{\mathbf{d}} = \arg \min_{\mathbf{d}_\mu} \Delta^2(\mathbf{d}_\mu, \mathbf{r}). \tag{2.45}$$

MLSE requires the evaluation of M^K squared Euclidean distances for the estimation of the data symbol vector $\hat{\mathbf{d}}$.

Maximum Likelihood Symbol-by-Symbol Estimation (MLSSE)

MLSSE minimizes the symbol error probability, which is equivalent to maximizing the conditional probability $P\{d_\mu^{(k)}|\mathbf{r}\}$ that $d_\mu^{(k)}$ was transmitted given the received sequence $\mathbf{r}$. The estimate of $d^{(k)}$ obtained by MLSSE is

$$\hat{d}^{(k)} = \arg \max_{d_\mu^{(k)}} P\{d_\mu^{(k)}|\mathbf{r}\}. \tag{2.46}$$

If the noise N_l is additive white Gaussian the most likely transmitted data symbol is

$$\hat{d}^{(k)} = \arg \max_{d_\mu^{(k)}} \sum_{\substack{\forall \mathbf{d}_\mu \text{ with same} \\ \text{realization of } d_\mu^{(k)}}} \exp\left(-\frac{1}{\sigma^2}\Delta^2(\mathbf{d}_\mu, \mathbf{r})\right). \tag{2.47}$$

The increased complexity with MLSSE compared to MLSE can be observed in the comparison of (2.47) with (2.45). An advantage of MLSSE compared to MLSE is that MLSSE inherently generates reliability information for detected data symbols that can be exploited in a subsequent soft decision channel decoder.

Block Linear Equalizer

The block linear equalizer is a sub-optimum, low complex multi-user detector which requires knowledge about the system matrix $\mathbf{A}$ in the receiver. Two criteria can be applied to use this knowledge in the receiver for data detection.

Zero Forcing Block Linear Equalizer

Joint detection applying a zero forcing block linear equalizer delivers at the output of the detector the soft decided data vector

$$\mathbf{v} = (\mathbf{A}^H\mathbf{A})^{-1}\mathbf{A}^H\mathbf{r} = (v^{(0)}, v^{(1)}, \ldots, v^{(K-1)})^T, \tag{2.48}$$

where $(\cdot)^H$ is the Hermitian transposition.

MMSE Block Linear Equalizer

An MMSE block linear equalizer delivers at the output of the detector the soft decided data vector

$$\mathbf{v} = (\mathbf{A}^H\mathbf{A} + \sigma^2\mathbf{I})^{-1}\mathbf{A}^H\mathbf{r} = (v^{(0)}, v^{(1)}, \ldots, v^{(K-1)})^T. \tag{2.49}$$

Hybrid combinations of block linear equalizers and interference cancellation schemes (see the next sub-section) are possible, resulting in block linear equalizers with decision feedback.

Interference Cancellation

The principle of interference cancellation is to detect and subtract interfering signals from the received signal before detection of the signal of interest. It can be applied to reduce intra-cell and inter-cell interference. Most detection schemes focus on intra-cell interference, which will be further discussed in this sub-section. Interference cancellation schemes

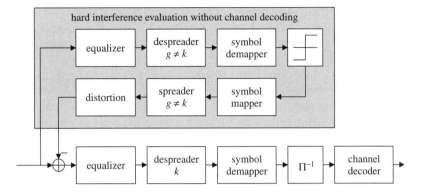

Figure 2-7 Hard interference cancellation scheme

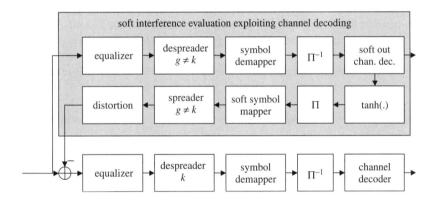

Figure 2-8 Soft interference cancellation scheme

can use signals for reconstruction of the interference either obtained at the detector output (see Figure 2-7) or at the decoder output (see Figure 2-8).

Both schemes can be applied in several iterations. Values and functions related to the iteration j are marked by an index $[j]$, where j may take on the values $j = 1, \ldots, J_{it}$ and J_{it} is the total number of iterations. The initial detection stage is indicated by the index $[0]$. Since the interference is detected more reliably at the output of the channel decoder than at the output of the detector, the scheme with channel decoding included in the iterative process outperforms the other scheme. Interference cancellation distinguishes between parallel and successive cancellation techniques. Combinations of parallel and successive interference cancellations are also possible.

Parallel Interference Cancellation
The principle of parallel interference cancellation is to detect and subtract all interfering signals in parallel before detection of the wanted signal. Parallel interference cancellation is suitable for systems where the interfering signals have similar power. In the

initial detection stage, the data symbols of all K active users are detected in parallel by single-user detection, i.e.

$$\hat{d}^{(k)[0]} = Q\{\mathbf{c}^{(k)*}\,\mathbf{G}^{(k)[0]}\,\mathbf{r}^T\}, \, k = 0, \ldots, K-1, \tag{2.50}$$

where $\mathbf{G}^{(k)[0]}$ denotes the equalization coefficients assigned to the initial stage. The following detection stages work iteratively by using the decisions of the previous stage to reconstruct the interfering contribution in the received signal. The obtained interference is subtracted, i.e. cancelled from the received signal, and the data detection is performed again with reduced multiple access interference. Thus, the second and further detection stages apply

$$\hat{d}^{(k)[j]} = Q\left\{\mathbf{c}^{(k)*}\mathbf{G}^{(k)[j]}\left(\mathbf{r} - \sum_{\substack{g \neq k \\ g=0}}^{K-1} \mathbf{H}^{(g)}\,d^{(g)[j-1]}\,\mathbf{c}^{(g)}\right)^T\right\}, \quad j = 1, \ldots, J_{it}, \tag{2.51}$$

where, except for the final stage, the detection has to be applied for all K users.

Parallel interference cancellation can be applied with different detection strategies in the iterations. Starting with EGC in each iteration [17, 18] various combinations have been proposed [6, 25, 30]. Promising results are obtained with MMSE equalization adapted in the first iteration to the actual system load and in all further iterations to MMSE equalization adapted to the single-user case [24]. The application of MRC seems theoretically to be of advantage for the second and further detection stages, since MRC is the optimum detection technique in the multiple access interference free case, i.e. in the single-user case. However, if one or more decision errors are made, MRC has a poor performance [25].

Successive Interference Cancellation
Successive interference cancellation detects and subtracts the interfering signals in the order of the interfering signal power. First, the strongest interferer is cancelled, before the second strongest interferer is detected and subtracted, i.e.

$$\hat{d}^{(k)[j]} = Q\{\mathbf{c}^{(k)*}\mathbf{G}^{(k)[j]}\,(\mathbf{r} - \mathbf{H}^{(g)}(d^{(g)[j-1]}\mathbf{c}^{(g)}))^T\}, \tag{2.52}$$

where g is the strongest interferer in the iteration $j, j = 1, \ldots, J_{it}$. This procedure is continued until a predefined stop criterion applies. Successive interference cancellation is suitable for systems with large power variations between the interferers [6].

Soft Interference Cancellation
Interference cancellation can use reliability information about the detected interference in the iterative process. These schemes can be without [41] and with [21, 28] channel decoding in the iterative process, and are termed soft interference cancellation. If reliability information about the detected interference is taken into account in the cancellation scheme, the performance of the iterative scheme can be improved since error propagation can be reduced compared to schemes with hard decided feedback. The block diagram of an MC-CDMA receiver with soft interference cancellation is illustrated in Figure 2-8.

The data of the desired user k are detected by applying interference cancellation with reliability information. Before detection of user k's data in the lowest path of Figure 2-8 with an appropriate single-user detection technique, the contributions of the $K - 1$ interfering users g, $g = 0, \ldots, K - 1$ with $g \neq k$, are detected with single-user detection and subtracted from the received signal. The principle of parallel or successive interference cancellation or combinations of both can be applied within a soft interference cancellation scheme.

In the following, we focus on the contribution of the interfering user g with $g \neq k$. The soft decided values $\mathbf{w}^{(g)[j]}$ are obtained after single-user detection, symbol de-mapping, and de-interleaving. The corresponding log-likelihood ratios (LLRs) for channel decoding are given by the vector $\mathbf{l}^{(g)[j]}$. LLRs are the optimum soft decided values which can be exploited in a Viterbi decoder (see Section 2.1.8). From the subsequent soft-in/soft-out channel decoder, besides the output of the decoded source bits, reliability information in the form of LLRs of the coded bits can be obtained. These LLRs are given by the vector

$$\mathbf{l}_{out}^{(g)\,[j]} = (\Gamma_{0,out}^{(g)\,[j]}, \Gamma_{1,out}^{(g)\,[j]}, \ldots, \Gamma_{L_b-1,out}^{(g)\,[j]})^T. \qquad (2.53)$$

In contrast to the LLRs of the coded bits at the input of the soft-in/soft-out channel decoder, the LLRs of the coded bits at the output of the soft-in/soft-out channel decoder

$$\Gamma_{\kappa,out}^{(g)\,[j]} = \ln \left(\frac{P\{b_\kappa^{(g)} = +1|\mathbf{w}^{(g)\,[j]}\}}{P\{b_\kappa^{(g)} = -1|\mathbf{w}^{(g)\,[j]}\}} \right), \quad \kappa = 0, \ldots, L_b - 1, \qquad (2.54)$$

are the estimates of all the other soft decided values in the sequence $\mathbf{w}^{(g)[j]}$ about this coded bit, and not only of one received soft decided value $w_\kappa^{(g)[j]}$. For brevity, the index κ is omitted since the focus is on the LLR of one coded bit in the sequel. To avoid error propagation, the average value of coded bit $b^{(g)}$ is used, which is the so-called soft bit $w_{out}^{(g)[j]}$ [21]. The soft bit is defined as

$$\begin{aligned} w_{out}^{(g)[j]} &= E\{b^{(g)}|\mathbf{w}^{(g)[j]}\} \\ &= (+1)P\{b^{(g)} = +1|\mathbf{w}^{(g)[j]}\} + (-1)P\{b^{(g)} = -1|\mathbf{w}^{(g)[j]}\}. \end{aligned} \qquad (2.55)$$

With Equation (2.54), the soft bit results in

$$w_{out}^{(g)\,[j]} = \tanh\left(\frac{\Gamma_{out}^{(g)\,[j]}}{2}\right). \qquad (2.56)$$

The soft bit $w_{out}^{(g)\,[j]}$ can take on values in the interval $[-1, +1]$. After interleaving, the soft bits are soft symbol mapped such that the reliability information included in the soft bits is not lost. The obtained complex-valued data symbols are spread with the user-specific spreading code and each chip is pre-distorted with the channel coefficient assigned to the sub-carrier on which the chip has been transmitted. The total reconstructed multiple access interference is subtracted from the received signal $\mathbf{r}$. After canceling the interference, the data of the desired user k are detected using single-user detection. However, in contrast to the initial detection stage, in further stages the equalizer coefficients given by the matrix $\mathbf{G}^{(k)[j]}$ and the LLRs given by the vector $\mathbf{l}^{(k)[j]}$ after soft interference cancellation are adapted to the quasi multiple access interference-free case.

2.1.6   Pre-Equalization

If information about the actual channel is *a priori* known at the transmitter, pre-equalization can be applied at the transmitter such that the signal at the receiver appears nondistorted and an estimation of the channel at the receiver is not necessary. Information about the channel state can, for example, be made available in TDD schemes if the TDD slots are short enough such that the channel of an uplink and a subsequent downlink slot can be considered as constant and the transceiver can use the channel state information obtained from previously received data.

An application scenario of pre-equalization in a TDD mobile radio system would be that the terminal station sends pilot symbols in the uplink which are used in the base station for channel estimation and detection of the uplink data symbols. The estimated channel state is used for pre-equalization of the downlink data to be transmitted to the terminal station. Thus, no channel estimation is necessary in the terminal station which reduces its complexity. Only the base station has to estimate the channel, i.e. the complexity can be shifted to the base station.

A further application scenario of pre-equalization in a TDD mobile radio system would be that the base station sends pilot symbols in the downlink to the terminal station, which performs channel estimation. In the uplink, the terminal station applies pre-equalization with the intention to get quasi-orthogonal user signals at the base station receiver antenna. This results in a high spectral efficiency in the uplink, since MAI can be avoided. Moreover, complex uplink channel estimation is not necessary.

The accuracy of pre-equalization can be increased by using prediction of the channel state in the transmitter where channel state information from the past is filtered.

Pre-equalization is performed by multiplying the symbols on each sub-channel with an assigned pre-equalization coefficient before transmission [10, 23, 37, 45, 47]. The selection criteria for the equalization coefficients is to compensate the channel fading as far as possible, such that the signal at the receiver antenna seems to be only affected by AWGN. In Figure 2-9, an OFDM transmitter with pre-equalization is illustrated which results with a spreading operation in an MC-SS transmitter.

2.1.6.1   Downlink

In a multi-carrier system in the downlink (e.g. SS-MC-MA) the pre-equalization operation is given by

$$\bar{s} = \overline{G}\,s,\tag{2.57}$$

where the source symbols S_l before pre-equalization are represented by the vector s and $\overline{G}$ is the diagonal $L \times L$ pre-equalization matrix with elements $\overline{G}_{l,l}$. In the case of spreading,

Figure 2-9   OFDM or MC-SS transmitter with pre-equalization

L corresponds to the spreading code length and in the case of OFDM (OFDMA, MC-TDMA), L is equal to the number of sub-carriers N_c. The pre-equalized sequence $\bar{s}$ is fed to the OFDM operation and transmitted.

In the receiver, the signal after inverse OFDM operation results in

$$\mathbf{r} = \mathbf{H}\bar{\mathbf{s}} + \mathbf{n}$$
$$= \mathbf{H}\overline{\mathbf{G}}\mathbf{s} + \mathbf{n} \tag{2.58}$$

where $\mathbf{H}$ represents the channel matrix with the diagonal components $H_{l,l}$ and $\mathbf{n}$ represents the noise vector. It can be observed from Equation (2.58) that by choosing

$$\overline{G}_{l,l} = \frac{1}{H_{l,l}} \tag{2.59}$$

the influence of the fading channel can be compensated and the signal is only disturbed by AWGN. In practice, this optimum technique cannot be realized since this would require transmission with very high power on strongly faded sub-channels. Thus, in the following section we focus on pre-equalization with a power constraint where the total transmission power with pre-equalization is equal to the transmission power without pre-equalization [37].

The condition for pre-equalization with power constraint is

$$\sum_{l=0}^{L-1} |\overline{G}_{l,l}\, S_l|^2 = \sum_{l=0}^{L-1} |S_l|^2. \tag{2.60}$$

When assuming that all symbols S_l are transmitted with the same power, the condition for pre-equalization with power constraint becomes

$$\sum_{l=0}^{L-1} |\overline{G}_{l,l}|^2 = \sum_{l=0}^{L-1} |G_{l,l}\, C|^2 = L, \tag{2.61}$$

where $G_{l,l}$ is the pre-equalization coefficient without power constraint and C is a normalizing factor that keeps the transmit power constant. The factor C results in

$$C = \sqrt{\frac{L}{\sum_{l=0}^{L-1} |G_{l,l}|^2}}. \tag{2.62}$$

By applying the equalization criteria introduced in Section 2.1.5.1, the following pre-equalization coefficients are obtained.

Maximum Ratio Transmission (MRT)

$$\overline{G}_{l,l} = H_{l,l}^* \sqrt{\frac{L}{\sum_{n=0}^{L-1} |H_{n,n}|^2}}. \tag{2.63}$$

Equal Gain Transmision (EGT)

$$\overline{G}_{l,l} = \frac{H_{l,l}^*}{|H_{l,l}|}. \tag{2.64}$$

Zero Forcing (ZF)

$$\overline{G}_{l,l} = \frac{H_{l,l}^*}{|H_{l,l}|^2} \sqrt{\frac{L}{\sum_{n=0}^{L-1} \frac{1}{|H_{n,n}|^2}}}. \tag{2.65}$$

Quasi Minimum Mean Square Error (MMSE) Pre-Equalization

$$\overline{G}_{l,l} = \frac{H_{l,l}^*}{|H_{l,l}|^2 + \sigma^2} \sqrt{\frac{L}{\sum_{n=0}^{L-1} \left| \frac{H_{n,n}^*}{|H_{n,n}|^2 + \sigma^2} \right|^2}}. \tag{2.66}$$

We call this technique quasi MMSE pre-equalization, since this is an approximation. The optimum technique requires a very high computational complexity, due to the power constraint condition.

As with the single-user detection techniques presented in Section 2.1.5.1, controlled pre-equalization can be applied. Controlled pre-equalization applies zero forcing pre-equalization on sub-carriers where the amplitude of the channel coefficients exceeds a predefined threshold a_{th}. All other sub-carriers apply equal gain transmission for pre-equalization.

2.1.6.2   Uplink

In an MC-CDMA uplink scenario, pre-equalization is performed in the terminal station of user k according to

$$\overline{\mathbf{s}}^{(k)} = \overline{\mathbf{G}}^{(k)} \mathbf{s}^{(k)}. \tag{2.67}$$

The received signal at the base station after inverse OFDM operation results in

$$\mathbf{r} = \sum_{k=0}^{K-1} \mathbf{H}^{(k)} \overline{\mathbf{s}}^{(k)} + \mathbf{n}$$

$$= \sum_{k=0}^{K-1} \mathbf{H}^{(k)} \overline{\mathbf{G}}^{(k)} \mathbf{s}^{(k)} + \mathbf{n}. \tag{2.68}$$

The pre-equalization techniques presented in Equations (2.63) to (2.66) are applied in the uplink individually for each terminal station; i.e. $\overline{G}_{l,l}^{(k)}$ and $H_{l,l}^{(k)}$ have to be applied instead of $\overline{G}_{l,l}$ and $H_{l,l}$ respectively.

Finally, knowledge about the channel in the transmitter can be exploited, not only to perform pre-equalization but also to apply adaptive modulation per sub-carrier in order to increase the capacity of the system (see Chapter 4).

2.1.7   Combined Equalization

With combined equalization [11, 12] channel state information is available at both the transmitter and receiver. This enables pre-equalization to be applied at the transmitter together with post-equalization at the receiver. In the following, three different combined equalization techniques are analyzed. The first technique is referred to as MRT–MRC combined equalization and is based on the combination of the MRT principle at the transmitter and the MRC principle at the receiver. The second technique is called selection diversity combined equalization and is the optimal single-user combined-equalization technique in the sense of BER minimization. The third technique is actually a class of techniques that represent a certain tradeoff between MRT–MRC combined equalization and SD combined equalization.

2.1.7.1   MRT–MRC Combined Equalization

Both MRC at the receiver and MRT at the transmitter are optimal in the single-user case. The assigned pre-equalization coefficients for MRT are given in Equation (2.63). In order to maximize the SNR at the receiver, the post-equalization coefficients must be adapted to the additional distortion caused by pre-equalization. The post-equalization coefficients result in

$$G_{l,l} = |H_{l,l}|^2 \sqrt{\frac{L}{\sum\limits_{n=0}^{L-1} |H_{n,n}|^2}}. \tag{2.69}$$

2.1.7.2   Selection Diversity Combined Equalization

The optimal combined equalization technique in the single-user case is based on the application of the selection diversity criterion at the transmitter and receiver. Selection diversity is the optimal technique in the sense of SNR maximization for the case of combined equalization within MC-CDMA. Selection diversity redistributes the transmission power such that only a single sub-carrier bears the complete transmission signal. The pre-equalization coefficients are given by

$$\overline{G}_{l,l} \begin{cases} \sqrt{L} & l = \arg\max_n (|H_{n,n}|), \quad n = 0, 1, \ldots, L-1 \\ 0 & \text{otherwise} \end{cases} \tag{2.70}$$

while the selection diversity post-equalization coefficients are chosen according to

$$G_{l,l} \begin{cases} \dfrac{H_{l,l}^*}{|H_{l,l}|}. & l = \arg\max_n (|H_{n,n}|), \quad n = 0, 1, \ldots, L-1 \\ 0 & \text{otherwise} \end{cases} \tag{2.71}$$

The function arg max $y(x)$ returns the value x that maximizes the function $y(x)$. The factor $\sqrt{L}$ is chosen in order to keep the transmit power the same, as in the case without pre-equalization. The phase correction of the fading coefficients can be performed on any of the two sides, transmitter or receiver. Here, phase correction is performed at the receiver. A formal proof of the optimality of the selection diversity combined equalization can be found in Reference [12].

2.1.7.3 Combined Equalization Based on Generalized Pre-Equalization

Since selection diversity combined equalization is only optimal in the single-user case, generalized pre-equalization [12] has been proposed as the combined equalization technique for the multi-user case. The coefficients of generalized pre-equalization are given by

$$\overline{G}_{l,l} = |H_{l,l}|^p H_{l,l}^* \sqrt{\frac{L}{\displaystyle\sum_{n=0}^{L-1} |H_{n,n}|^{2p+2}}}. \tag{2.72}$$

Generalized pre-equalization is a unified approach for pre-equalization and comprises several well-known pre-equalization techniques, such as MRT ($p = 0$), EGT ($p = -1$), and pre-equalization zero-forcing ($p = -2$). Setting $p > 0$ allows more power to be invested on strong sub-carriers and less on weak sub-carriers compared to MRT. The post-equalization coefficients results in

$$G_{l,l} = |H_{l,l}|^{p+2} \sqrt{\frac{L}{\displaystyle\sum_{n=0}^{L-1} |H_{n,n}|^{2p+2}}}. \tag{2.73}$$

For $p \to \infty$, generalized pre-equalization becomes equivalent to selection diversity combined equalization.

In the single-user case, the problem of BER minimization is equivalent to SNR maximization as the only impairment is SNR-degradation caused by the fading channel. In contrast to that, in the multi-user case, in addition to SNR-degradation, multiple access interference occurs. Therefore, the BER minimization is more complex as it depends on both SNR and multiple access interference at the receiver. In the BER minimization concept proposed in Reference [11], pre-equalization is employed to distribute the transmit power efficiently among the available subcarriers rather than as a method to cancel the multiple access interference. Generalized pre-equalization is applied at the transmitter, where p is a design parameter that is chosen to achieve a reasonable tradeoff between efficient allocation of the transmit power with respect to SNR maximization and multiple access interference mitigation capabilities. Residual interference will be present at the receiver when $p \neq -2$. Thus, some form of post-equalization is required to counteract the detrimental effect of multiple access interference. An efficient solution to cope with multiple access interference is to apply soft interference cancellation (see Section 2.1.5.2) at the receiver.

2.1.8 Soft Channel Decoding

Channel coding with bit interleaving is an efficient technique to combat degradations due to fading, noise, interference, and other channel impairments. The basic idea of channel coding is to introduce controlled redundancy into the transmitted data that is exploited at the receiver to correct channel-induced errors by means of forward error correction (FEC). Binary convolutional codes are chosen as channel codes in current mobile radio, digital

broadcasting, WLAN, and WLL systems, since very simple decoding algorithms based
on the Viterbi algorithm exist that can achieve a soft decision decoding gain. Moreover,
convolutional codes are used as component codes for Turbo codes, which have become
part of 3 G mobile radio standards. A detailed channel coding description is given in
Chapter 4.

Many of the convolutional codes that have been developed for increasing the reliability
in the transmission of information are effective when errors caused by the channel are
statistically independent. Signal fading due to time-variant multi-path propagation often
causes the signal to fall below the noise level, thus resulting in a large number of errors
called burst errors. An efficient method for dealing with burst error channels is to interleave
the coded bits in such a way that the bursty channel is transformed into a channel with
independent errors. Thus, a code designed for independent errors or short bursts can be
used. Code bit interleaving has become an extremely useful technique in 2 G and 3 G
digital cellular systems, and can, for example, be realized as a block, diagonal, or random
interleaver.

A block diagram of channel encoding and user-specific spreading in an MC-CDMA
transmitter assigned to user k is shown in Figure 2-10. The block diagram is the same
for up- and downlinks. The input sequence of the convolutional encoder is represented
by the source bit vector

$$\mathbf{a}^{(k)} = (a_0^{(k)}, a_1^{(k)}, \ldots, a_{L_a-1}^{(k)})^T \qquad (2.74)$$

of length L_a. The code word is the discrete time convolution of $\mathbf{a}^{(k)}$ with the impulse
response of the convolutional encoder. The memory M_c of the code determines the complexity of the convolutional decoder, given by 2^{M_c} different memory realizations, also
called states, for binary convolutional codes. The output of the channel encoder is a coded
bit sequence of length L_b, which is represented by the coded bit vector

$$\mathbf{b}^{(k)} = (b_0^{(k)}, b_1^{(k)}, \ldots, b_{L_b-1}^{(k)})^T. \qquad (2.75)$$

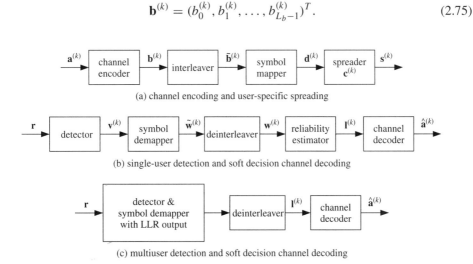

(a) channel encoding and user-specific spreading

(b) single-user detection and soft decision channel decoding

(c) multiuser detection and soft decision channel decoding

Figure 2-10   Channel encoding and decoding in MC-CDMA systems

The channel code rate is defined as the ratio

$$R = \frac{L_a}{L_b}.$$
(2.76)

The interleaved coded bit vector $\tilde{\mathbf{b}}^{(k)}$ is passed to a symbol mapper, where $\tilde{\mathbf{b}}^{(k)}$ is mapped into a sequence of L_d complex-valued data symbols, i.e.

$$\mathbf{d}^{(k)} = (d_0^{(k)}, d_1^{(k)}, \dots, b_{L_d-1}^{(k)})^T.$$
(2.77)

A data symbol index κ, $\kappa = 0, \dots, L_d - 1$, is introduced to distinguish the different data symbols $d_\kappa^{(k)}$ assigned to $\mathbf{d}^{(k)}$. Each data symbol is multiplied with the spreading code $\mathbf{c}^{(k)}$ according to Equation (2.3) and processed as described in Section 2.1.

With single-user detection, the L_d soft decided values at the output of the detector are given by the vector

$$\mathbf{v}^{(k)} = (v_0^{(k)}, v_1^{(k)}, \dots, v_{L_d-1}^{(k)})^T.$$
(2.78)

The L_d complex-valued, soft decided values of $\mathbf{v}^{(k)}$ assigned to the data symbols of $\mathbf{d}^{(k)}$ are mapped on to L_b real-valued, soft decided values represented by $\tilde{\mathbf{w}}^{(k)}$ assigned to the coded bits of $\tilde{\mathbf{b}}^{(k)}$. The output of the symbol de-mapper after de-interleaving is written as the vector

$$\mathbf{w}^{(k)} = (w_0^{(k)}, w_1^{(k)}, \dots, w_{L_b-1}^{(k)})^T.$$
(2.79)

Based on the vector $\mathbf{w}^{(k)}$, LLRs of the detected coded bits are calculated. The vector

$$\mathbf{l}^{(k)} = (\Gamma_0^{(k)}, \Gamma_1^{(k)}, \dots, \Gamma_{L_b-1}^{(k)})^T$$
(2.80)

of length L_b represents the LLRs assigned to the transmitted coded bit vector $\mathbf{b}^{(k)}$. Finally, the sequence $\mathbf{l}^{(k)}$ is soft decision-decoded by applying the Viterbi algorithm. At the output of the channel decoder, the detected source bit vector

$$\hat{\mathbf{a}}^{(k)} = (\hat{a}_0^{(k)}, \hat{a}_1^{(k)}, \dots, \hat{a}_{L_a-1}^{(k)})^T$$
(2.81)

is obtained.

Before presenting the coding gains of different channel coding schemes applied in MC-CDMA systems, the calculation of LLRs in fading channels is given generally for multi-carrier modulated transmission systems. Based on this introduction, the LLRs for MC-CDMA systems are derived. The LLRs for MC-CDMA systems with single-user detection and with joint detection are in general applicable for the up- and downlink. In the uplink, only the user index $^{(k)}$ has to be assigned to the individual channel fading coefficients of the corresponding users.

2.1.8.1 Log-Likelihood Ratio for OFDM Systems

The LLR is defined as

$$\Gamma = \ln \left(\frac{p(w \mid b = +1)}{p(w \mid b = -1)} \right),$$

(2.82)

which is the logarithm of the ratio between the likelihood functions $p(w|b = +1)$ and $p(w|b = -1)$. The LLR can take on values in the interval $[-\infty, +\infty]$. With flat fading $H_{l,l}$ on the sub-carriers and in the presence of AWGN, the log-likelihood ratio for OFDM systems results in

$$\Gamma = \frac{4| H_{l,l} |}{\sigma^2} w.$$

(2.83)

2.1.8.2 Log-Likelihood Ratio for MC-CDMA Systems

Since in MC-CDMA systems a coded bit $b^{(k)}$ is transmitted in parallel on L sub-carriers, where each sub-carrier may be affected by both independent fading and multiple access interference, the LLR for OFDM systems is not applicable for MC-CDMA systems. The LLR for MC-CDMA systems is presented in the next sub-section.

Single-User Detection

A received MC-CDMA data symbol after single-user detection results in the soft decided value

$$v^{(k)} = \sum_{l=0}^{L-1} C_l^{(k)*} G_{l,l} \left(H_{l,l} \sum_{g=0}^{K-1} d^{(g)} C_l^{(g)} + N_l \right)$$

$$= d^{(k)} \underbrace{\sum_{l=0}^{L-1} \left| C_l^{(k)} \right|^2 G_{l,l} H_{l,l}}_{\text{desired symbol}} + \underbrace{\sum_{\substack{g \neq k \\ g=0}}^{K-1} d^{(g)} \sum_{l=0}^{L-1} G_{l,l} H_{l,l} C_l^{(g)} C_l^{(k)*}}_{\text{MAI}} + \underbrace{\sum_{l=0}^{L-1} N_l G_{l,l} C_l^{(k)*}}_{\text{noise}}.$$

(2.84)

Since a frequency interleaver is applied, the L complex-valued fading factors $H_{l,l}$ affecting $d^{(k)}$ can be assumed to be independent. Thus, for sufficiently long spreading codes, the multiple access interference (MAI) can be considered to be additive zero-mean Gaussian noise according to the central limit theorem. The noise term can also be considered as additive zero-mean Gaussian noise. The attenuation of the transmitted data symbol $d^{(k)}$ is the magnitude of the sum of the equalized channel coefficients $G_{l,l} H_{l,l}$ of the L sub-carriers used for the transmission of $d^{(k)}$, weighted with $|C_l^{(k)}|^2$. The symbol de-mapper delivers the real-valued soft decided value $w^{(k)}$. According to Equation (2.83), the LLR for MC-CDMA systems can be calculated as

$$\Gamma^{(k)} = \frac{2 \left| \sum_{l=0}^{L-1} |C_l^{(k)}|^2 G_{l,l} H_{l,l} \right|}{\sigma_{MAI}^2 + \sigma_{noise}^2} w^{(k)}.$$

(2.85)

Since the variances are assigned to real-valued noise, $w^{(k)}$ is multiplied by a factor of 2. When applying Walsh–Hadamard codes as spreading codes, the property can be exploited that the product $C_l^{(g)} C_l^{(k)*}$, $l = 0, \ldots, L-1$, in half of the cases equals -1 and in the other half equals $+1$ if $g \neq k$. Furthermore, when assuming that the realizations $b^{(k)} = +1$ and $b^{(k)} = -1$ are equally probable, the LLR for MC-CDMA systems with single-user detection results in [27, 29]

$$\Gamma^{(k)} = \frac{2 \left| \sum\limits_{l=0}^{L-1} G_{l,l} H_{l,l} \right|}{(K-1) \left(\dfrac{1}{L} \sum\limits_{l=0}^{L-1} |G_{l,l} H_{l,l}|^2 - \left| \dfrac{1}{L} \sum\limits_{l=0}^{L-1} G_{l,l} H_{l,l} \right|^2 \right) + \dfrac{\sigma^2}{2} \sum\limits_{l=0}^{L-1} |G_{l,l}|^2} \, w^{(k)}. \quad (2.86)$$

When MMSE equalization is used in MC-CDMA systems, Equation (2.86) can be approximated by [27]

$$\Gamma^{(k)} = \frac{4}{L\sigma^2} \sum_{l=0}^{L-1} |H_{l,l}| w^{(k)}, \quad (2.87)$$

since the variance of $G_{l,l} H_{l,l}$ reduces such that only the noise remains relevant.

The gain with soft decision decoding compared to hard decision decoding in MC-CDMA systems with single-user detection depends on the spreading code length and is in the order of 4 dB for small L (e.g. $L = 8$) and reduces to 3 dB with increasing L (e.g. $L = 64$). This shows the effect that the spreading averages the influence of the fading on a data symbol. When using LLRs instead of the soft decided information $w^{(k)}$, the performance further improves up to 1 dB [26].

Maximum Likelihood Detection

The LLR for coded MC-CDMA mobile radio systems with joint detection based on MLSSE is given by

$$\Gamma^{(k)} = \ln \left(\frac{p(\mathbf{r} \mid b^{(k)} = +1)}{p(\mathbf{r} \mid b^{(k)} = -1)} \right) \quad (2.88)$$

and is inherently delivered in the symbol-by-symbol estimation process presented in Section 2.1.5.2. The set of all possible transmitted data vectors $\mathbf{d}_\mu$ where the considered coded bit $b^{(k)}$ of user k is equal to $+1$ is denoted by $D_+^{(k)}$. The set of all possible data vectors where $b^{(k)}$ is equal to -1 is denoted by $D_-^{(k)}$. The LLR for MC-CDMA systems with MLSSE results in [27, 29]

$$\Gamma^{(k)} = \ln \left(\frac{\sum\limits_{\forall \mathbf{d}_\mu \in D_+^{(k)}} \exp \left(-\dfrac{1}{\sigma^2} \Delta^2 (\mathbf{d}_\mu, \mathbf{r}) \right)}{\sum\limits_{\forall \mathbf{d}_\mu \in D_-^{(k)}} \exp \left(-\dfrac{1}{\sigma^2} \Delta^2 (\mathbf{d}_\mu, \mathbf{r}) \right)} \right), \quad (2.89)$$

where $\Delta^2(\mathbf{d}_\mu, \mathbf{r})$ is the squared Euclidean distance according to Equation (2.44).

For coded MC-CDMA systems with joint detection based on MLSE, the sequence estimation process cannot provide reliability information on the detected, coded bits. However, an approximation for the LLR with MLSE is given by [19]

$$\Gamma^{(k)} \approx \frac{1}{\sigma^2} (\Delta^2 (\mathbf{d}_{\mu-}, \mathbf{r}) - \Delta^2 (\mathbf{d}_{\mu+}, \mathbf{r})). \tag{2.90}$$

The indices μ_- and μ_+ mark the smallest squared Euclidean distances $\Delta^2(\mathbf{d}_{\mu-}, \mathbf{r})$ and $\Delta^2(\mathbf{d}_{\mu+}, \mathbf{r})$, where $b^{(k)}$ is equal to -1 and $b^{(k)}$ is equal to $+1$ respectively.

Interference Cancellation

MC-CDMA receivers using interference cancellation exploit the LLRs derived for single-user detection in each detection stage, where in the second and further stages the term representing the multiple access interference in the LLRs can approximately be set to zero.

2.1.9 Flexibility in System Design

The MC-CDMA signal structure introduced in Section 2.1.1 enables the realization of powerful receivers with low complexity due to the avoidance of ISI and ICI in the detection process. Moreover, the spreading code length L has not necessarily to be equal to the number of sub-carriers N_c in an MC-CDMA system, which enables a flexible system design and can further reduce the complexity of the receiver. The three MC-CDMA system modifications presented in the following are referred to as M-Modification, Q-Modification, and $M\&Q$-Modification [18, 19, 26]. These modifications can be applied in the up- and downlinks of a mobile radio system.

2.1.9.1 Parallel Data Symbols (M-Modification)

As depicted in Figure 2-11, the M-Modification increases the number of sub-carriers N_c while maintaining constant the overall bandwidth B, the spreading code length L, and the

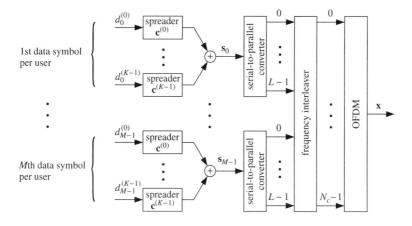

Figure 2-11 M-Modification

maximum number of active users K. The OFDM symbol duration increases and the loss in spectral efficiency due to the guard interval decreases with the M-Modification. Moreover, the tighter sub-carrier spacing enables one to guarantee flat fading per sub-channel in propagation scenarios with a small coherence bandwidth. With the M-Modification, each user transmits simultaneously $M > 1$ data symbols per OFDM symbol.

The total number of sub-carriers of the modified MC-CDMA system is

$$N_c = ML. \tag{2.91}$$

Each user exploits the total of N_c sub-carriers for data transmission. The OFDM symbol duration (including the guard interval) increases to

$$T_s' = T_g + MLT_c, \tag{2.92}$$

where it can be observed that the loss in spectral efficiency due to the guard interval decreases with increasing M. The maximum number of active users is still $K = L$.

The data symbol index $m, m = 0, \ldots, M - 1$, is introduced in order to distinguish the M simultaneously transmitted data symbols $d_m^{(k)}$ of user k. The number M is upper-limited by the coherence time $(\Delta t)_c$ of the channel. To exploit frequency diversity optimally, the components of the sequences $\mathbf{s}_m, m = 0, \ldots, M - 1$, transmitted in the same OFDM symbol, are interleaved over the frequency. The interleaving is carried out prior to OFDM.

2.1.9.2   Parallel User Groups (Q-Modification)

With an increasing number of active users K, the number of required spreading codes and, thus, the spreading code length L increase. Since L and K determine the complexity of the receiver, both values have to be kept as small as possible. The Q-Modification introduces an OFDMA component (see Chapter 3) on sub-carrier level and with that reduces the receiver complexity by reducing the spreading code length per user, while maintaining constant the maximum number of active users K and the number of sub-carriers N_c. The MC-CDMA transmitter with Q-Modification is shown in Figure 2-12

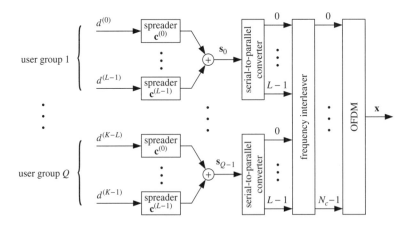

Figure 2-12 Q-Modification

where Q different user groups transmit simultaneously in one OFDM symbol. Each user group has a specific set of sub-carriers for transmission, which avoids interference between different user groups. Assuming that each user group applies spreading codes of length L, the total number of sub-carriers is

$$N_c = QL,$$ (2.93)

where each user exploits a subset of L sub-carriers for data transmission. Depending on the coherence bandwidth $(\Delta f)_c$ of the channel, it can be sufficient to apply spreading codes with $L \ll N_c$ to obtain the full diversity gain [20, 26].

To exploit the frequency diversity of the channel optimally, the components of the spread sequences $s_q, q = 0, \ldots, Q - 1$, transmitted in the same OFDM symbol are interleaved over the frequency. The interleaving is carried out prior to OFDM. The OFDM symbol duration (including the guard interval) is

$$T_s' = T_g + QLT_c.$$ (2.94)

Only one set of L spreading codes of length L is required within the whole MC-CDMA system. This set of spreading codes can be used in each subsystem. An adaptive sub-carrier allocation can also increase the capacity of the system [2, 13].

2.1.9.3 *M&Q*-Modification

The $M\&Q$-Modification combines the flexibility of the M- and the Q-Modification. The transmission of M data symbols per user and, additionally, the splitting of the users in Q independent user groups according to the $M\&Q$-Modification is illustrated in Figure 2-13.

The total number of sub-carriers used is

$$N_c = MQL,$$ (2.95)

where each user only exploits a subset of ML sub-carriers for data transmission due to the OFDMA component introduced by Q-Modification. The total OFDM symbol duration (including the guard interval) results in

$$T_s' = T_g + MQLT_c.$$ (2.96)

A frequency interleaver scrambles the information of all sub-systems prior to OFDM to guarantee an optimum exploitation of the frequency diversity offered by the mobile radio channel.

The M-, Q-, and $M\&Q$-Modifications are also suitable for the uplink of an MC-CDMA mobile radio system. For Q- and $M\&Q$-Modifications in the uplink only the inputs of the frequency interleaver of the user group of interest are connected in the transmitter; all other inputs are set to zero.

Finally, it should be noted that an MC-CDMA system with its basic implementation or with any of the three modifications presented in this section could support an additional TDMA component in the up- and downlink, since the transmission is synchronized on OFDM symbols.

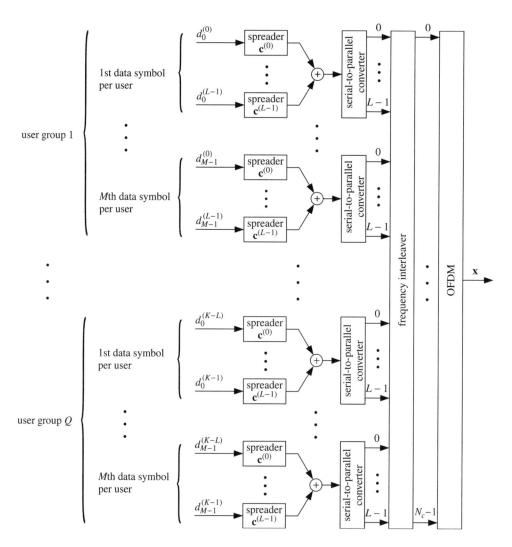

Figure 2-13 $M\&Q$-Modification

2.1.10 *Performance Analysis*

2.1.10.1 **System Parameters**

The parameters of the MC-CDMA system analyzed in this section are summarized in Table 2-2. Orthogonal Walsh–Hadamard codes are used for spreading. The spreading code length in a subsystem is $L = 8$. Unless otherwise stated, cases with fully loaded systems are considered. QPSK, 8-PSK, and 16-QAM with Gray encoding are applied for data symbol mapping. Moreover, the guard interval of the reference system is chosen such that ISI and ICI are eliminated. The mobile radio channel is implemented as an uncorrelated Rayleigh fading channel, described in detail in Section 1.1.6.

Table 2-2   MC-CDMA system parameters

Parameter	Value/characteristics
Spreading codes	Walsh–Hadamard codes
Spreading code length L	8
System load	Fully loaded
Symbol mapping	QPSK, 8-PSK, 16-QAM
FEC codes	Convolutional codes with memory 6
FEC code rate R and FEC decoder	4/5, 2/3, 1/2, 1/3 with Viterbi decoder
Channel estimation and synchronization	Perfect
Mobile radio channel	Uncorrelated Rayleigh fading channel

The performance of the MC-CDMA reference system presented in this section is applicable to any MC-CDMA system with an arbitrary transmission bandwidth B, an arbitrary number of sub-systems Q, and an arbitrary number of data symbols M transmitted per user in an OFDM symbol, resulting in an arbitrary number of sub-carriers. The number of sub-carriers within a sub-system has to be 8, while the amplitudes of the channel fading have to be Rayleigh-distributed and have to be uncorrelated on the sub-carriers of a sub-system due to appropriate frequency interleaving. The loss in SNR due to the guard interval is not taken into account in the results. The intention is that the loss in SNR due to the guard interval can be calculated individually for each specified guard interval. Therefore, the results presented can be adapted to any guard interval.

2.1.10.2   Synchronous Downlink

The BER versus the SNR per bit for the single-user detection techniques MRC, EGC, ZF, and MMSE equalization in an MC-CDMA system without FEC coding is depicted in Figure 2-14. The results show that with a fully loaded system the MMSE equalization outperforms the other single-user detection techniques. ZF equalization restores the orthogonality between the user signals and avoids multiple access interference. However, it introduces noise amplification. EGC avoids noise amplification but does not counteract the multiple access interference caused by the loss of the orthogonality between the user signals, resulting in a high error floor. The worst performance is obtained with MRC, which additionally enhances the multiple access interference. As reference, the matched filter bound (lower bound) for the MC-CDMA system is given. Analytical approaches to evaluate the performance of MC-CDMA systems with MRC and EGC are given in Reference [55], with ZF equalization in Reference [51] and with MMSE equalization in [25].

Figure 2-15 shows the BER versus the SNR per bit for the multi-user detection techniques with parallel interference cancellation, MLSE, and MLSSE applied in an MC-CDMA system without FEC coding. The performance of parallel interference cancellation with adapted MMSE equalization is presented for two detection stages. The significant performance improvements with parallel interference cancellation are obtained after the

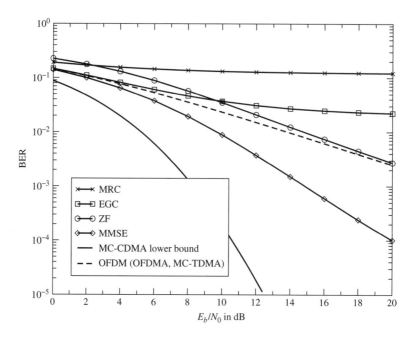

Figure 2-14 BER versus SNR for MC-CDMA with different single-user detection techniques: fully loaded system; no FEC coding; QPSK; Rayleigh fading

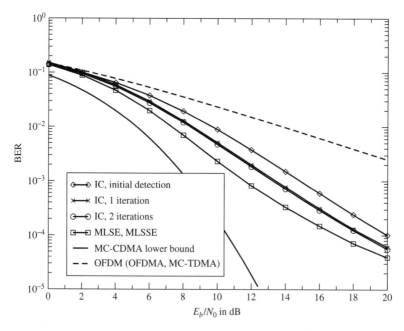

Figure 2-15 BER versus SNR for MC-CDMA with different multi-user detection techniques: fully loaded system; no FEC coding; QPSK; Rayleigh fading

first iteration. The optimum joint detection techniques MLSE and MLSSE perform almost identically and outperform the other detection techniques. The SNR-degradation with the optimum detection techniques compared to the matched filter bound (lower bound) is caused by the superposition of orthogonal Walsh–Hadamard codes, resulting in sequences of length L that can contain up to $L-1$ zeros. Sequences with many zeros perform worse in a fading channel due to the reduced diversity gain. These diversity losses can be reduced by applying rotated constellations, as described in Section 2.1.4.4. An upper bound of the BER for MC-CDMA systems applying joint detection with MLSE and MLSSE for the uncorrelated Rayleigh channel is derived in Reference [19] and for the uncorrelated Rice fading channel in Reference [25]. Analytical approaches to determine the performance of MC-CDMA systems with interference cancellation are shown in References [25] and [30].

The FEC coded BER versus the SNR per bit for single-user detection with MRC, EGC, ZF, and MMSE equalization in MC-CDMA systems is presented in Figure 2-16. It can be observed that rate 1/2 coded OFDM (OFDMA, MC-TDMA) systems slightly outperform rate 1/2 coded MC-CDMA systems with MMSE equalization when considering cases with full system load in a single cell. Furthermore, the performance of coded MC-CDMA systems with simple EGC requires only an SNR of about 1 dB higher to reach the BER of 10^{-3} compared to more complex MC-CDMA systems with MMSE equalization. With a fully loaded system, the single-user detection technique MRC is not of interest in practice.

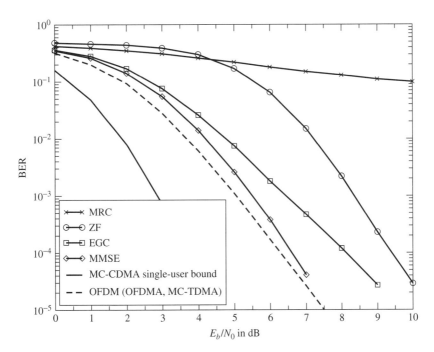

Figure 2-16   FEC coded BER versus SNR for MC-CDMA with different single-user detection techniques: fully loaded system; channel code rate $R = 1/2$; QPSK; Rayleigh fading

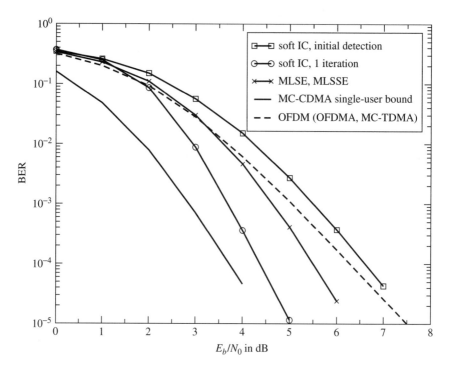

Figure 2-17   FEC coded BER versus SNR for MC-CDMA with different multi-user detection techniques: fully loaded system; channel code rate $R = 1/2$; QPSK; Rayleigh fading

The FEC coded BER versus the SNR per bit for multiuser detection with soft IC, MLSE, MLSSE, and single-user detection with MMSE equalization is shown in Figure 2-17 for code rate 1/2. Coded MC-CDMA systems with the soft IC detection technique outperform coded OFDM (OFDMA, MC-TDMA) systems and MC-CDMA systems with MLSE/MLSSE. The performance of the initial stage with soft IC is equal to the performance with MMSE equalization. Promising results are obtained with soft IC already after the first iteration.

The FEC coded BER versus the SNR per bit for different symbol mapping schemes in MC-CDMA systems with soft IC and in OFDM (OFDMA, MC-TDMA) systems is shown in Figure 2-18 for code rate 2/3. Coded MC-CDMA systems with the soft IC detection technique outperform coded OFDM (OFDMA, MC-TDMA) systems for all symbol mapping schemes at lower BERs due to the steeper slope obtained with MC-CDMA.

Finally, the spectral efficiency of MC-CDMA with soft IC and of OFDM (OFDMA, MC-TDMA) versus the SNR is shown in Figure 2-19. The results are given for the code rates 1/3, 1/2, 2/3, and 4/5 and are shown for a BER of 10^{-4}. The curves in Figure 2-19 show that MC-CDMA with soft IC can outperform OFDM (OFDMA, MC-TDMA).

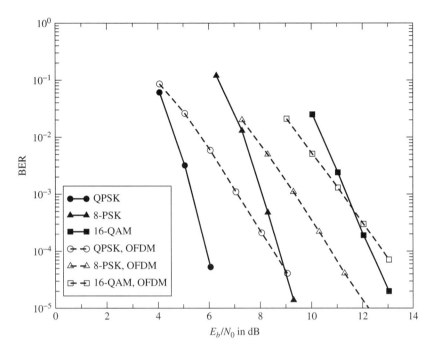

Figure 2-18   FEC coded BER versus SNR for MC-CDMA with different symbol mapping schemes: fully loaded system; channel code rate $R = 2/3$; Rayleigh fading

Figure 2-19 represents the most important results regarding spectral/power efficiency in a cellular system, which are in favor of MC-CDMA schemes. These curves lead to the following conclusions:

- For a given coverage, the transmitted data rate can be augmented by at least 40 % compared to MC-TDMA or OFDMA.
- Alternately, for a given data rate, about 2.5 dB can be gained in SNR. The 2.5 dB extension in power will give a higher coverage for an MC-CDMA system, or battery live extension.

2.1.10.3   Synchronous Uplink

The parameters used for the synchronous uplink are the same as for the downlink presented in the previous section. Orthogonal spreading codes outperform other codes such as Gold codes in the synchronous MC-CDMA uplink scenario, which motivates the choice of Walsh–Hadamard codes also in the uplink. Each user has an uncorrelated Rayleigh fading channel. Due to the loss of orthogonality of the spreading codes at the receiver antenna, MRC is the optimum single-user detection technique in the uplink (see

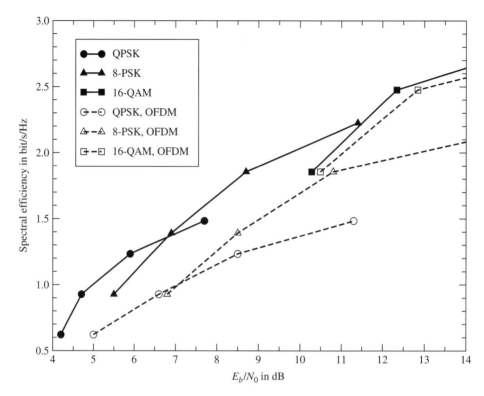

Figure 2-19   Spectral efficiency of MC-CDMA and OFDM (OFDMA, MC-TDMA): fully loaded system; Rayleigh fading; $BER = 10^{-4}$

Section 2.1.5.1). The performance of an MC-CDMA system with different loads and MRC in the synchronous uplink is shown in Figure 2-20. It can be observed that due to the loss of orthogonality between the user signals in the uplink only moderate numbers of active users can be handled with single-user detection.

The performance of MC-CDMA in the synchronous uplink can be significantly improved by applying multi-user detection techniques. Various concepts have been investigated in the literature. In the uplink, the performance of MLSE and MLSSE closely approximates the single-user bound (one user curve in Figure 2-20) since here the Walsh–Hadamard codes do not superpose orthogonally and the maximum diversity can be exploited [47]. The performance degradation of a fully loaded MC-CDMA system with MLSE/MLSSE compared to the single-user bound is about 1 dB in SNR.

Moreover, suboptimum multi-user detection techniques have also been investigated for MC-CDMA in the uplink, which benefit from reduced complexity in the receiver. Interference cancellation schemes are analyzed in References [1] and [32] and joint detection schemes in References [5] and [49].

To take advantage of MC-CDMA with nearly orthogonal user separation at the receiver antenna, the pre-equalization techniques presented in Section 2.1.6 can be applied. The parameters of the MC-CDMA system under investigation are presented in Table 2-3.

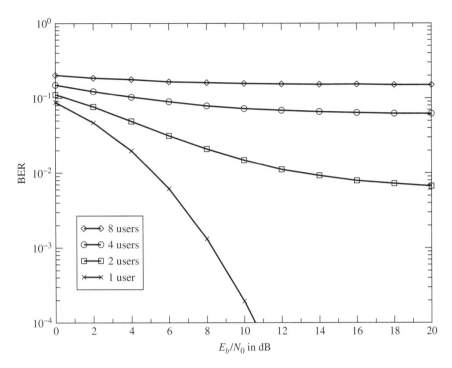

Figure 2-20   BER versus SNR for MC-CDMA in the synchronous uplink: MRC; no FEC coding; QPSK; $L = 8$; Rayleigh fading

Table 2-3   Parameters of the MC-CDMA uplink system with pre-equalization

Parameter	Value/characteristics
Bandwidth	20 MHz
Carrier frequency	5.2 GHz
Number of sub-carriers	256
Spreading codes	Walsh–Hadamard codes
Spreading code length L	16
Symbol mapping	QPSK
FEC coding	None
Mobile radio channel	Indoor fading channel with $T_g > \tau_{max}$
Maximum Doppler frequency	26 Hz

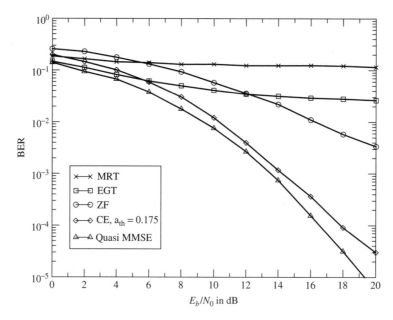

Figure 2-21   BER versus SNR for MC-CDMA with different pre-equalization techniques in the uplink: fully loaded system; no FEC coding

In Figure 2-21, the BER versus the SNR for an MC-CDMA system with different pre-equalization techniques in the uplink is shown. The system is fully loaded. It can be observed that with pre-equalization a fully loaded MC-CDMA system can achieve promising results in the uplink. Quasi MMSE pre-equalization and controlled pre-equalization with a threshold of $a_{th} = 0.175$ outperform other pre-equalization techniques.

When assuming that the information about the uplink channel for pre-equalization is only available at the beginning of each transmission frame, the performance of the system degrades with increasing frame duration due to the time variation of the channel. A typical scenario would be that at the beginning of each frame a feedback channel provides the transmitter with the required channel state information. Of importance for the selection of a proper frame duration is that it is smaller than the coherence time of the channel. The influence of the frame length for an MC-CDMA system with controlled pre-equalization is shown in Figure 2-22. The Doppler frequency is 26 Hz and the OFDM symbol duration is 13.6 μs.

In Figure 2-23, the performance of an MC-CDMA system with controlled pre-equalization and an update of the channel coefficients at the beginning of each OFDM frame is shown for different system loads. An OFDM frame consists of 200 OFDM symbols.

In the following, the performance of different combined equalization techniques in a multi-user MC-CDMA uplink is shown. The system is fully loaded, i.e. L users are active per subsystem. Walsh–Hadamard codes of length $L = 8$ are used. At the receiver, a soft interference canceller with three iterations is applied. The ETSI BRAN Channel E [36] is taken as the propagation channel. Figure 2-24 illustrates the BER versus SNR performance of a fully loaded rate 1/2 coded uplink MC-CDMA system.

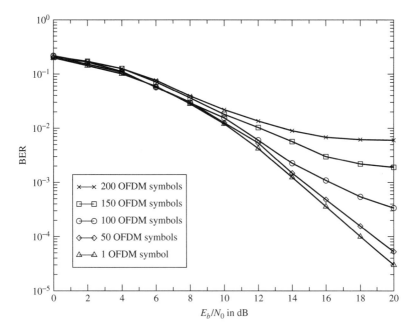

Figure 2-22 BER versus SNR for MC-CDMA with controlled pre-equalization and different frame lengths in the uplink: fully loaded system; no FEC coding

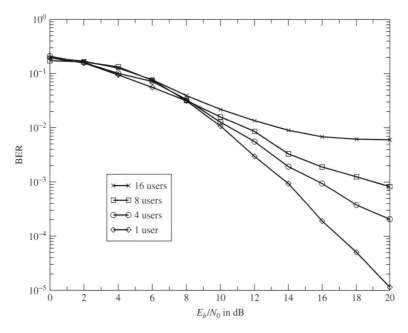

Figure 2-23 BER versus SNR for MC-CDMA with controlled pre-equalization and different system loads in the uplink: frame length is equal to 200 OFDM symbols; no FEC coding

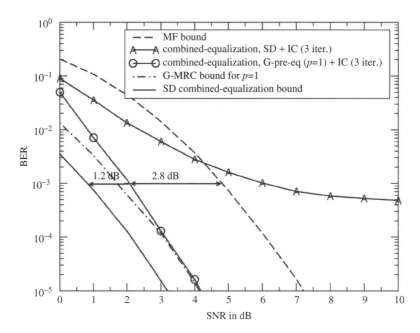

Figure 2-24 FEC coded BER versus SNR for MC-CDMA with combined equalization: fully loaded system; channel code rate $R = 1/2$; QPSK; $L = 8$; Rayleigh fading

At the transmitter, generalized pre-equalization (G-pre-eq) is applied with $p = 1$. The corresponding selection diversity (SD) and generalized equalization single-user bounds are given as references. As further reference, the matched filter performance is shown. Although the system is fully loaded, the performance gap towards the corresponding selection diversity combined equalization bounds is only around 1 dB at a BER of 10^{-3}. Compared to the conventional matched filter bound, a gain of nearly 3 dB is obtained at a BER of 10^{-3}. Additionally, the performance of a fully loaded coded uplink MC-CDMA system that applies selection diversity together with soft interference cancellation is shown.

2.2 MC-DS-CDMA

2.2.1 Signal Structure

The MC-DS-CDMA signal is generated by serial-to-parallel converting data symbols into N_c sub-streams and applying DS-CDMA on each individual sub-stream. With MC-DS-CDMA, each data symbol is spread in bandwidth within its sub-channel, but in contrast to MC-CDMA or DS-CDMA not over the whole transmission bandwidth for $N_c > 1$. An MC-DS-CDMA system with one sub-carrier is identical to a single-carrier DS-CDMA system. MC-DS-CDMA systems can be distinguished in systems where the sub-channels are narrowband and the fading per sub-channel appears flat and in systems with broadband sub-channels where the fading is frequency-selective per sub-channel. The fading over the whole transmission bandwidth can be frequency-selective in both cases. The complexity

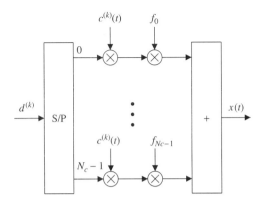

Figure 2-25   MC-DS-CDMA transmitter

of the receiver with flat fading per sub-channel is comparable to that of an MC-CDMA receiver, when OFDM is assumed for multi-carrier modulation. As soon as the fading per sub-channel is frequency-selective and ISI occurs, more complex detectors have to be applied. MC-DS-CDMA is of special interest for the asynchronous uplink of mobile radio systems, due to its close relation to asynchronous single-carrier DS-CDMA systems. On the one hand, a synchronization of users can be avoided, but, on the other hand, the spectral efficiency of the system decreases due to asynchronism.

Figure 2-25 shows the generation of a multi-carrier direct sequence spread spectrum signal. The data symbol rate is $1/T_d$. A sequence of N_c complex-valued data symbols $d_n^{(k)}, n = 0, \ldots, N_c - 1$, of user k is serial-to-parallel converted into N_c sub-streams. The data symbol rate on each sub-stream becomes $1/(N_c T_d)$. Within a single sub-stream, a data symbol is spread with the user-specific spreading code

$$c^{(k)}(t) = \sum_{l=0}^{L-1} c_l^{(k)} p_{T_c}(t - lT_s) \qquad (2.97)$$

of length L. The pulse form of the chips is given by $p_{Tc}(t)$. For a description of the MC-DS-CDMA signal, the continuous time representation is chosen, since MC-DS-CDMA systems are of interest for the asynchronous uplink. Here, OFDM might not necessarily be the best choice as multi-carrier modulation technique. The duration of a chip within a sub-stream is

$$T_c = T_s = \frac{N_c T_d}{L}. \qquad (2.98)$$

With multi-carrier direct sequence spread spectrum, each data symbol is spread over L multi-carrier symbols, each of duration T_s. The complex-valued sequence obtained after spreading is given by

$$x^{(k)}(t) = \sum_{n=0}^{N_c-1} d_n^{(k)} c^{(k)}(t) e^{j2\pi f_n t}, \quad 0 \le t < LT_s. \qquad (2.99)$$

The nth sub-carrier frequency is

$$f_n = \frac{(1 + \alpha)n}{T_s},$$

(2.100)

where $0 \leq \alpha \leq 1$. The choice of α depends on the chosen chip form $p_{Tc}(t)$ and is typically chosen such that the N_c parallel sub-channels are disjoint. In the case of OFDM, α is equal to 0 and $p_{Tc}(t)$ has a rectangular form.

A special case of MC-DS-CDMA systems is obtained when the sub-carrier spacing is equal to $1/(N_c T_s)$. The tight sub-carrier spacing allows the use of longer spreading codes to reduce multiple access interference; however, it results in an overlap of the signal spectra of the sub-carriers and introduces ICI. This special case of MC-DS-CDMA is referred to as multi-tone CDMA (MT-CDMA) [52]. An MT-CDMA signal is generated by first modulating a block of N_c data symbols on N_c sub-carriers applying OFDM before spreading the resulting signal with a code of length $N_c L$, where L is the spreading code length of conventional MC-DS-CDMA. Due to the N_c times increased sub-channel bandwidth with MT-CDMA, each sub-channel is broadband and more complex receivers are required.

2.2.2 Downlink Signal

In the synchronous downlink, the signals of K users are superimposed in the transmitter. The resulting transmitted MC-DS-CDMA signal is

$$x(t) = \sum_{k=0}^{K-1} x^{(k)}(t).$$

(2.101)

The signal received at a terminal station is given by

$$y(t) = x(t) \otimes h(t) + n(t).$$

(2.102)

Since the downlink can be synchronized, ISI and ICI can be avoided by choosing an appropriate guard interval. Moreover, if narrowband sub-channels are achieved in the system design, a low-complex implementation of an MC-DS-CDMA system exploiting the advantages of OFDM can be realized.

2.2.3 Uplink Signal

In the uplink, the MC-DS-CDMA signal transmitted by user k is $x^{(k)}(t)$. The channel output assigned to user k is given by the convolution of $x^{(k)}(t)$ with the channel impulse response $h^{(k)}(t)$, i.e.

$$y^{(k)}(t) = x^{(k)}(t) \otimes h^{(k)}(t).$$

(2.103)

The received signal of all K users at the base station including the additive noise signal $n(t)$ results in

$$y(t) = \sum_{k=0}^{K-1} y^{(k)}(t - \tau^{(k)}) + n(t).$$

(2.104)

The user-specific delay relative to the first arriving signal is given by $\tau^{(k)}$. If all users are synchronized, then $\tau^{(k)} = 0$ for all K users.

As soon as each sub-channel can be considered as narrowband, i.e. the sub-channel bandwidth is smaller than the coherence bandwidth $(\Delta f)_c$, the fading per sub-channel is frequency nonselective and low complex detection techniques compared to broadband sub-channels can be realized. Narrowband sub-channels are achieved by choosing a sufficiently large number of sub-carriers relative to the bandwidth B. A rough approximation for the minimum number of sub-carriers is given by

$$N_c \geq \tau_{\max} B. \qquad (2.105)$$

The overall transmission bandwidth is given by B and $\tau_{\max}$ is the maximum delay of the mobile radio channel.

2.2.4 Spreading

Since MC-DS-CDMA is of interest for the asynchronous uplink, spreading codes such as PN or Gold codes described in Section 2.1.4.1 are of interest for this scenario. As for asynchronous single-carrier DS-CDMA systems, good auto- and cross-correlation properties are required. In the case of a synchronous downlink, orthogonal codes are preferable.

2.2.5 Detection Techniques

MC-DS-CDMA systems with broadband sub-channels can be split into N_c classical broadband DS-CDMA systems. Thus, single- and multi-user detection techniques known for DS-CDMA can be applied in each data stream, which are described in Section 1.3.1.2 and in Reference [53]. Analyses for MC-DS-CDMA systems with broadband sub-channels and pre-rake diversity combining techniques have been carried out in References [22] and [38].

MC-DS-CDMA is typically applied for asynchronous uplink scenarios. The assigned single-user detector for MC-DS-CDMA with narrowband sub-channels can be realized by a spreading code correlator on each sub-channel. Figure 2-26 shows the single-user detector with N_c correlators.

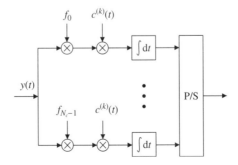

Figure 2-26   MC-DS-CDMA correlation detector

Modified MC-DS-CDMA systems transmit the same spread data symbol in parallel on p, $0 < p \leq N_c$, sub-carriers in order to achieve an additional diversity gain. However, this reduces the spectral efficiency of the system by the factor of p. In the receiver, the detected data of the p sub-channels can be combined with EGC or MRC [31, 48].

In the case of synchronous uplink and downlink transmissions with narrowband sub-channels, the same detection techniques described in Section 2.1.5 for MC-CDMA can be applied for MC-DS-CDMA.

2.2.6 Performance Analysis

The BER performance of an MC-DS-CDMA system in the synchronous and asynchronous uplink is analyzed in this section. The transmission bandwidth is 5 MHz and the carrier frequency is 5 GHz. Compared to the MC-CDMA parameters, longer spreading codes of length $L = 31$ with Gold codes and $L = 32$ with Walsh–Hadamard codes are chosen with MC-DS-CDMA due to spreading in the time direction. The number of sub-carriers is identical to the spreading code length, i.e. $N_c = L$. QPSK symbol mapping is chosen. No FEC coding scheme is applied. The mobile radio channels of the individual users are modeled by the COST 207 rural area (RA) channel model and the Doppler frequency is 1.15 kHz (250 km/h). A high Doppler frequency is of advantage for MC-DS-CDMA, since it offers high time diversity.

2.2.6.1 Synchronous Uplink

Figure 2-27 shows the BER versus the SNR per bit for an MC-DS-CDMA system with different spreading codes and detection techniques. The number of active users is 8. It can be observed that Walsh–Hadamard codes outperform Gold codes in the synchronous uplink. Moreover, the single-user detection technique MRC outperforms MMSE equalization. All curves show a quite high error floor due to multiple access interference.

The influence of the system load is presented in Figure 2-28 for a system with Walsh–Hadamard codes and MRC. The MC-DS-CDMA system can only handle moderate numbers of users in the synchronous uplink due to the limitation by the multiple access inference.

To overcome the limitation with single-user detection techniques in the uplink, more complex multi-user detection techniques have to be used, which have been analyzed, for example, in Reference [34] for MC-DS-CDMA in the synchronous uplink.

2.2.6.2 Asynchronous Uplink

The BER performance of an asynchronous MC-DS-CDMA in the uplink is shown in Figure 2-29 for different spreading codes and system loads. MRC has been chosen as the detection technique, since it is the optimum single-user detection technique in the uplink (see Section 2.1.5.1). It can be observed that Gold codes outperform Walsh–Hadamard codes due to better cross-correlation properties, which are of importance in the asynchronous case. The considered MC-DS-CDMA system shows a high error floor due to multiple access interference, which is higher than in the synchronous case.

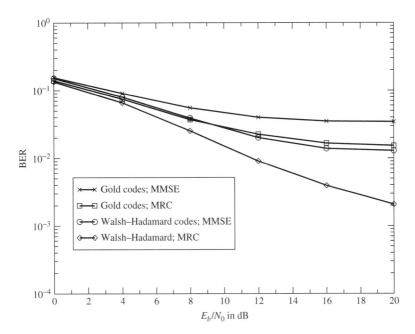

Figure 2-27   BER versus SNR for MC-DS-CDMA with different spreading codes and detection techniques: synchronous uplink; $K = 8$ users; QPSK; COST 207 RA fading channel

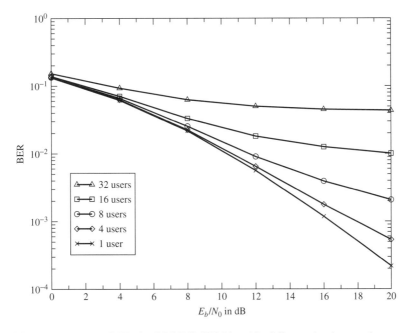

Figure 2-28   BER versus SNR for MC-DS-CDMA with different loads: synchronous uplink; Walsh–Hadamard codes; MRC; QPSK; COST 207 RA fading channel

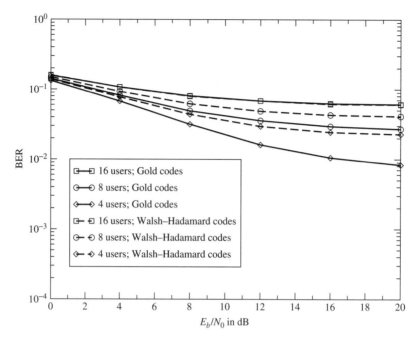

Figure 2-29   BER versus SNR for MC-DS-CDMA with different loads and spreading codes: asynchronous uplink; MRC; QPSK; COST 207 RA fading channel

These results show that MC-DS-CDMA in the asynchronous uplink needs more complex multi-user detectors to handle larger numbers of users. Investigations with interference cancellation have been carried out in Reference [17].

References

[1] Akther M. S., Asenstorfer J., Alexander P. D., and Reed M. C., "Performance of multi-carrier CDMA with iterative detection," in *Proc. IEEE International Conference on Universal Personal Communications (ICUPC '98)*, Florence, Italy, pp. 131–135, Oct. 1998.

[2] Al-Susa E. and Cruickshank D., "An adaptive orthogonal multicarrier multiuser CDMA technique for a broadband mobile communication system," in *Proc. International Workshop on Multi-Carrier Spread Spectrum and Related Topics (MC-SS 2001)*, Oberpfaffenhofen, Germany, pp. 45–52, Sept. 2001.

[3] Atarashi H., Maeda N., Abeta S., and Sawahashi M., "Broadband packet wireless access based on VSF-OFCDM and MC/DS-CDMA," in *Proc. IEEE International Symposium on Personal, Indoor and Mobile Radio Communications (PIMRC 2002)*, Lisbon, Portugal, pp. 992–997, Sept. 2002.

[4] Aue V. and Fettweis G., "Multi-carrier spread spectrum modulation with reduced dynamic range," in *Proc. IEEE Vehicular Technology Conference (VTC'96)*, Atlanta, USA, pp. 914–917, May/June 1996.

[5] Bader F., Zazo S., and Borrallo J. M., "Decorrelation MUD for MC-CDMA in an uplink transmission mode," in *Proc. International Workshop on Multi-Carrier Spread Spectrum and Related Topics (MC-SS 2001)*, Oberpfaffenhofen, Germany, pp. 173–180, Sept. 2001.

[6] Baudais J.-Y., Helard J.-F., and Citerne J., "An improved linear MMSE detection technique for multi-carrier CDMA systems: comparison and combination with interference cancellation schemes," *European Transactions on Telecommunications (ETT)*, vol. 11, pp. 547–554, Nov./Dec. 2000.

[7] Brüninghaus K. and Rohling H., "On the duality of multi-carrier spread spectrum and single-carrier transmission," in *Proc. International Workshop on Multi-Carrier Spread Spectrum (MC-SS '97)*, Oberpfaffenhofen, Germany, pp. 187–194, April 1997.

[8] Bury A., *Efficient Multi-Carrier Spread Spectrum Transmission*, Düsseldorf: VDI-Verlag, Fortschritt-Berichte VDI, series 10, no. 685, 2001, PhD thesis.

[9] Bury A. and Lindner J., "Comparison of amplitude distributions for Hadamard spreading and Fourier spreading in multi-carrier code division multiplexing," in *Proc. IEEE Global Telecommunications Conference (GLOBECOM 2000)*, San Francisco, USA, pp. 857–860, Nov./Dec. 2000.

[10] Cosovic I. and Kaiser S., "Link adaptation for uplink MC-CDMA systems," *European Transactions on Telecommunications (ETT)*, vol. 17, pp. 671–683, Nov./Dec. 2006.

[11] Cosovic I. and Kaiser S., "A unified analysis of diversity exploitation in multicarrier CDMA," *IEEE Transactions on Vehicular Technology*, vol. 56, pp. 2051–2062, July 2007.

[12] Cosovic I., Schnell M., and Springer A., "Combined equalization for coded uplink MC-CDMA in Rayleigh fading channels," *IEEE Transactions on Communications*, vol. 53, pp. 1609–1615, Oct. 2005.

[13] Costa E., Haas H., and Schulz E., "Optimization of capacity assignment in MC-CDMA transmission systems," in *Proc. International Workshop on Multi-Carrier Spread Spectrum and Related Topics (MC-SS 2001)*, Oberpfaffenhofen, Germany, pp. 217–224, Sept. 2001.

[14] Dekorsy A. and Kammeyer K.-D., "A new OFDM-CDMA uplink concept with M-ary orthogonal modulation," *European Transactions on Telecommunications (ETT)*, vol. 10, pp. 377–389, July/Aug. 1999.

[15] Dekorsy A. and Kammeyer K.-D., "Serial code concatenation with complex valued Walsh–Hadamard codes applied to OFDM-CDMA," in *Proc. International Workshop on Multi-Carrier Spread Spectrum and Related Topics (MC-SS 2001)*, Oberpfaffenhofen, Germany, pp. 131–138, Sept. 2001.

[16] Egle J., Reinhardt M., and Lindner J., "Equalization and coding for extended MC-CDMA over time and frequency selective channels," in *Proc. International Workshop on Multi-Carrier Spread Spectrum (MC-SS '97)*, Oberpfaffenhofen, Germany, pp. 127–134, April 1997.

[17] Fang L. and Milstein L. B., "Successive interference cancellation in multicarrier DS/CDMA," *IEEE Transactions on Communications*, vol. 48, pp. 1530–1540, Sept. 2000.

[18] Fazel K., "Performance of CDMA/OFDM for mobile communication system," in *Proc. IEEE International Conference on Universal Personal Communications (ICUPC '93)*, Ottawa, Canada, pp. 975–979, Oct. 1993.

[19] Fazel K. and Papke L., "On the performance of convolutionally-coded CDMA/OFDM for mobile communication system," in *Proc. IEEE International Symposium on Personal, Indoor and Mobile Radio Communications (PIMRC '93)*, Yokohama, Japan, pp. 468–472, Sept. 1993.

[20] Fujii M., Shimizu R., Suzuki S., Itami M., and Itoh K., "A study on downlink capacity of FD-MC/CDMA for channels with frequency selective fading," in *Proc. International Workshop on Multi-Carrier Spread Spectrum and Related Topics (MC-SS 2001)*, Oberpfaffenhofen, Germany, pp. 139–146, Sept. 2001.

[21] Hagenauer J., "Forward error correcting for CDMA systems," in *Proc. IEEE International Symposium on Spread Spectrum Techniques and Applications (ISSSTA '96)*, Mainz, Germany, pp. 566–569, Sept. 1996.

[22] Jarot S. P. W. and Nakagawa M., "Investigation on using channel information of MC-CDMA for pre-rake diversity combining in TDD/CDMA system," in *Proc. International Workshop on Multi-Carrier Spread Spectrum and Related Topics (MC-SS 2001)*, Oberpfaffenhofen, Germany, pp. 265–272, Sept. 2001.

[23] Jeong D. G. and Kim M. J., "Effects of channel estimation error in MC-CDMA/TDD systems," in *Proc. IEEE Vehicular Technology Conference (VTC 2000-Spring)*, Tokyo, Japan, pp. 1773–1777, May 2000.

[24] Kaiser S., "On the performance of different detection techniques for OFDM-CDMA in fading channels," in *Proc. IEEE Global Telecommunications Conference (GLOBECOM '95)*, Singapore, pp. 2059–2063, Nov. 1995.

[25] Kaiser S., "Analytical performance evaluation of OFDM-CDMA mobile radio systems," in *Proc. European Personal and Mobile Communications Conference (EPMCC '95)*, Bologna, Italy, pp. 215–220, Nov. 1995.

[26] Kaiser S., *Multi-Carrier CDMA Mobile Radio Systems – Analysis and Optimization of Detection, Decoding, and Channel Estimation*, Düsseldorf: VDI-Verlag, Fortschritt-Berichte VDI, series 10, no. 531, 1998, PhD thesis.

[27] Kaiser S., "OFDM code division multiplexing in fading channels," *IEEE Transactions on Communications*, vol. 50, pp. 1266–1273, Aug. 2002.

[28] Kaiser S. and Hagenauer J., "Multi-carrier CDMA with iterative decoding and soft-interference cancellation," in *Proc. IEEE Global Telecommunications Conference (GLOBECOM '97)*, Phoenix, USA, pp. 6–10, Nov. 1997.

[29] Kaiser S. and Papke L., "Optimal detection when combining OFDM-CDMA with convolutional and Turbo channel coding," in *Proc. IEEE International Conference on Communications (ICC '96)*, Dallas, USA, pp. 343–348, June 1996.

[30] Kalofonos D. N. and Proakis J. G., "Performance of the multistage detector for a MC-CDMA system in a Rayleigh fading channel," in *Proc. IEEE Global Telecommunications Conference (GLOBECOM '96)*, London, UK, pp. 1784–1788, Nov. 1997.

[31] Kondo S. and Milstein L. B., "Performance of multi-carrier DS-CDMA systems," *IEEE Transactions on Communications*, vol. 44, pp. 238–246, Feb. 1996.

[32] Kühn V., "Combined MMSE-PIC in coded OFDM-CDMA systems," in *Proc. IEEE Global Telecommunications Conference (GLOBECOM 2001)*, San Antonio, USA, Nov. 2001.

[33] Kühn V., Dekorsy A., and Kammeyer K.-D., "Channel coding aspects in an OFDM-CDMA system," in *Proc. ITG Conference on Source and Channel Coding*, Munich, Germany, pp. 31–36, Jan. 2000.

[34] Liu H. and Yin H., "Receiver design in multi-carrier direct-sequence CDMA communications," *IEEE Transactions on Communications*, vol. 49, pp. 1479–1487, Aug. 2001.

[35] Maxey J. J. and Ormondroyd R. F., "Multi-carrier CDMA using convolutional coding and interference cancellation over fading channels," in *Proc. International Workshop on Multi-Carrier Spread Spectrum (MC-SS '97)*, Oberpfaffenhofen, Germany, pp. 89–96, April 1997.

[36] J. Medbo, "Channel models for HIPERLAN/2 in different indoor scenarios," *ETSI EP BRAN 3ERI085B*, Mar. 1998.

[37] Mottier D. and Castelain D., "SINR-based channel pre-compensation for uplink multi-carrier CDMA systems," in *Proc. IEEE International Symposium on Personal, Indoor and Mobile Radio Communications (PIMRC 2002)*, Lisbon, Portugal, Sept. 2002.

[38] Nakagawa M. and Esmailzadeh R., "Time division duplex-CDMA," in *Proc. International Workshop on Multi-Carrier Spread Spectrum and Related Topics (MC-SS 2001)*, Oberpfaffenhofen, Germany, pp. 13–21, Sept. 2001.

[39] Nobilet S., Helard J.-F., and Mottier D., "Spreading sequences for uplink and downlink MC-CDMA systems: PAPR and MAI minimization," *European Transactions on Telecommunications (ETT)*, vol. 13, pp. 465–474, Sept./Oct. 2002.

[40] Ochiai H. and Imai H., "Performance of OFDM-CDMA with simple peak power reduction," *European Transactions on Telecommunications (ETT)*, vol. 10, pp. 391–398, July/Aug. 1999.

[41] Ochiai H. and Imai H., "Performance of downlink MC-CDMA with simple interference cancellation," in *Proc. International Workshop on Multi-Carrier Spread Spectrum and Related Topics (MC-SS'99)*, Oberpfaffenhofen, Germany, pp. 211–218, Sept. 1999.

[42] Persson A., Ottosson T., and Ström E., "Time-frequency localized CDMA for downlink multi-carrier systems," in *Proc. IEEE International Symposium on Spread Spectrum Techniques and Applications (ISSSTA 2002)*, Prague, Czech Republic, pp. 118–122, Sept. 2002.

[43] Popovic B. M., "Spreading sequences for multicarrier CDMA systems," *IEEE Transactions on Communications*, vol. 47, pp. 918–926, June 1999.

[44] Proakis J., *Digital Communications*, New York: McGraw-Hill, 1995.

[45] Pu Z., You X., Cheng S., and Wang H., "Transmission and reception of TDD multicarrier CDMA signals in mobile communications system," in *Proc. IEEE Vehicular Technology Conference (VTC '99-Spring)*, Houston, USA, pp. 2134–2138, May 1999.

[46] Reinhardt M., Egle J., and Lindner J., "Transformation methods, coding and equalization for time- and frequency-selective channels," *European Transactions on Telecommunications (ETT)*, vol. 11, pp. 555–565, Nov./Dec. 2000.

[47] Schnell M., *Systeminhärente Störungen bei "Spread Spectrum"-Vielfachzugriffsverfahren für die Mobilfunkübertragung*, Düsseldorf: VDI-Verlag, Fortschritt-Berichte VDI, series 10, no. 505, 1997, PhD thesis.

[48] Sourour E. A. and Nakagawa M., "Performance of orthogonal multi-carrier CDMA in a multipath fading channel," *IEEE Transactions on Communications*, vol. 44, pp. 356–367, Mar. 1996.

[49] Steiner B., "Uplink performance of a multicarrier-CDMA mobile radio system concept," in *Proc. IEEE Vehicular Technology Conference (VTC '97)*, Phoenix, USA, pp. 1902–1906, May 1997.

[50] Stirling-Gallacher R. A. and Povey G. J. R., "Different channel coding strategies for OFDM-CDMA," in *Proc. IEEE Vehicular Technology Conference (VTC '97)*, Phoenix, USA, pp. 845–849, May 1997.

[51] Tomba L. and Krzymien W. A., "Downlink detection schemes for MC-CDMA systems in indoor environments," *IEICE Transactions on Communications*, vol. E79-B, pp. 1351–1360, Sept. 1996.

[52] Vandendorpe L., "Multitone spread spectrum multiple access communications system in a multipath Rician fading channel," *IEEE Transactions on Vehicular Technology*, vol. 44, pp. 327–337, May 1995.

[53] Verdu S., *Multiuser Detection*, Cambridge: Cambridge University Press, 1998.

[54] Viterbi A. J., "Very low rate convolutional codes for maximum theoretical performance of spread spectrum multiple-access channels," *IEEE Journal on Selected Areas in Communications*, vol. 8, pp. 641–649, May 1990.

[55] Yee N., Linnartz J.-P. and Fettweis G., "Multi-carrier CDMA in indoor wireless radio networks," in *Proc. IEEE International Symposium on Personal, Indoor and Mobile Radio Communications (PIMRC '93)*, Yokohama, Japan, pp. 109–113, Sept. 1993.

[56] Yip K.-W. and Ng T.-S., "Tight error bounds for asynchronous multi-carrier CDMA and their applications," *IEEE Communications Letters*, vol. 2, pp. 295–297, Nov. 1998.

3

Hybrid Multiple Access Schemes

3.1 Introduction

The simultaneous transmission of multiple data streams over the same medium can be achieved with different multiplexing schemes. Most communications systems, such as GSM, LTE, DECT, and IEEE 802.11a, use multiplexing based on either time division, frequency division, or a combination of both. Space division multiplexing is applied to increase the user capacity of the system further. The simplest scheme of space division multiplexing is antenna sectorization at the base station, where often antennas with $120°/90°$ beams are used. Multiplexing schemes using code division have gained significant interest and have become part of wireless standards such as WCDMA/UMTS, HSPA, IS-95, CDMA-2000, and WLAN.

Time Division Multiplexing

The separation of different data streams with time division multiplexing is carried out by assigning each stream exclusively a certain period of time, i.e. time slot, for transmission. After each time slot, the next data stream transmits in the subsequent time slot. The number of slots assigned to each user can be supervised by the medium access controller (MAC). A MAC frame determines a group of time slots in which all data streams transmit once. The duration of the different time slots can vary according to the requirements of the different data streams.

If the different data streams belong to different users, the access scheme is called time division multiple access (TDMA). Time division multiplexing can be used with both time division duplex (TDD) and frequency division duplex (FDD). However, it is often used in communication systems with TDD transmission, where up- and downlinks are separated by the assignment of different time slots. It is adopted in several wireless LAN and WLL systems including IEEE 802.11a as well as IEEE 802.16x and HIPERMAN.

Frequency Division Multiplexing

With frequency division multiplexing, the different data streams are separated by exclusively assigning each stream a certain fraction of the frequency band for transmission.

In contrast to time division multiplexing, each stream can continuously transmit within its sub-band. The efficiency of frequency division multiplexing strongly depends on the minimum separation of the sub-bands to avoid adjacent channel interference. OFDM is an efficient frequency division multiplexing schemes, which offers minimum spacing of the sub-bands without interference from adjacent channels in the synchronous case.

In multiple access schemes, where different data streams belong to different users, the frequency division multiplexing scheme is known as frequency division multiple access (FDMA). Frequency division multiplexing is often used in communication systems with FDD, where up- and downlinks are separated by the assignment of different frequency bands for each link. FDD is used, for example, in mobile radio systems such as GSM, IS-95, and WCDMA/UMTS FDD mode.

Code Division Multiplexing
Multiplexing of different data streams can be carried out by multiplying the data symbols of a data stream with a spreading code exclusively assigned to this data stream before superposition with the spread data symbols of the other data streams. All data streams use the same bandwidth at the same time in code division multiplexing. Depending on the application, the spreading codes should as far as possible be orthogonal to each other in order to reduce interference between different data streams.

Multiple access schemes where the user data are separated by code division multiplexing are referred to as code division multiple access (CDMA). They are used, for example, in mobile radio systems, WCDMA/UMTS, HSPA, IS-95, and CDMA-2000.

Space Division Multiplexing
The spatial dimension can also be used for multiplexing of different data streams by transmitting the data streams over different, non-overlapping transmission channels. Space division multiplexing can be achieved using beam-forming or sectorization. The use of space division multiplexing for multiple access is termed space division multiple access (SDMA).

Hybrid Multiplexing Schemes
The above listed multiplexing schemes are often combined resulting in hybrid schemes in communication systems like GSM, where TDMA and FDMA are applied, or UMTS, where CDMA, TDMA, and FDMA are used. These hybrid combinations can additionally increase the user capacity and flexibility of the system. For example, the combination of MC-CDMA with DS-CDMA or TDMA offers the possibility to overload the limited MC-CDMA scheme. The idea is to load the orthogonal MC-CDMA scheme up to its limits and, in the case of additional users, to superimpose other nonorthogonal multiple access schemes. For small numbers of overload and using efficient interference cancellation schemes nearly all additional multiple access interference caused by the system overlay can be canceled [33].

In this chapter, different hybrid multiple access concepts will be presented and compared to each other.

3.2   Multi-Carrier FDMA

The concept of the combination of spread spectrum and frequency hopping with multi-carrier transmission opened the door for alternative hybrid multiple access solutions such as: *OFDMA* [28], *OFDMA with CDM* (*SS-MC-MA*) [18], distributed DFT-spread OFDM (*Interleaved FDMA*) [35], and localized DFT-spread OFDM. All of these schemes are discussed in the following.

3.2.1   *Orthogonal Frequency Division Multiple Access (OFDMA)*

3.2.1.1   Basic Principle

Orthogonal frequency division multiple access (OFDMA) consists of assigning one or several sub-carrier frequencies to each user (terminal station) with the constraint that the sub-carrier spacing is equal to the OFDM frequency spacing $1/T_s$ (see References [28] and [30] to [32]).

To introduce the basic principle of OFDMA we will make the following assumptions:

-   One sub-carrier is assigned per user (the generalization for several sub-carriers per user is straightforward).
-   In addition, the only source of disturbance is AWGN.

The signal of user $k, k = 0, 1, \ldots, K - 1$, where $K = N_c$, has the form

$$s^{(k)}(t) = \mathrm{Re}\{d^{(k)}(t)e^{j2\pi f_k t}e^{j2\pi f_c t}\}, \tag{3.1}$$

with

$$f_k = \frac{k}{T_s} \tag{3.2}$$

and f_c representing the carrier frequency. Furthermore, we assume that the frequency f_k is permanently assigned to user k, although in practice a frequency assignment could be made upon request. Therefore, an OFDMA system with, for example, $N_c = 1024$ sub-carriers and adaptive sub-carrier allocation is able to handle up to thousands of simultaneous users.

In the following, we consider a permanent channel assignment scheme in which the number of sub-carriers is equal to the number of users. Under this assumption the modulator of the terminal station of user k has the form of an unfiltered modulator with a rectangular pulse (e.g. unfiltered QPSK) and carrier $f_k + f_c$. The transmitted data symbols are given by

$$d^{(k)}(t) = \sum_{i=-\infty}^{+\infty} d_i^{(k)} \mathrm{rect}(t - iT_s), \tag{3.3}$$

where $d_i^{(k)}$ designates the data symbol transmitted by user k during the ith symbol period and rect(t) is a rectangular pulse shape that spans the time interval $[0, T_s]$.

The received signal of all K users before down-conversion at the base station in the presence of only noise (in the absence of multi-path) can be written as

$$q(t) = \sum_{k=0}^{K-1} s^{(k)}(t) + n(t), \tag{3.4}$$

where $n(t)$ is an additive noise term. After demodulation at the base station using a local oscillator with carrier frequency f_c, we obtain

$$r(t) = \sum_{k=0}^{K-1} r^{(k)}(t) + w(t), \tag{3.5}$$

where $r^{(k)}(t)$ is the complex envelope of the kth user signal and $w(t)$ is the baseband equivalent noise. This expression can also be written as

$$r(t) = \sum_{i=-\infty}^{\infty} \sum_{k=0}^{K-1} d_i^{(k)}(t) e^{j2\pi f_k t} + w(t), \tag{3.6}$$

where we explicitly find in this expression the information part $d_i^{(k)}(t)$.

The demodulated signal is sampled at a sampling rate of N_c/T_s and a block of N_c regularly spaced signal samples is generated per symbol period T_s. Over the ith symbol period, we generate an N_c-point sequence

$$r_{n,i} = \sum_{k=0}^{K-1} d_i^{(k)} e^{j2\pi kn/N_c} + w_{n,i}, \quad n = 0, \ldots, N_c - 1. \tag{3.7}$$

It is simple to verify that except for a scaling factor $1/N_c$, the above expression is a noisy version of the IDFT of the sequence $d_i^{(k)}$, $k = 0, \ldots, K - 1$. This indicates that the data symbols can be recovered using an N_c-point DFT after sampling. In other words, the receiver at the base station is an OFDM receiver.

As illustrated in Figure 3-1, in the simplest OFDMA scheme (one sub-carrier per user) each user signal is a single-carrier signal. At the base station (of, for example, fixed wireless access or interactive DVB-T) the received signal, being the sum of K users' signals, acts as an OFDM signal due to its multi-point to point nature. Unlike conventional FDMA, which requires K demodulators to handle K simultaneous users, OFDMA requires only a single demodulator, followed by an N_c-point DFT.

Hence, the basic components of an OFDMA transmitter at the terminal station are FEC channel coding, mapping, sub-carrier assignment, and a single-carrier modulator (or multi-carrier modulator in the case where several sub-carriers are assigned per user).

Since OFDMA is preferably used for the uplink in a multi-user environment, low order modulation such as QPSK with Gray mapping is preferred. However, basically high order modulation (e.g. 16-QAM or 64-QAM) can also be employed.

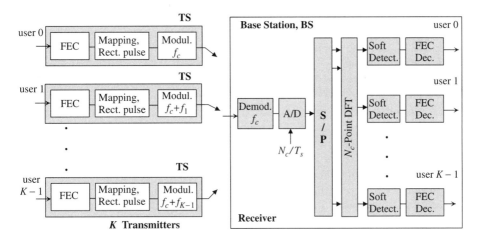

Figure 3-1 Basic principle of OFDMA

The sub-carrier assignment can be fixed or dynamic. In practice, in order to increase the system robustness (to exploit frequency diversity) a dynamic assignment of sub-carriers (i.e. frequency hopping) for each user is preferable. This approach is similar to M- or Q-Modification in MC-CDMA (see Section 2.1.9). For pulse shaping, rectangular shaping is usually used, which results for K users in an OFDM-type signal at the receiver side.

In summary, where only one sub-carrier is assigned to a user, the modulator for the user could be a single-carrier modulator. If several carriers are used for a given terminal station, the modulator is a multi-carrier (OFDM) modulator.

A very accurate clock and carrier synchronization is essential for an OFDMA system to ensure orthogonality between the K modulated signals originating from different terminal stations. This can be achieved, for instance, by transmitting synchronization signals from the base stations to all terminal stations. Each terminal station modulator derives its carrier frequency and symbol timing from these common downlink signals.

At the base station the main components of the receiver are the demodulator (including synchronization functions), FFT, and channel decoder (with soft decisions). Since in the case of a synchronous system the clock and carrier frequencies are available at the base station (see Section 3.2.1.2), simple carrier and clock recovery circuits are sufficient in the demodulator to extract this information from the received signal [30]. This fact can greatly simplify the OFDM demodulator.

3.2.1.2 Synchronization Sensitivity

As mentioned before, OFDMA requires an accurate carrier spacing between different users and precise symbol clock frequency. Hence, in a synchronous system, the OFDMA transmitter is synchronized (clock and frequency) to the base station downlink signal, received by all terminal stations [3, 4, 11].

In order to avoid time drift, the symbol clock of the terminal station is locked to the downlink reference clock and on some extra time synchronization messages (e.g. regular time stamps), transmitted periodically from the base station to all terminal stations. The reference clock in the base station requires a quite high accuracy [3]. Furthermore, the terminal station can synchronize the transmit sub-carriers in phase and frequency to the received downlink channel.

Since the clock and carrier frequencies are available at the reception side in the base station, no complex carrier and clock recovery circuits are necessary in the demodulator to extract this information from the received signal [30]. This simplifies the OFDMA demodulator. Although the carrier frequency is locally available, there are phase differences between different user signals and local references. These phase errors can be compensated, for instance, by a phase equalizer, which takes the form of a complex multiplier bank with one multiplier per sub-carrier. This phase equalization is not necessary if the transmitted data are differentially encoded and demodulated.

Regarding the sensitivity to the oscillator's phase noise, the OFDMA technique will have the same sensitivity as a conventional OFDM system. Therefore, low noise oscillators are needed, particularly if the number of sub-carriers is high or high order modulation is used.

If each terminal station is fixed positioned (e.g. return channel of DVB-T), the ranging procedure (i.e. measuring the delay and power of individual signals) and adjusting the phase and the transmit power of the transmitters can be done at the installation and later on periodically in order to cope with drifts, which may be due to weather or aging variations and other factors. The ranging information can be transmitted periodically from the base station to all terminal stations within a given frame format [3, 4, 11].

Phase alignment of different users through ranging is typically not perfect. Residual misalignment can be compensated for by using a larger guard interval (cyclic extension).

3.2.1.3 Pulse Shaping

In the basic version of OFDMA, one sub-carrier is assigned to each user. The spectrum of each user is quite narrow, which makes OFDMA more sensitive to narrowband interference. In this section, another variant is described that may lead to increased robustness against narrowband interference.

With rectangular pulse shaping, OFDMA has a $sinc^2(f)$ shaped spectrum with overlapping sub-channels (see Figure 3-2(a)). The consequence of this is that a narrowband interferer will affect not only one sub-carrier but several sub-carriers [31]. The robustness of OFDMA to band-limited interference can be increased if the bandwidth of individual sub-channels is strictly limited so that either adjacent sub-channels do not overlap or each sub-channel spectrum only overlaps with two adjacent sub-channels. The non-overlapping concept is illustrated in Figure 3-2(b). As long as the bandwidth of one sub-channel is smaller than $1/T_s$, the narrowband interferer will only affect one sub-channel. As shown in Figure 3-2(b), the orthogonality between sub-channels is guaranteed, since there is no overlapping between the spectra of adjacent sub-channels. Here a Nyquist pulse shaping is needed for ISI-free transmission on each sub-carrier, comparable to a conventional single-carrier transmission scheme. This requires oversampling of the received signal and DFT operations at a higher rate than N_c/T_s. In other words, the increased robustness to narrowband interference is achieved at the expense of increased complexity.

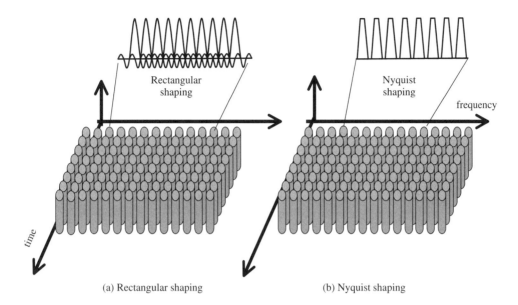

(a) Rectangular shaping   (b) Nyquist shaping

Figure 3-2   Example of OFDMA with band-limited spectra

The Nyquist shaping function $g(t)$ can be implemented with a time-limited square root raised cosine pulse with a roll-off factor α,

$$
g(t) =
\begin{cases}
\dfrac{\sin\left[\dfrac{\pi t}{T_s'}(1-\alpha)\right] + \dfrac{4\alpha t}{T_s'}\cos\left[\dfrac{\pi t}{T_s'}(1-\alpha)\right]}{\dfrac{\pi t}{T_s'}\left[1-\left(\dfrac{4\alpha t}{T_s'}\right)^2\right]} & \text{for } t \in \{-4T_s', 4T_s'\} \\[4ex]
0 & \text{otherwise}
\end{cases}
\tag{3.8}
$$

The relationship between T_s', T_s, and α (roll-off factor) provides the property of the individual separated spectra, where $T_s' = (1+\alpha)T_s$.

3.2.1.4 Frequency Hopping OFDMA

The application of frequency hopping (FH) in an OFDMA system is straightforward. Rather than assigning a fixed particular frequency to a given user, the base station assigns a hopping pattern [2, 12, 28, 36]. In the following it is assumed that N_c sub-carriers are available and that the frequency hopping sequence is periodic and uniformly distributed over the signal bandwidth.

Suppose that the frequency sequence $(f_0, f_7, f_{14}, \ldots)$ is assigned to the first user, the sequence $(f_1, f_8, f_{15}, \ldots)$ to the second user, and so on. The frequency assignment to N_c users can be written as

$$
f(n, k) = f_{k+(7n \bmod N_c)}, \quad k = 0, \ldots, N_c - 1,
\tag{3.9}
$$

where $f(n, k)$ designates the sub-carrier frequency assigned to user k at symbol time n.

OFDMA with frequency hopping has a close relationship with MC-CDMA. We know that MC-CDMA is based on spreading the signal bandwidth using direct sequence spreading with processing gain P_G. In OFDMA, frequency assignments can be specified with a code according to a frequency hopping (FH) pattern, where the number of hops can be slow. Both schemes employ OFDM for chip transmission.

3.2.1.5   General OFDMA Transceiver

A general conceptual block diagram of an OFDMA transceiver for the uplink of a multi-user cellular system is illustrated in Figure 3-3. The terminal station is synchronized to the base station. The transmitter of the terminal station extracts from the demodulated downlink received data MAC messages with information about sub-carrier allocation, frequency hopping pattern, power control messages and timing, and further clock and frequency synchronization information. Synchronization of the terminal station is achieved by using the MAC control messages to perform time synchronization and using frequency information issued from the terminal station demodulator (the recovered base station system clock). The MAC control messages are processed by the MAC management block to instruct the terminal station modulator on the transmission resources assigned to it and to tune the access performed to the radio frequency channel. The pilot symbols are inserted to ease the channel estimation task at the base station.

At the base station, the received signals issued by all terminal stations are demodulated by the use of an FFT as a conventional OFDM receiver, assisted by the MAC layer management block.

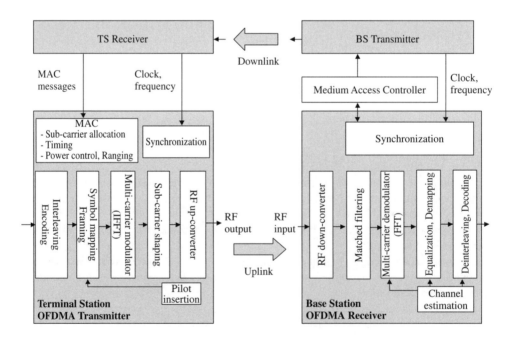

Figure 3-3   General OFDMA conceptual transceiver

It should be emphasized that the transmitter and the receiver structure of an OFDMA system is quite similar to an OFDM system. The same components, like FFT, channel estimation, equalization and soft channel decoding, can be used for both systems.

In order to offer a variety of multi-media services requiring different data rates, the OFDMA scheme needs to be flexible in terms of data rate assignment. This can be achieved by assigning the required number of sub-carriers according to the bandwidth request of a given user. This method of assignment is part of an MAC protocol at the base station.

Note that if the number of assigned sub-carriers is an integer power of two, the inverse FFT can be used at the terminal station transmitter, which will be equivalent to a conventional OFDM transmitter. OFDMA is adopted in IEEE802.16e/WiMAX as the access scheme for the uplink. This is due to its low power consumption and its high flexibility.

3.2.2 OFDMA with Code Division Multiplexing: SS-MC-MA

The extension of OFDMA by code division multiplexing (CDM) results in a multiple access scheme referred to as spread spectrum multi-carrier multiple access (SS-MC-MA) [18, 19]. It applies OFDMA for user separation and additionally uses CDM on data symbols belonging to the same user. The CDM component is introduced in order to achieve additional diversity gains. Like MC-CDMA, SS-MC-MA exploits the advantages given by the combination of the spread spectrum technique and multi-carrier modulation. The SS-MC-MA scheme is similar to the MC-CDMA transmitter with M-Modification. Both transmitters are identical except for the mapping of the user data to the sub-systems. In SS-MC-MA systems, one user maps L data symbols to one sub-system, which this user exclusively uses for transmission. Different users use different sub-systems in SS-MC-MA systems. In MC-CDMA systems, M data symbols per user are mapped to M different sub-systems, where each sub-system is shared by different users. The principle of SS-MC-MA is illustrated for a downlink transmitter in Figure 3-4.

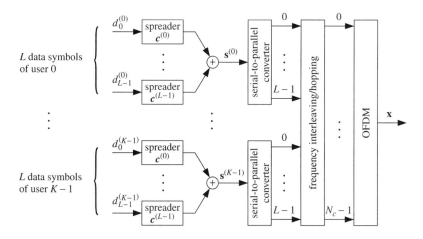

Figure 3-4   SS-MC-MA downlink transmitter

The SS-MC-MA and MC-CDMA systems have the following similarities:

- SS-MC-MA and MC-CDMA systems exploit frequency diversity by spreading each data symbol over L sub-carriers.
- Per subsystem, the same data detection techniques can be applied with SS-MC-MA and MC-CDMA systems.
- ISI and ICI can be avoided in SS-MC-MA and MC-CDMA systems, resulting in simple data detection techniques.

Their main differences are:

- In SS-MC-MA systems, CDM is used for the simultaneous transmission of the data of one user on the same sub-carriers, whereas in MC-CDMA systems, CDM is used for the transmission of the data of different users on the same sub-carriers. Therefore, SS-MC-MA is an OFDMA scheme on the sub-carrier level whereas MC-CDMA is a CDMA scheme.
- MC-CDMA systems have to cope with multiple access interference, which is not present in SS-MC-MA systems. Instead of multiple access interference, SS-MC-MA systems have to cope with self-interference caused by the superposition of signals from the same user.
- In SS-MC-MA systems, each sub-carrier is exclusively used by one user, enabling low complex channel estimation, especially for the uplink. In MC-CDMA systems, the channel estimation in the uplink has to cope with the superposition of signals from different users, which are faded independently on the same sub-carriers, increasing the complexity of the uplink channel estimation.

After this comparative introduction of SS-MC-MA, the uplink transmitter and the assigned receiver are described in detail in this section.

Figure 3-5 shows an SS-MC-MA uplink transmitter with channel coding for the data of user k. The vector

$$\mathbf{d}^{(k)} = (d_0^{(k)}, d_1^{(k)}, \ldots, d_{L-1}^{(k)})^T \tag{3.10}$$

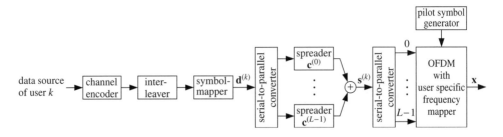

Figure 3-5   SS-MC-MA transmitter of user k

represents one block of L parallel converted data symbols of user k. Each data symbol is multiplied with another orthogonal spreading code of length L. The $L \times L$ matrix

$$\mathbf{C} = (\mathbf{c}_0, \mathbf{c}_1, \ldots, \mathbf{c}_{L-1}) \tag{3.11}$$

represents the L different spreading codes $c_l, l = 0, \ldots, , L - 1$, used by user k. The spreading matrix $\mathbf{C}$ can be the same for all users. The modulated spreading codes are synchronously added, resulting in the transmission vector

$$\mathbf{s}^{(k)} = \mathbf{C}\mathbf{d}^{(k)} = (S_0^{(k)}, S_1^{(k)}, \ldots, S_{L-1}^{(k)})^T. \tag{3.12}$$

To increase the robustness of SS-MC-MA systems, less than L data modulated spreading codes can be added in one transmission vector $\mathbf{s}^{(k)}$.

Comparable to frequency interleaving in MC-CDMA systems, the SS-MC-MA transmitter performs a user-specific frequency mapping such that subsequent chips of $\mathbf{s}^{(k)}$ are interleaved over the whole transmission bandwidth. The user-specific frequency mapping assigns each user exclusively its L sub-carriers, avoiding multiple access interference. The Q-Modification introduced in Section 2.1.9.2 for MC-CDMA systems is inherent in SS-MC-MA systems. The M-Modification can, as in MC-CDMA systems, be applied to SS-MC-MA systems by assigning a user more than one sub-system.

OFDM with guard interval is applied in SS-MC-MA systems in the same way as in MC-CDMA systems. In order to perform coherent data detection at the receiver and to guarantee robust time and frequency synchronization, pilot symbols are multiplexed in the transmitted data.

An SS-MC-MA receiver with coherent detection of the data of user k is shown in Figure 3-6. After inverse OFDM with user-specific frequency de-mapping and extraction of the pilot symbols from the symbols with user data, the received vector

$$\mathbf{r}^{(k)} = \mathbf{H}^{(k)}\mathbf{s}^{(k)} + \mathbf{n}^{(k)} = (R_0^{(k)}, R_1^{(k)}, \ldots, R_{L-1}^{(k)})^T \tag{3.13}$$

with the data of user k is obtained. The $L \times L$ diagonal matrix $\mathbf{H}^{(k)}$ and the vector $\mathbf{n}^{(k)}$ of length L describe the channel fading and noise respectively on the sub-carriers exclusively used by user k.

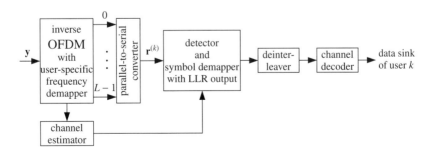

Figure 3-6   SS-MC-MA receiver of user k

Any of the single-user or multi-user detection techniques presented for MC-CDMA systems in Section 2.1.5 can be applied for the detection of the data of a single user per sub-system in SS-MC-MA systems. However, SS-MC-MA systems offer (especially in the downlink) the advantage that with multi-symbol detection (equivalent to multi-user detection in MC-CDMA systems) in one estimation step simultaneously L data symbols of a single user are estimated. Compared to MC-CDMA systems, the complexity per data symbol of multi-symbol detection in SS-MC-MA systems reduces by a factor of L in the downlink. With multi-symbol detection, LLRs can inherently be obtained from the detection algorithm, which may also include the symbol de-mapping. After de-interleaving and decoding of the LLRs, the detected source bits of user k are obtained.

A future mobile radio system may use MC-CDMA in the downlink and SS-MC-MA in the uplink. This combination achieves for both links a high spectral efficiency and flexibility. Furthermore, in both links the same hardware can be used, only the user data have to be mapped differently [16]. Alternatively, a modified SS-MC-MA scheme with flexible resource allocation can achieve a high throughput in the downlink [24].

SS-MC-MA can cope with a certain amount of asynchronism. It has been shown in References [21] and [22] that it is possible to avoid any additional measures for uplink synchronization in cell radii up to several kilometers. The principle is to apply a synchronized downlink and each user transmits in the uplink directly after receiving its data without any additional time correction. A guard time shorter than the maximum time difference between the user signals is used, which increases the spectral efficiency of the system. Thus, SS-MC-MA can be applied with a low complex synchronization in the uplink.

Moreover, the SS-MC-MA scheme can be modified such that with not fully loaded systems, the additional available resources are used for more reliable transmission [6, 7]. With a full load, these BER performance improvements can only be obtained by reducing the spectral efficiency of the system.

3.2.3 Distributed DFT-Spread OFDM: Interleaved FDMA (IFDMA)

Interleaved FDMA (IFDMA) belongs to the class of DFT-spread access schemes. More specifically, IFDMA represents distributed DFT-spread OFDM. The multiple access scheme IFDMA is based on the principle of FDMA where no multiple access interference occurs [34, 35]. The signal is designed in such a way that the transmitted signal can be considered as multi-carrier signal where each user is exclusively assigned a sub-set of sub-carriers. The sub-carriers of the different users are interleaved. It is an inherent feature of the IFDMA signal that the sub-carriers of a user are equally spaced over the transmission bandwidth B, which guarantees a maximum exploitation of the available frequency diversity. The signal design of IFDMA is performed in the time domain and the resulting signal has the advantage of a low PAPR.

The transmission of IFDMA signals in multi-path channels results in ISI, which can efficiently be compensated by frequency domain equalization [5]. Compared to OFDMA, an IFDMA scheme is less flexible, since it does not support adaptive sub-carrier allocation and sub-carrier loading. This disadvantage can to some extend be overcame by localized DFT-spread OFDM, introduced in Section 3.2.4.

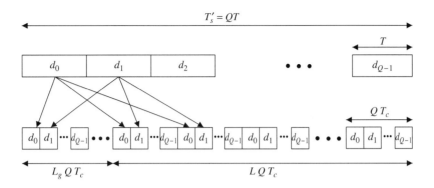

Figure 3-7   IFDMA signal design with guard interval

The IFDMA signal design is illustrated in Figure 3-7. A block of Q data symbols

$$\mathbf{d}^{(k)} = (d_0^{(k)}, d_1^{(k)}, \ldots, d_{Q-1}^{(k)})^T \tag{3.14}$$

assigned to user k is used for the construction of one IFDMA symbol. The duration of a data symbol is T and the duration of an IFDMA symbol is

$$T_s' = QT. \tag{3.15}$$

In order to limit the effect of ISI to one IFDMA symbol, a guard interval consisting of a cyclic extension of the symbol is included between adjacent IFDMA symbols, comparable to the guard interval in multi-carrier systems. Each IFDMA symbol of duration T_s' includes the guard interval of duration

$$T_g = L_g Q T_c. \tag{3.16}$$

An IFDMA symbol is obtained by compressing each of the Q symbols from symbol duration T to chip duration T_c, i.e.

$$T_c = \frac{T}{L_g + L}, \tag{3.17}$$

and repeating the resulting compressed block $(L_g + L)$ times. Thus, the transmission bandwidth is spread by the factor

$$P_G = L_g + L. \tag{3.18}$$

The compressed vector of user k can be written as

$$\mathbf{s}^{(k)} = \frac{1}{L_g + L} \left(\underbrace{\mathbf{d}^{(k)^T}, \mathbf{d}^{(k)^T}, \ldots, \mathbf{d}^{(k)^T}}_{(L_g+L)\text{copies}} \right)^T. \tag{3.19}$$

The transmission signal $\mathbf{x}^{(k)}$ is constructed by element-wise multiplication of the compressed vector $\mathbf{s}^{(k)}$ with a user-dependent phase vector $\mathbf{c}^{(k)}$ of length $(L_g + L)Q$ having the components

$$c_l^{(k)} = e^{-j2\pi lk/(QL)}, \quad l = 0, \ldots, (L_g + L)Q - 1. \tag{3.20}$$

The element-wise multiplication of the two vectors $\mathbf{s}^{(k)}$ and $\mathbf{c}^{(k)}$ ensures that each user is assigned a set of sub-carriers orthogonal to the sub-carrier sets of all other users. Each sub-carrier set contains Q sub-carriers and the number of active users is restricted to

$$K \leq L. \tag{3.21}$$

The IFDMA receiver has to perform an equalization to cope with the ISI which is present with IFDMA in multi-path channels. For low numbers of Q, the optimum maximum likelihood sequence estimation can be applied with reasonable complexity whereas for higher numbers of Q, less complex sub-optimum detection techniques such as linear equalization or decision feedback equalization are required to deal with the ISI.

Due to its low PAPR, a practical application of IFDMA can be an uplink where power-efficient terminal stations are required that benefit from the quasi constant envelope and more complex receivers that have to cope with ISI being part of the base station.

3.2.4 Localized DFT-Spread OFDM

Localized DFT-spread OFDM differs from distributed DFT-spread OFDM only in the mapping of the sub-carriers. While with distributed DFT-spread OFDM the sub-carriers of the users are distributed over the whole available bandwidth, with localized DFT-spread OFDM blocks of adjacent sub-carriers are mapped to a user. Figure 3-8 illustrates the differences in sub-carrier mapping between both schemes.

Localized DFT-spread OFDM has the advantage that flexible channel dependent scheduling in frequency can be implemented. Moreover, it is more robust to frequency errors between signals from different users. The disadvantage compared to distributed DFT-spread OFDM is that the frequency diversity achieved per user is reduced. Both localized and distributed DFT-spread OFDM have the same low PAPR. Due to the low PAPR and the possibility of flexible user scheduling in frequency, localized DFT-spread OFDM has been chosen as the access technology for the uplink of LTE (see Chapter 5). The receiver for localized DFT-spread OFDM can efficiently be realized by frequency domain equalization [5].

3.3 Multi-Carrier TDMA

The combination of OFDM and TDMA is referred to as MC-TDMA or OFDM-TDMA. Due to its well understood TDMA component, MC-TDMA is part of several high rate standards; e.g. WLAN-EEE 802.11 a/g and WLL-IEEE 802.16a/ETSI HIPERMAN [4, 10, 11] (see Chapter 5).

MC-TDMA transmission is done in a frame manner like in a TDMA system. One time frame within MC-TDMA has K time slots (or bursts), each allocated to one of the K

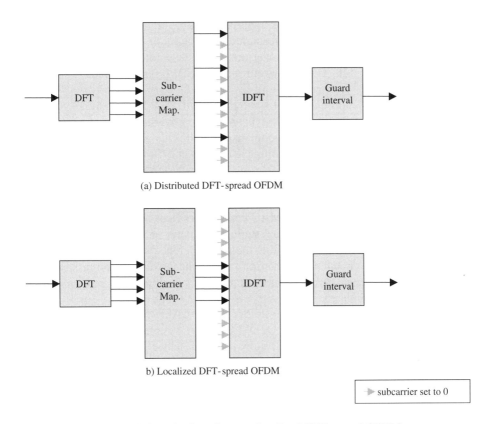

(a) Distributed DFT-spread OFDM

b) Localized DFT-spread OFDM

▶ subcarrier set to 0

Figure 3-8   Distributed versus localized DFT-spread OFDM

terminal stations. One time slot/burst consists of one or several OFDM symbols. The allocation of time slots to the terminal stations is controlled by the base station medium access controller (MAC). Multiple access interference can be avoided when ISI between adjacent OFDM symbols can be prevented by using a sufficiently long guard interval or with a timing advance control mechanism.

Adaptive coding and modulation is usually applied in conjunction with MC-TDMA systems, where the coding and modulation can be easily adapted per transmitted burst.

The main advantages of MC-TDMA are in guaranteeing a high peak data rate, in its multiplexing gain (bursty transmission), in the absence of multiple access interference, and in simple receiver structures that can be designed, for instance, by applying differential modulation in the frequency direction. In the case of coherent demodulation a quite robust OFDM burst synchronization is needed, especially for the uplink. A frequency synchronous system where the terminal station transmitter is frequency-locked to the received signal in the downlink or is spending a high amount of overhead transmitted per burst could remedy this problem.

Besides the complex synchronization mechanism required for an OFDM system, the other disadvantage of MC-TDMA is that diversity can only be exploited by using additional measures like channel coding or applying multiple transmit/receive antennas. As a TDMA system, the instantaneous transmitted power in the terminal station is high, which requires more powerful high power amplifiers than for FDMA or OFDMA systems. Furthermore, the MC-TDMA system as an OFDM system needs a high output power back-off.

As shown in Figure 3-9, the terminal station of an MC-TDMA system is synchronized to the base station in order to reduce the synchronization overhead. The transmitter of the terminal station extracts from the demodulated downlink data such as MAC messages, burst allocation, power control and timing advance, and further clock and frequency synchronization information. In other words, the synchronization of the terminal station is achieved using the MAC control messages to perform time synchronization and using frequency information issued from the terminal station downlink demodulator (the recovered base station system clock). MAC control messages are processed by the MAC management block to instruct the terminal station modulator on the transmission resources assigned to it and to tune the access. Here, the pilot/reference symbols are inserted at the transmitter side to ease the burst synchronization and channel estimation tasks at the base station. At the base station, the received burst issued by each terminal station is detected and multi-carrier demodulated.

It should be emphasized that the transmitter and receiver structure of an MC-TDMA system is quite similar to that of an OFDM/OFDMA system. The same components, such as FFT, channel estimation, equalization, and soft channel decoding, can be used for both, except that for an MC-TDMA system a burst synchronization is required, equivalent

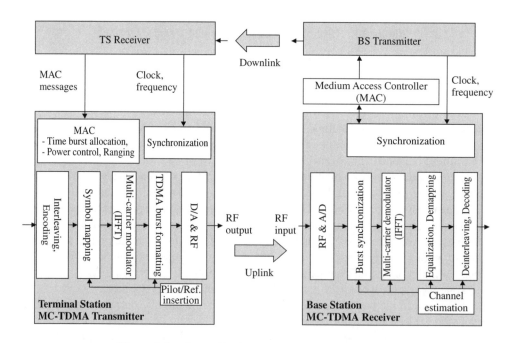

Figure 3-9   General MC-TDMA conceptual transceiver

to a single-carrier TDMA system. Furthermore, a frequency synchronous system would simplify the MC-TDMA receiver synchronization tasks.

Combining OFDMA and MC-TDMA achieves a flexible multi-user system with high throughput [9].

3.4   Ultra Wideband Systems

The technique for generating an ultra wideband (UWB) signal has existed for more than three decades [27], which is better known to the radar community as a *baseband carrier less short pulse* [1]. A classical way to generate a UWB signal is to spread the data with a code with a very large processing gain, i.e. 50 to 60 dB, resulting in a transmitted bandwidth of several GHz. Multiple access can be realized by classical CDMA, where for each user a given spreading code is assigned. However, the main problem of such a technique is its implementation complexity.

As the power spectral density of the UWB signal is extremely low, the transmitted signal appears as a negligible white noise for other systems. In the increasingly crowded spectrum, the transmission of the data as a noise-like signal can be considered a main advantage for the UWB systems. However, its drawbacks are the small coverage and the low data rate for each user

In References [25] and [37] an alternative approach compared to classical CDMA is proposed for generating a UWB signal that does require sine-wave generation. It is based on time-hopping spread spectrum. The key advantages of this method are the ability to resolve multi-paths and the low complexity technology availability for its implementation.

3.4.1   *Pseudo-Random PPM UWB Signal Generation*

The idea of generating a UWB signal by transmitting ultra-short Gaussian monocycles with controlled pulse-to-pulse intervals can be found in Reference [25]. The monocycle is a wideband signal with center frequency and bandwidth dependent on the monocycle duration. In the time domain, a Gaussian monocycle is derived by the first derivative of the Gaussian function, given by

$$s(t) = 6a\sqrt{\frac{e\pi}{3}}\frac{t}{\tau}e^{-6\pi\left(\frac{t}{\tau}\right)^2}, \tag{3.22}$$

where a is the peak amplitude of the monocycle and τ is the monocycle duration. In the frequency domain, the monocycle spectrum is given by

$$S(f) = -j\frac{2f\tau^2}{3}\sqrt{\frac{e\pi}{2}}e^{-\frac{\pi}{6}(f\tau)^2}, \tag{3.23}$$

with center frequency and bandwidth approximately equal to $1/\tau$.

In Figure 3-10, a Gaussian monocycle with $\tau = 0.5$ ns duration is illustrated. This monocycle will result in a center frequency of 2 GHz with 3 dB bandwidth of approximately 2 GHz (from 1 to 3.16 GHz). For data transmission, pulse position modulation (PPM) can be used, which varies the precise timing of transmission of a monocycle about the nominal position. By shifting each monocycle's actual transmission time over a large time frame

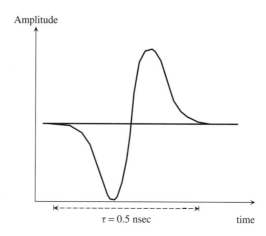

Figure 3-10   Gaussian monocycle with duration of 0.5 ns

in accordance with a specific PN code, i.e. performing time hopping (see Figure 3-11), this pseudo-random time modulation makes the UWB spectrum a pure white noise in the frequency domain. In the time domain each user will have a unique PN time-hopping code, hence resulting in a time-hopping multiple access.

A single data bit is generally spread over multiple monocycles, i.e. pulses. The duty cycle of each transmitted pulse is about $0.5-1\%$. Hence, the processing gain obtained by this technique is the sum of the duty cycle (about $20-23$ dB) and the number of pulses used per data bit. As an example, if we consider a transmission with 10^6 pulses per second with a duty cycle of 0.5% and with a pulse duration of 0.5 ns ($B = 2$ GHz bandwidth) for 8 kbit/s transmitted data, the resulting processing gain is 54 dB, which is significantly high. The ultra wideband signal generated above can be seen as a combination of spread spectrum with pulse position modulation.

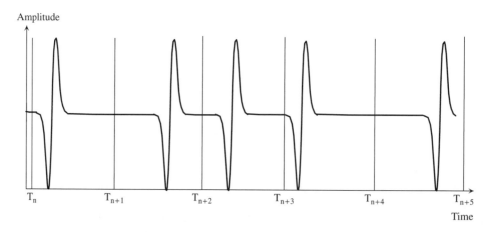

Figure 3-11   PN time modulation with five pulses

3.4.2 UWB Transmission Schemes

A UWB transmission scheme for a multi-user environment is illustrated in Figure 3-12, where for each user a given time-hopping pattern, i.e. PN code, is assigned. The transmitter is quite simple. It does not include any amplifier or any IF generation. The signal of the transmitted data after pulse position modulation according to the user's PN code is emitted directly at the Tx antenna. A critical point of the transmitter is the antenna, which may act as a filter.

The receiver components are similar to the transmitter. A rake receiver as in a conventional DS-CDMA system might be required to cope with multi-path propagation. The baseband signal processing extracts the modulated signal and controls both signal acquisition and tracking.

The main application fields of UWB could be short range (e.g. indoor) multi-user communications, radar systems, and location determination/positioning. UWB may have a potential application in the automotive industry.

3.5 Comparison of Hybrid Multiple Access Schemes

A multitude of performance comparisons have been carried out between MC-CDMA and DS-CDMA as well as between the multi-carrier multiple access schemes MC-CDMA, MC-DS-CDMA, SS-MC-MA, OFDMA, and MC-TDMA. It has been shown that MC-CDMA can significantly outperform DS-CDMA with respect to BER performance and bandwidth efficiency in the synchronous downlink [8, 13, 14]. The reason for the better performance with MC-CDMA is that it can avoid ISI and ICI, allowing an efficient and simple user signal separation. The results of these comparisons are the motivation to consider MC-CDMA as a potential candidate for future mobile radio systems, which should outperform 3G systems based on DS-CDMA.

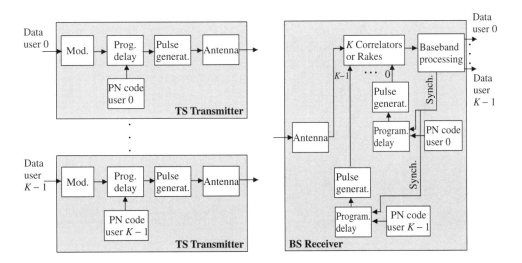

Figure 3-12   Multi-user UWB transmission scheme

The design of a future air interface for broadband mobile communications requires a comprehensive comparison between the various multi-carrier based multiple access schemes. In Section 2.1.10, the performance of MC-CDMA, OFDMA, and MC-TDMA has been compared in a Rayleigh fading channel for scenarios with and without FEC channel coding, where different symbol mapping schemes have also been taken into account. It can generally be said that MC-CDMA outperforms the other multiple access schemes but requires additional complexity for signal spreading and detection. The reader is referred to Section 2.1.10 and to References [15, 17, 23, 26], and [29] to compare the performance of the various schemes directly.

In the following, we show a performance comparison between MC-CDMA and OFDMA for the downlink and between SS-MC-MA and OFDMA for the uplink. The transmission bandwidth is 2 MHz and the carrier frequency is 2 GHz. The guard interval exceeds the maximum delay of the channel. The mobile radio channels are chosen according to the COST 207 models. Simulations are carried out with a bad urban (BU) profile and a velocity of 3 km/h of the mobile user and with a hilly terrain (HT) profile and a velocity of 150 km/h of the mobile user. QPSK is chosen for symbol mapping. All systems are fully loaded and synchronized.

In Figure 3-13, the BER versus the SNR per bit for MC-CDMA and OFDMA systems with different channel code rates in the downlink is shown. The number of sub-carriers is 512. Perfect channel knowledge is assumed in the receiver. The results for MC-CDMA are obtained with soft interference cancellation [20] after the first iteration. It can be observed

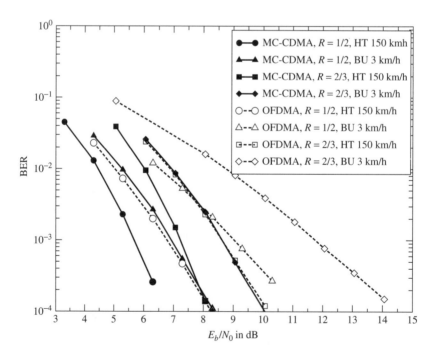

Figure 3-13   BER versus SNR for MC-CDMA and OFDMA in the downlink: QPSK; fully loaded system

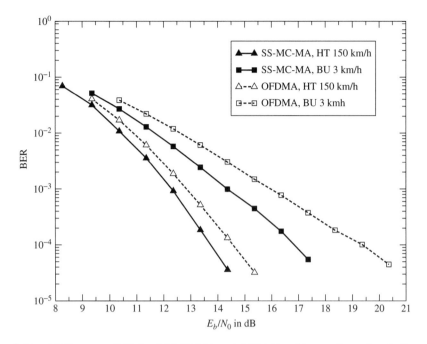

Figure 3-14   BER versus SNR for SS-MC-MA and OFDMA with one-dimensional pilot symbol aided channel estimation in the uplink: $R = 2/3$; QPSK; fully loaded system

that MC-CDMA outperforms OFDMA. The SNR gain with MC-CDMA compared to OFDMA strongly depends on the propagation scenario and code rate.

Figure 3-14 shows the BER versus the SNR per bit of an SS-MC-MA system and an OFDMA system in the uplink. The number of sub-carriers is 256. Both systems apply one-dimensional channel estimation which requires an overhead on pilot symbols of 22.6 %. The channel code rate is 2/3. The SS-MC-MA system applies maximum likelihood detection. The performance of SS-MC-MA can be further improved by applying soft interference cancellation in the receiver. The SS-MC-MA system outperforms OFDMA in the uplink, but it requires more complex receivers. The SS-MC-MA system and the OFDMA system would improve in performance by about 1 dB in the downlink due to the reduced overhead with two-dimensional channel estimation.

References

[1] Bennett C. L. and Ross G. F., "Time-domain electromagnetics and its applications," *Proceedings IEEE*, vol. 66, pp. 299–318, March 1978.

[2] Chen Q., Sousa E. S., and Pasupathy S., "Multi-carrier CDMA with adaptive frequency hopping for mobile radio systems," *IEEE Journal on Selected Areas in Communications*, vol. 14, pp. 1852–1858, Dec. 1996.

[3] ETSI DVB-RCT (TS 301 958), "Interaction channel for digital terrestrial television (RCT) incorporating multiple access OFDM," Sophia Antipolis, France, March 2001.

[4] ETSI HIPERMAN (TS 102 177), "High performance metropolitan area network, Part 1: physical layer," Sophia Antipolis, France, 2004.

[5] Falconer D., Ariyavisitakul, S. L., Benyamin-Seeyar A., and Eidson, B. "Frequency domain equalization for single-carrier broadband wireless systems," *IEEE Communications Magazine*, vol. 40, pp. 58–66, April 2002.

[6] Giannakis G. B., Anghel P. A., Wang Z., and Scaglione A., "Generalized multi-carrier CDMA for MUI/ISI-resilient uplink transmissions irrespective of frequency-selective multi-path," in *Proc. International Workshop on Multi-Carrier Spread Spectrum and Related Topics (MC-SS'99)*, Oberpfaffenhofen, Germany, pp. 25–33, Sept. 1999.

[7] Giannakis G. B., Stamoulis A., Wang Z., and Anghel A., "Load-adaptive MUI/ISI-resilient generalized multi-carrier CDMA with linear and DF receivers," *European Transactions on Telecommunications (ETT)*, vol. 11, pp. 527–537, Nov./Dec. 2000.

[8] Hara S. and Prasad R., "Overview of multi-carrier CDMA," *IEEE Communications Magazine*, vol. 35, pp. 126–133, Dec. 1997.

[9] Ibars C. and Bar-Ness Y., "Rate-adaptive coded multi-user OFDM for downlink wireless systems," in *Proc. International Workshop on Multi-Carrier Spread Spectrum and Related Topics (MC-SS 2001)*, Oberpfaffenhofen, Germany, pp. 199–207, Sept. 2001.

[10] IEEE-802.11 (P802.11a/D6.0), "LAN/MAN specific requirements – Part 2: wireless MAC and PHY specifications – high speed physical layer in the 5 GHz band," IEEE 802.11, May 1999.

[11] IEEE 802.16d, "Air interface for fixed broadband wireless access systems," IEEE 802.16, May 2004.

[12] Jankiraman M. and Prasad R., "Wideband multimedia solution using hybrid CDMA/OFDM/SFH techniques," in *Proc. International Workshop on Multi-Carrier Spread Spectrum and Related Topics (MC-SS '99)*, Oberpfaffenhofen, Germany, pp. 15–24, Sept. 1999.

[13] Kaiser S., "OFDM-CDMA versus DS-CDMA: performance evaluation in fading channels," in *Proc. IEEE International Conference on Communications (ICC '95)*, Seattle, USA, pp. 1722–1726, June 1995.

[14] Kaiser S., "On the performance of different detection techniques for OFDM-CDMA in fading channels," in *Proc. IEEE Global Telecommunications Conference (GLOBECOM '95)*, Singapore, pp. 2059–2063, Nov. 1995.

[15] Kaiser S., "Trade-off between channel coding and spreading in multi-carrier CDMA systems," in *Proc. IEEE International Symposium on Spread Spectrum Techniques and Applications (ISSSTA '96)*, Mainz, Germany, pp. 1366–1370, Sept. 1996.

[16] Kaiser S., *Multi-Carrier CDMA Mobile Radio Systems – Analysis and Optimization of Detection, Decoding, and Channel Estimation*, Düsseldorf: VDI-Verlag, Fortschritt-Berichte VDI, series 10, no. 531, 1998, PhD thesis.

[17] Kaiser S., "MC-FDMA and MC-TDMA versus MC-CDMA and SS-MC-MA: performance evaluation for fading channels," in *Proc. IEEE International Symposium on Spread Spectrum Techniques and Applications (ISSSTA '98)*, Sun City, South Africa, pp. 115–120, Sept. 1998.

[18] Kaiser S. and Fazel K., "A spread spectrum multi-carrier multiple-access system for mobile communications," in *Proc. International Workshop on Multi-Carrier Spread Spectrum (MC-SS '97)*, Oberpfaffenhofen, Germany, pp. 49–56, April 1997.

[19] Kaiser S. and Fazel K., "A flexible spread spectrum multi-carrier multiple-access system for multi-media applications," in *Proc. IEEE International Symposium on Personal, Indoor and Mobile Communications (PIMRC '97)*, Helsinki, Finland, pp. 100–104, Sept. 1997.

[20] Kaiser S. and Hagenauer J., "Multi-carrier CDMA with iterative decoding and soft-interference cancellation," in *Proc. IEEE Global Telecommunications Conference (GLOBECOM '97)*, Phoenix, USA, pp. 6–10, Nov. 1997.

[21] Kaiser S. and Krzymien W. A., "Performance effects of the uplink asynchronism in a spread spectrum multi-carrier multiple access system," *European Transactions on Telecommunications (ETT)*, vol. 10, pp. 399–406, July/Aug. 1999.

[22] Kaiser S., Krzymien W. A., and Fazel K., "SS-MC-MA systems with pilot symbol aided channel estimation in the asynchronous uplink," *European Transactions on Telecommunications (ETT)*, vol. 11, pp. 605–610, Nov./Dec. 2000.

[23] Lindner J., "On coding and spreading for MC-CDMA," in *Proc. International Workshop on Multi-Carrier Spread Spectrum and Related Topics (MC-SS '99)*, Oberpfaffenhofen, Germany, pp. 89–98, Sept. 1999.

[24] Novak R. and Krzymien W. A., "A downlink SS-OFDM-F/TA packet data system employing multi-user diversity," in *Proc. International Workshop on Multi-Carrier Spread Spectrum and Related Topics (MC-SS 2001)*, Oberpfaffenhofen, Germany, pp. 181–190, Sept. 2001.

[25] Petroff A. and Withington P., "Time modulated ultra-wideband (TM-UWB) overview," in *Proc. Wireless Symposium 2000*, San Jose, USA, Feb. 2000.

[26] Rohling H. and Grünheid R., "Performance comparison of different multiple access schemes for the downlink of an OFDM communication system," in *Proc. IEEE Vehicular Technology Conference (VTC '97)*, Phoenix, USA, pp. 1365–1369, May 1997.

[27] Ross G. F., "The transient analysis of certain TEM mode four-post networks," *IEEE Transactions on Microwave Theory and Techniques*, vol. 14, pp. 528–542, Nov. 1966.

[28] Sari H., "Orthogonal frequency-division multiple access with frequency hopping and diversity," in *Proc. International Workshop on Multi-Carrier Spread Spectrum (MC-SS '97)*, Oberpfaffenhofen, Germany, pp. 57–68, April 1997.

[29] Sari H., "A review of multi-carrier CDMA," in *Proc. International Workshop on Multi-Carrier Spread Spectrum and Related Topics (MC-SS 2001)*, Oberpfaffenhofen, Germany, pp. 3–12, Sept. 2001.

[30] Sari H. and Karam G., "Orthogonal frequency-division multiple access and its application to CATV networks," *European Transactions on Telecommunications (ETT)*, vol. 9, pp. 507–516, Nov./Dec. 1998.

[31] Sari H., Levy Y., and Karam G., "Orthogonal frequency-division multiple access for the return channel on CATV networks," in *Proc. International Conference on Telecommunications (ICT '96)*, Istanbul, Turkey, pp. 52–57, April 1996.

[32] Sari H., Levy Y., and Karam G., "OFDMA – a new multiple access technique and its application to interactive CATV networks," in *Proc. European Conference on Multimedia Applications, Services and Techniques (ECMAST '96)*, Louvain-la-Neuve, Belgium, pp. 117–127, May 1996.

[33] Sari H., Vanhaverbeke F., and Moeneclaey M., "Some novel concepts in multiplexing and multiple access," in *Proc. International Workshop on Multi-Carrier Spread Spectrum and Related Topics (MC-SS '99)*, Oberpfaffenhofen, Germany, pp. 3–12, Sept. 1999.

[34] Schnell M., De Broeck I., and Sorger U., "A promising new wideband multiple access scheme for future mobile communications systems," *European Transactions on Telecommunications (ETT)*, vol. 10, pp. 417–427, July/Aug. 1999.

[35] Sorger U., De Broeck I., and Schnell S., "Interleaved FDMA – a new spread spectrum multiple access scheme," in *Proc. IEEE International Conference on Communications (ICC '98)*, Atlanta, USA, pp. 1013–1017, June 1998.

[36] Tomba L. and Krzymien W. A., "An OFDM/SFH-CDMA transmission scheme for the uplink," in *Proc. International Workshop on Multi-Carrier Spread Spectrum (MC-SS '97)*, Oberpfaffenhofen, Germany, pp. 203–210, April 1997.

[37] Win M. Z. and Scholtz R. A., "Ultra-wideband bandwidth time-hopping spread spectrum impulse radio for wireless multiple access communications," *IEEE Transactions on Communications*, vol. 48, pp. 679–691, April 2000.

4

Implementation Issues

A general PHY block diagram of a multi-carrier transceiver employed in a cellular environment with a central base station and several terminal stations in a point to multi-point topology is depicted in Figure 4-1.

For the downlink, transmission occurs in the base station and reception in the terminal station and for the uplink, transmission occurs in the terminal station and reception in the base station. Although very similar in concept, note that in general the base station equipment handles more than one terminal station; hence, its architecture is more complex.

The transmission operation starts with a stream of data symbols (bits, bytes, or packets) sent from a higher protocol layer, i.e. the medium access control (MAC) layer. These data symbols are channel encoded, mapped into constellation symbols according to the designated symbol alphabet, spread (only in MC-SS), and optionally interleaved. The modulated symbols and the corresponding reference/pilot symbols are multiplexed to form a frame or a burst. The resulting symbols after framing or burst formatting are multiplexed and multi-carrier modulated by using OFDM and finally forwarded to the radio transmitter through a physical interface with digital-to-analogue (D/A) conversion.

The reception operation starts with receiving an analogue signal from the radio receiver. The analogue-to-digital converter (A/D) converts the analogue signal to the digital domain. After multi-carrier demodulation (IOFDM) and deframing, the extracted pilot symbols and reference symbols are used for channel estimation and synchronization. After optionally de-interleaving, de-spreading (only in the case of MC-SS) and de-mapping, the channel decoder corrects the channel errors to guarantee data integrity. Finally, the received data symbols (bits, bytes, or packets) are forwarded to the higher protocol layer for further processing.

Although the heart of an orthogonal multi-carrier transmission is the FFT/IFFT operation, synchronization and channel estimation process together with channel decoding play a major role. To ensure a low cost receiver (low cost local oscillator and RF components) and to guarantee a high spectral efficiency, robust digital synchronization and channel estimation mechanisms are needed. The throughput of an OFDM system not only depends on the used modulation constellation and FEC scheme but also on the amount of reference and pilot symbols spent to guarantee reliable synchronization and channel estimation.

Multi-Carrier and Spread Spectrum Systems Second Edition   K. Fazel and S. Kaiser
© 2008 John Wiley & Sons, Ltd

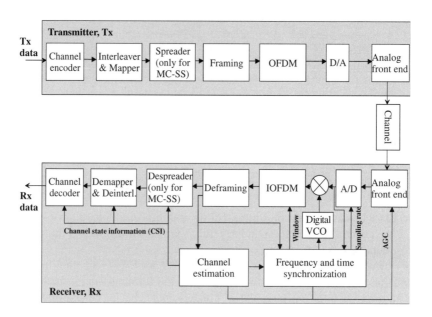

Figure 4-1   General PHY block diagram of a multi-carrier transceiver

In Chapter 2 the different de-spreading and detection strategies for MC-SS systems were analysed. It was shown that with an appropriate detection strategy, especially in full load conditions (where all users are active), a high system capacity can be achieved. In the performance analysis in Chapter 2 we assumed that the modem is perfectly synchronized and the channel is perfectly known at the receiver.

The principal goal of this chapter is to describe in detail the remaining components of a multi-carrier transmission scheme with or without spreading. The focus is given to multi-carrier modulation/demodulation, digital I/Q generation, sampling, channel coding/decoding, framing/de-framing, synchronization, and channel estimation mechanisms. Especially for synchronization and channel estimation units the effects of the transceiver imperfections (i.e. frequency drift, imperfect sampling time, phase noise) are highlighted. Finally, the effects of the amplifier non-linearity in multi-carrier transmission are analyzed.

4.1   Multi-Carrier Modulation and Demodulation

After symbol mapping (e.g. M-QAM) and spreading (in MC-SS), each block of N_c complex-valued symbols is serial-to-parallel (S/P) converted and submitted to the multi-carrier modulator, where the symbols are transmitted simultaneously on N_c parallel sub-carriers, each occupying a small fraction ($1/N_c$) of the total available bandwidth B.

Figure 4-2 shows the block diagram of a multi-carrier transmitter. The transmitted baseband signal is given by

$$s(t) = \frac{1}{N_c} \sum_{i=-\infty}^{+\infty} \sum_{n=0}^{N_c-1} d_{n,i} g(t - iT_s) \, e^{j2\pi f_n t}, \tag{4.1}$$

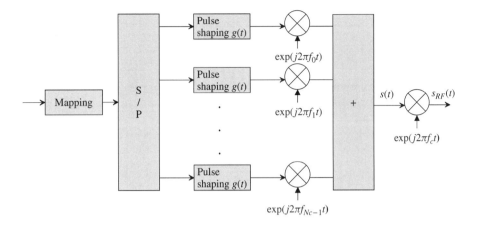

Figure 4-2 Block diagram of a multi-carrier transmitter

where N_c is the number of sub-carriers, $1/T_s$ is the symbol rate associated with each sub-carrier, $g(t)$ is the impulse response of the transmitter filters, $d_{n,i}$ is the complex constellation symbol, and f_n is the frequency of sub-carrier n. We assume that the sub-carriers are equally spaced, i.e.

$$f_n = \frac{n}{T_s}, \quad n = 0, \dots, N_c - 1. \tag{4.2}$$

The up-converted transmitted RF signal $S_{RF}(t)$ can be expressed by

$$S_{RF}(t) = \frac{1}{N_c} \, \mathrm{Re} \left\{ \sum_{i=-\infty}^{+\infty} \sum_{n=0}^{N_c-1} d_{n,i} g(t - iT_s) \, e^{j2\pi(f_n+f_c)t} \right\}$$

$$= \mathrm{Re}\{s(t) \, e^{j2\pi f_c t}\} \tag{4.3}$$

where f_c is the carrier frequency.

As shown in Figure 4-3, at the receiver side after down-conversion of the RF signal $r_{RF}(t)$, a bank of N_c matched filters is required to demodulate all sub-carriers. The received baseband signal after demodulation and filtering and before sampling at sub-carrier frequency f_m is given by

$$r_m(t) = [r(t) \, e^{-j2\pi f_m t}] \otimes h(t)$$

$$= \left[\sum_{i=-\infty}^{+\infty} \sum_{n=0}^{N_c-1} d_{n,i} g(t - iT_s) \, e^{j2\pi(f_n-f_m)t} \right] \otimes h(t), \tag{4.4}$$

where $h(t)$ is the impulse response of the receiver filter, which is matched to the transmitter filter (i.e. $h(t) = g^*(-t)$). The symbol $\otimes$ indicates the convolution operation. For simplicity, the received signal is given in the absence of fading and noise. After sampling

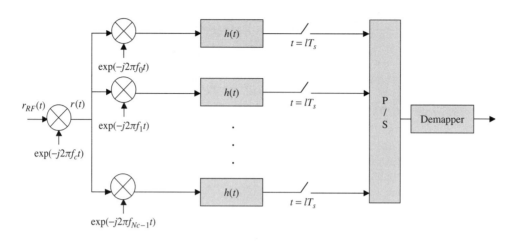

Figure 4-3 Block diagram of a multi-carrier receiver

at the optimum sampling time $t = lT_s$, the samples result in $r_m(lT_s) = d_{m,l}$, if the transmitter and the receiver of the multi-carrier transmission system fulfill both the ISI and ICI-free Nyquist conditions [70].

To fulfill these conditions, different pulse shaping filtering can be used:

Rectangular Band-Limited System
Each sub-carrier has a rectangular band-limited transmission filter with the impulse response

$$g(t) = \frac{\sin\left(\pi \dfrac{t}{T_s}\right)}{\pi \dfrac{t}{T_s}} = \text{sinc}\left(\pi \frac{t}{T_s}\right). \tag{4.5}$$

The spectral efficiency of the system is equal to the optimum value, i.e. the normalized value of 1 bit/s/Hz.

Rectangular Time-Limited System
Each sub-carrier has a rectangular time-limited transmission filter with the impulse response

$$g(t) = \text{rect}(t) = \begin{cases} 1 & 0 \leqslant t < T_s \\ 0 & \text{otherwise} \end{cases}. \tag{4.6}$$

The spectral efficiency of the system is equal to the normalized value $1/(1 + BT_s/N_c)$. For large N_c, it approaches the optimum normalized value of 1 bit/s/Hz.

Raised Cosine Filtering
Each sub-carrier is filtered by a time-limited ($t \in \{-kT_s', kT_s'\}$) square root of a raised cosine filter with the roll-off factor α and impulse response [70]

$$
g(t) = \frac{\sin\left[\dfrac{\pi t}{T_s'}(1-\alpha)\right] + \dfrac{k\alpha t}{T_s'}\cos\left[\dfrac{\pi t}{T_s'}(1+\alpha)\right]}{\dfrac{\pi t}{T_s'}\left[1-\left(\dfrac{k\alpha t}{T_s'}\right)^2\right]},
\tag{4.7}
$$

where $T_s' = (1+\alpha)T_s$ and k is the maximum number of samples that the pulse shall not exceed. The spectral efficiency of the system is equal to $1/(1+(1+\alpha)/N_c)$. For large N_c, it approaches the optimum normalized value of 1 bit/s/Hz.

4.1.1 Pulse Shaping in OFDM

OFDM employs a time-limited rectangular pulse shaping which leads to a simple digital implementation. OFDM without guard time is an optimum system, where for large numbers of sub-carriers its efficiency approaches the optimum normalized value of 1 bit/s/Hz.

The impulse response of the receiver filter is

$$
h(t) = \begin{cases} 1 & \text{if} -T_s < t \leqslant 0 \\ 0 & \text{otherwise} \end{cases}.
\tag{4.8}
$$

It can easily be shown that the condition of absence of ISI and ICI is fulfilled.

In the case of inserting a guard time T_g, the spectral efficiency of OFDM will be reduced to $1 - T_g/(T_s + T_g)$ for large N_c.

4.1.2 Digital Implementation of OFDM

By omitting the time index i in Equation (4.1), the transmitted OFDM baseband signal, i.e. one OFDM symbol with a guard time, is given by

$$
s(t) = \frac{1}{N_c}\sum_{n=0}^{N_c-1} d_n\, e^{j2\pi\frac{nt}{T_s}}, \quad -T_g \leqslant t < T_s,
\tag{4.9}
$$

where d_n is a complex-valued data symbol, T_s is the symbol duration, and T_g is the guard time between two consecutive OFDM symbols in order to prevent ISI and ICI in a multi-path channel. The sub-carriers are separated by $1/T_s$.

Note that for burst transmission, i.e. burst formatting, a pre-/post-fix of duration T_a can be added to the original OFDM symbol of duration $T_s' = T_s + T_g$ so that the total OFDM symbol duration becomes

$$
T' = T_s + T_g + T_a.
\tag{4.10}
$$

The pre-/post-fix can be designed such that it has good correlation properties in order to perform channel estimation or synchronization. One possibility for the pre-/post-fix is to extend the OFDM symbol by a specific PN sequence with good correlation properties. At the receiver, as a guard time, the pre-/post-fix is skipped and the OFDM symbol is rebuilt as described in Section 4.5.

From the above expression we note that the transmitted OFDM symbol can be generated by using an inverse complex FFT operation (IFFT), where the de-multiplexing is done by an FFT operation. In the complex digital domain this operation leads to an IDFT operation with N_c points at the transmitter side and a DFT with N_c points at the receiver side (see Figure 4-4). Note that for the guard time and pre-/post-fix L_g samples are inserted after the IDFT operation at the transmitter side and removed before the DFT at the receiver side.

Highly repetitive structures based on elementary operations such as butterflies for the FFT operation can be applied if N_c is of the power of 2 [1]. Depending on the transmission media and the carrier frequency f_c, the actual OFDM transmission systems employ from 64 up to 2048 (2k) sub-carriers. In the DVB-T standard [19], up to 8192 (8k) sub-carriers are chosen to combat long echoes in a single frequency network operation.

The complexity of the FFT operation (multiplications and additions) depends on the number of FFT points N_c. It can be approximated by $(N_c/2)\log N_c$ operations [1]. Furthermore, large numbers of FFT points, resulting in long OFDM symbol durations T_s', make the system more sensitive to the time variance of the channel (Doppler effect) and more vulnerable to the oscillator phase noise (technological limitation). However, on the other hand, a large symbol duration increases the spectral efficiency due to a decrease of the guard interval loss.

Therefore, for any OFDM realization a tradeoff between the number of FFT points, the sensitivity to the Doppler and phase noise effects, and the loss due to the guard interval has to be found.

4.1.3 Virtual Sub-Carriers and DC Sub-Carrier

By employing large numbers of sub-carriers in OFDM transmission, a high frequency resolution in the channel bandwidth can be achieved. This enables a much easier

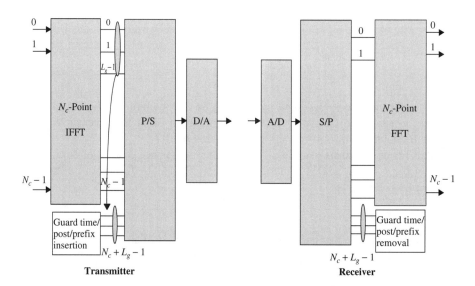

Figure 4-4   Digital implementation of OFDM

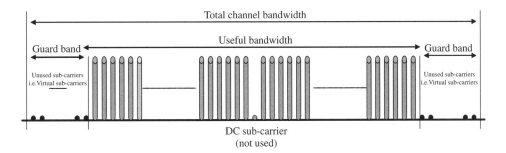

Figure 4-5   Virtual sub-carriers used for filtering

implementation and design of the filters. If the number of FFT points is slightly higher than that required for data transmission, a simple filtering can be achieved by putting at both sides of the spectrum null sub-carriers (guard bands), called virtual sub-carriers (see Figure 4-5). Furthermore, in order to avoid the DC problem, a null sub-carrier can be put in the middle of the spectrum; i.e. the DC sub-carrier is not used.

4.1.4   D/A and A/D Conversion, I/Q Generation

The digital implementation of multi-carrier transmission at the transmitter and the receiver sides requires digital-to-analogue (D/A) and analogue-to-digital (A/D) conversion and methods for modulating and demodulating a carrier with a complex OFDM time signal.

4.1.4.1   D/A and A/D Conversion and Sampling Rate

The main advantage of an OFDM transmission and reception is its digital implementation using digital FFT processing. Therefore, at the transmission side the digital signal after digital IFFT processing is converted to the analogue domain with a D/A converter, ready for IF/RF upconversion and vice versa at the receiver side.

The number of bits reserved for the D/A and A/D conversions depends on many parameters: (a) the accuracy needed for a given constellation, (b) required Tx/Rx dynamic ranges (e.g. the difference between the maximum received power and the receiver sensitivity), and (c) the used sampling rate, i.e. complexity. It should be noticed that at the receiver side, due to a higher disturbance, a more accurate converter is required. In practice, in order to achieve a good tradeoff between complexity, performance, and implementation loss typically for a 64-QAM transmission, D/A converters with 8 bits or higher should be used, and 10 bits or higher are recommended for the receiver A/D converters. However, for low order modulation, these constraints can be relaxed.

The sampling rate is a crucial parameter. To avoid any problem with aliasing, the sampling rate f_{samp} should be at least twice the maximum frequency of the signal. This requirement is theoretically satisfied by choosing the sampling rate [1]

$$f_{samp} = 1/T_{samp} = N_c/T_s = B. \tag{4.11}$$

However, in order to provide a better channel selectivity in the receiver regarding adjacent channel interference, a higher sampling rate than the channel bandwidth might be used, i.e. $f_{samp} > N_c/T_s$.

4.1.4.2   I/Q Generation

At least two methods exist for modulating and demodulating a carrier (I and Q generation) with a complex OFDM time signal. These are described below.

Analogue quadrature method
This is a conventional solution in which the in-phase carrier component I is fed by the real part of the modulating signal and the quadrature component Q is fed by the imaginary part of the modulating signal [70].

The receiver applies the inverse operations using an I/Q demodulator (see Figure 4-6). This method has two drawbacks for an OFDM transmission, especially for large numbers of sub-carriers and high order modulation (e.g. 64-QAM): (a) due to imperfections in the RF components, it is difficult at moderate complexity to avoid cross-talk between the I and Q signals and, hence, to maintain an accurate amplitude and phase matching between the I and Q components of the modulated carrier across the signal bandwidth; this imperfection may result in high received baseband signal degradation, i.e. interference; and (b) it requires two A/D converters.

A low cost front-end may result in I/Q mismatching, emanating from the gain mismatch between the I and Q signals and from nonperfect quadrature generation. These problems can be solved in the digital domain.

Digital FIR filtering method
The second approach is based on employing digital techniques in order to shift the complex time domain signal up in frequency and produce a signal with no imaginary components that is fed to a single modulator. Similarly, the receiver requires a single demodulator. However, the A/D converter has to work at double sampling frequency (see Figure 4-7).

The received analogue signal can be written as

$$r(t) = I(t)\,\cos(\pi t/T_{samp}) + Q(t)\,\sin(\pi t/T_{samp}), \tag{4.12}$$

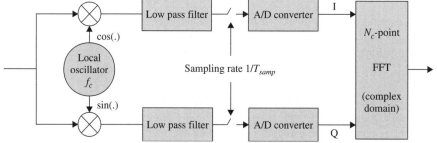

Figure 4-6   Conventional I/Q generation with two analogue demodulators

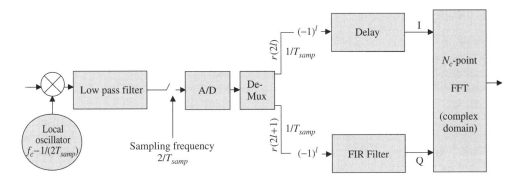

Figure 4-7   Digital I/Q generation using FIR filtering with a single analogue demodulator

where T_{samp} is the sampling period of each I and Q component. By doubling the sampling rate to $2/T_{samp}$ we get the sampled signal

$$r(l) = I(l)\cos(\pi l/2) + Q(l)\sin(\tau l/2). \tag{4.13}$$

This stream can be separated into two sub-streams with rate $1/T_{samp}$ by taking the even and odd samples:

$$r(2l) = I(2l)\cos(\pi l) + Q(2l)\sin(\pi l)$$

$$r(2l+1) = I(2l+1)\cos(\pi(2l+1)/2) + Q(2l+1)\sin(\pi(2l+1)/2). \tag{4.14}$$

It is straightforward to show that the desired output I and Q components are related to $r(2l)$ and $r(2l+1)$ by

$$I(l) = (-1)^l r(2l) \tag{4.15}$$

and the $Q(l)$ outputs are obtained by delaying $(-1)^l r(2l+1)$ by $T_{samp}/2$, i.e. passing the $(-1)^l r(2l+1)$ samples through an interpolator filter (FIR). The $I(l)$ components have to be delayed as well to compensate the FIR filtering delay.

In other words, at the transmission side this method consists (at the output of the complex digital IFFT processing) of filtering the Q channel with an FIR interpolator filter to implement a 1/2 sample time shift. Both I and Q streams are then oversampled by a factor of 2. By taking the even and odd components of each stream, only one digital stream at twice the sampling frequency is formed. This digital signal is converted to analogue and used to modulate the RF carrier. At the reception side, the inverse operation is applied. The incoming analogue signal is down-converted and centered on a frequency $f_{samp}/2$, filtered, and converted to digital by sampling at twice the sampling frequency (i.e. $2f_{samp}$). It is de-multiplexed into the two streams $r(2l)$ and $r(2l+1)$ at rate $f_{samp} = 1/T_{samp}$. The I and Q channels are multiplied by $(-1)^l$ to ensure transposition of the spectrum of the signal into baseband [1]. The Q channel is filtered using the same FIR interpolator filter as the transmitter while the I components are delayed by a corresponding amount so that the I and Q components can be delivered simultaneously to the digital FFT processing unit.

4.2 Synchronization

Reliable receiver synchronization is one of the most important issues in multi-carrier communication systems, and is especially demanding in fading channels when coherent detection of high order modulation schemes is employed.

A general block diagram of a multi-carrier receiver synchronization unit is depicted in Figure 4-8. The incoming signal in the analogue front end unit is first down-converted, performing the complex demodulation to baseband time domain digital I and Q signals of the received OFDM signal. The local oscillator(s) of the analogue front end has/have to work with sufficient accuracy. Therefore, the local oscillator(s) is/are continuously adjusted by the frequency offset estimated in the synchronization unit. In addition, before the FFT operation a fine frequency offset correction signal might be required to reduce the ICI.

Furthermore, the sampling rate of the A/D clock needs to be controlled by the time synchronization unit as well, in order to prevent any frequency shift after the FFT operation that may result in an additional ICI. The correct positioning of the FFT window is another important task of the timing synchronization.

The remaining task of the OFDM synchronization unit is to estimate the phase and amplitude distortion of each sub-carrier, where this function is performed by the *channel estimation* core (see Section 4.3). These estimated channel state information (CSI) values are used to derive for each demodulated symbol the *reliability information* that is directly applied for de-spreading and/or for channel decoding.

An automatic gain control (AGC) of the incoming analogue signal is also needed to adjust the gain of the received signal to its desired values.

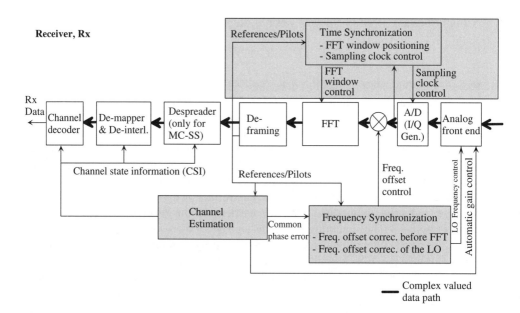

Figure 4-8 General block diagram of a multi-carrier synchronization unit

The performance of any synchronization and channel estimation algorithm is determined by the following parameters:

- *Minimum SNR* under which the operation of synchronization is guaranteed
- *Acquisition time* and *acquisition range* (e.g. maximum tolerable deviation range of timing offset, local oscillator frequency)
- *Overhead* in terms of reduced data rate or power excess
- *Complexity*, regarding implementation aspects and
- *Robustness* and *accuracy* in the presence of multi-path and interference disturbances.

In a wireless cellular system with a point to multi-point topology, the base station acts as a central control of the available resources among several terminal stations. Signal transmission from the base station towards the terminal station in the downlink is often done in a continuous manner. However, the uplink transmission from the terminal station towards the base station might be different and can be performed in a bursty manner.

In case of a continuous downlink transmission, both acquisition and tracking algorithms for synchronization can be applied [24], where all fine adjustments to counteract time-dependent variations (e.g. local oscillator frequency offset, Doppler, timing drift, common phase error) are carried out in the tracking mode. Furthermore, in the case of a continuous transmission, non-pilot-aided algorithms (blind synchronization) might be considered.

However, the situation is different for a bursty transmission. All synchronization parameters for each burst have to be derived with required accuracy within the limited time duration. Two ways exist to achieve simple and accurate burst synchronization:

- enough reference and pilot symbols are appended to each burst or
- the terminal station is synchronized to the downlink, where the base station will continuously broadcast synchronization signaling to all terminal stations.

The first solution requires a significant amount of overhead, which leads to a considerable loss in uplink spectral efficiency. The second solution is widely adopted in burst transmission. Here all terminal stations synchronize their transmit frequency and clock to the received base station signal. The time-advance variation (moving vehicle) between the terminal station and the base station can be adjusted through a closed loop by transmitting regular ranging messages individually from the base station to each terminal station. Hence, the *burst receiver* at the base station does not need to regenerate the terminal station clock and carrier frequency; it only has to estimate the channel. Note that in FDD the uplink carrier frequency has only to be shifted.

In time- and frequency-synchronous multi-carrier transmission the receiver at the base station needs to detect the start position of an OFDM symbol or frame and to estimate the channel state information from some known pilot symbols inserted in each OFDM symbol. If the coherence time of the channel exceeds an OFDM symbol, the channel estimation can estimate the time variation as well. This strategy, which will be considered in the following, simplifies a burst receiver.

To summarize, in the next sections we make the following assumptions:

- The terminal stations are frequency/time-synchronized to the base station.
- The Doppler variation is slow enough to be considered constant during one OFDM symbol of duration T_s'.
- The guard interval duration T_g is larger than the channel impulse response.

4.2.1 General

The synchronization algorithms employed for multi-carrier demodulation are based either on the analysis of the received signal (*non-pilot-aided*, i.e. blind synchronization) [12, 13, 38] or on the processing of special dedicated data time and/or frequency multiplexed with the transmitted data, i.e. *pilot-aided* synchronization [13, 24, 25, 60, 81]. For instance, in non-pilot-aided synchronization some of these algorithms exploit the intrinsic redundancy present in the guard time (cyclic extension) of each OFDM symbol. Maximum likelihood estimation of parameters can also be applied, exploiting the guard time redundancy [78] or using some dedicated transmitted reference symbols [60].

As shown in Figure 4-8, there are three main synchronization tasks around the FFT: (a) timing recovery, (b) carrier frequency recovery, and (c) carrier phase recovery. In this part, we concentrate on the first two items, since the carrier phase recovery is closely related to the channel estimation (see Section 4.3). Hence, the two main synchronization parameters that have to be estimated are: (a) time positioning of the FFT window including the sampling rate adjustment that can be controlled in a two-stage process, coarse- and fine-timing control, and (b) the possible large frequency difference between the receiver and transmitter local oscillators that has to be corrected to a very high accuracy.

As known from DAB [16], DVB-T [19], and other standards, the transmission is usually performed in a *frame-by-frame* basis. An example of an OFDM frame is depicted in Figure 4-9, where each frame consists of a so-called null symbol (without signal power) transmitted at the frame beginning, followed by some known reference symbols and data symbols. Furthermore, within data symbols some reference pilots are scattered in time and frequency. The null symbol may serve two important purposes: interference and noise estimation, and coarse timing control. The coarse timing control may use the

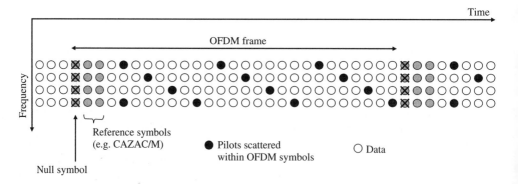

Figure 4-9 Example of an OFDM frame

null symbol as a means of quickly establishing frame synchronization prior to fine time synchronization.

Fine timing control can be achieved by time [81] or frequency domain processing [14] using the reference symbols. These symbols have good partial auto-correlation properties. The resulting signal can either be used to directly control the fine positioning of the FFT window or to alter the sampling rate of the A/D converters. In addition, for time synchronization the properties of the guard time can be exploited [38, 78].

If the frequency offset is smaller than half the sub-carrier spacing a maximum likelihood frequency estimation can be applied by exploiting the reference symbols [60] or the guard time redundancy [78]. In the case where the frequency offset exceeds several sub-carrier spacings, a frequency offset estimation technique using again the OFDM reference symbol as above for timing can be used [63, 81]. These reference symbols allow coarse and fine adjustment of the local oscillator frequency in a two-step process. Here, frequency domain processing can be used. The more such special reference symbols are embedded into the OFDM frame, the faster the acquisition time and the higher the accuracy. Finally, a *common phase error* (CPE) estimation can be performed that partially counters the effect of phase noise of the local oscillator [74]. The common phase error estimation may exploit pilot symbols in each OFDM symbol (see Section 4.7.1.3), which can also be used for channel estimation.

In the following, after examining the effects of synchronization imperfections on multi-carrier transmission, we will detail the maximum likelihood estimation algorithms and other time and frequency synchronization techniques that are commonly employed.

4.2.2 Effects of Synchronization Errors

Large timing and frequency errors in multi-carrier systems cause an increase of ISI and ICI, resulting in high performance degradations.

Let us assume that the receiver local oscillator frequency f_c (see Figure 4-3) is not perfectly locked to the transmitter frequency. The baseband received signal after down-conversion is

$$r(t) = s(t)\, e^{j2\pi f_{error} t} + n(t), \tag{4.16}$$

where f_{error} is the frequency error and $n(t)$ the complex-valued AWGN.

The above signal in the absence of fading after demodulation and filtering (i.e. convolution) at sub-carrier m can be written as [73]

$$r_m(t) = [s(t)\, e^{j2\pi f_{error} t} + n(t)]\, e^{-j2\pi f_m t} \otimes h(t)$$

$$= \left[\sum_{i=-\infty}^{+\infty} \sum_{n=0}^{N_c-1} d_{n,i}\, g(t - iT_s)\, e^{-j2\pi \left(\frac{n-m}{T_s} \right) t}\, e^{j2\pi f_{error} t} \right] \otimes h(t) + n'(t), \tag{4.17}$$

where $h(t)$ is the impulse response of the receiver filter and $n'(t)$ is the filtered noise. Let us assume that the sampling clock has a static error τ_{error}. The sample at instant lT_s of the received signal at sub-carrier m is made up of four terms as follows:

$$r_m(lT_s + \tau_{error}) = d_{m,l}A_m(\tau_{error})\, e^{j2\pi f_{error}lT_s} + ISI_{m,l} + ICI_{m,l} + n'(lT_s + \tau_{error}), \quad (4.18)$$

where the first term corresponds to the transmitted data $d_{m,i}$ which is attenuated and phase shifted. The second and third terms are the ISI and ICI, given by

$$ISI_{m,l} = \sum_{\substack{i=-\infty \\ i \neq l}}^{+\infty} d_{m,i}A_m[(l-i)T_s + \tau_{error}]\, e^{j2\pi f_{error}iT_s}, \quad (4.19)$$

$$ICI_{m,l} = \sum_{i=-\infty}^{+\infty} \sum_{\substack{n=0 \\ n \neq m}}^{N_c-1} d_{n,i}A_n[(l-i)T_s + \tau_{error}]\, e^{j2\pi f_{error}iT_s}, \quad (4.20)$$

where

$$A_n(t) = \left(g(t)\, e^{j2\pi f_{error}t}\, e^{j2\pi \frac{(n-m)t}{T_s}} \right) \otimes h(t), \quad (4.21)$$

$g(t)$ is the impulse response of the transmitter filter, and $A_n(lT_s)$ represents the sampled components of Equation (4.21), i.e. samples after convolution.

4.2.2.1 Analysis of the SNR in the Presence of a Frequency Error

Here we consider only the effect of a frequency error; i.e. we put $\tau_{error} = 0$ in the above expressions. For simplicity, the guard time is omitted. Then Equation (4.21) becomes [73]

$$A_n(t) = \begin{cases} e^{j\pi f_{error}t}\, e^{j\pi \frac{n-m}{T_s}t}\, \text{sinc}\left[\left(f_{error} + \frac{n-m}{T_s} \right)(T_s - t) \right]\left(1 - \frac{t}{T_s} \right) & 0 < t \leqslant T_s \\ 0 & \text{otherwise} \end{cases}.$$

(4.22)

After sampling at instant lT_s, at sub-carrier $m = n$, $A_m(0) = \text{sinc}(f_{error}T_s)$ and $A_m(lT_s) = 0$. However, for $m \neq n$, $A_m(0) = \text{sinc}(f_{error}T_s + n - m)$ and $A_m(lT_s) = 0$. Therefore, it can be shown that the received data after the FFT operation at time $t = 0$ and sub-carrier m can be written as [73]

$$r_m = d_m \text{sinc}(f_{error}T_s)\, e^{j2\pi f_{error}T_s} + ICI_m + n', \quad (4.23)$$

by omitting the time index. Note that the frequency error does not introduce any ISI.

Equation (4.23) shows that a frequency error creates besides the ICI a reduction of the received signal amplitude and a phase rotation of the symbol constellation on each sub-carrier. For large numbers of sub-carriers, the ICI can be modeled as AWGN. The resulting SNR can be written as

$$SNR_{ICI} \approx \frac{|d_m|^2 \mathrm{sinc}^2(f_{error}T_s)}{\sum\limits_{\substack{n=0 \\ n\neq m}}^{N_c-1} |d_n|^2 \mathrm{sinc}^2(n-m+f_{error}T_s) + P_N}, \tag{4.24}$$

where P_N is the power of the noise n'. If E_s is the average received energy of the individual sub-carriers and $N_0/2$ is the noise power spectral density of the AWGN, then

$$\frac{E_s}{N_0} = \frac{|d_m|^2}{P_N} \tag{4.25}$$

and the SNR can be expressed as

$$SNR_{ICI} \approx \frac{E_s}{N_0} \frac{\mathrm{sinc}^2(f_{error}T_s)}{1 + \frac{E_s}{N_0} \sum\limits_{\substack{n=0 \\ n\neq m}}^{N_c-1} \mathrm{sinc}^2(n-m+f_{error}T_s)}. \tag{4.26}$$

This equation shows that a frequency error can cause a significant loss in SNR. Furthermore, the SNR depends on the number of sub-carriers.

4.2.2.2 Analysis of the SNR in the Presence of a Clock Error

Here, we consider only the effect of a clock error, i.e. $f_{error} = 0$ in the above expressions. If the clock error is within the guard time, i.e. $|\tau_{error}| \leqslant T_g$ (or early synchronization), the timing error is absorbed and hence there is no ISI and no ICI. It results only in a phase shift at a given sub-carrier which can be compensated for by the channel estimation (see Section 4.3).

However, if the timing error exceeds the guard time, i.e. $|\tau_{error}| > T_g$ (or late synchronization), both ISI and ICI appear. As Equations (4.18) to (4.21) show, the clock error also introduces an amplitude reduction and a phase rotation which is proportional to the sub-carrier index. In a similar manner as above, the expression of the SNR can be derived as [73]

$$SNR_{ICI+ISI} \approx \frac{|d_m|^2(1 - \tau_{error}/T_s)^2}{(\tau_{error}/T_s)^2 \left(1 + 2 \sum\limits_{\substack{n=0 \\ n\neq m}}^{N_c-1} |d_n|^2 \mathrm{sinc}^2[(n-m)\tau_{error}/T_s] \right) + P_N}$$

$$\approx \frac{\frac{E_s}{N_0}(1 - \tau_{error}/T_s)^2}{\frac{E_s}{N_0}(\tau_{error}/T_s)^2 \left(1 + 2 \sum\limits_{\substack{n=0 \\ n\neq m}}^{N_c-1} \mathrm{sinc}^2[(n-m)\tau_{error}/T_s] \right) + 1}. \tag{4.27}$$

It can be observed again that a clock error exceeding the guard time will introduce a reduction in SNR.

4.2.2.3 Requirements on OFDM Frequency and Clock Accuracy

Figures 4-10 and 4-11 show the simulated SNR degradations in dB for different bit error rates (BERs) versus the frequency error and timing error for QPSK respectively. These diagrams show that an OFDM system is sensitive to frequency and to clock errors. In order to keep an acceptable performance degradation (loss smaller than 0.5–1 dB) the error after frequency synchronization and time synchronization should not exceed the following limits [73]:

$$\tau_{error}/T_s < 0.01$$
$$f_{error}T_s < 0.02. \tag{4.28}$$

Thus, the error relative to the sampling period should fulfill $\tau_{error}/(N_cT_{samp}) < 0.01$ and the error relative to the sub-carrier spacing shall not be greater than 2 % of the sub-carrier spacing, where the latter is usually a quite difficult condition.

It should be noticed that for dimensioning the length of the guard time, the time synchronization inaccuracy should be taken into account. As long as the sum of the timing offset and the maximum multi-path propagation delay is smaller than the guard time, the only effect is a phase rotation that can be estimated by the channel estimator (see Section 4.3) and compensated for by the channel equalizer (see Section 4.5).

4.2.3 Maximum Likelihood Parameter Estimation

Let us consider a frequency error f_{error} and a timing error τ_{error}. The joint maximum likelihood estimates $\hat{f}_{error}$ of the frequency error and $\hat{\tau}_{error}$ of the timing error are obtained by the maximization of the log-likelihood function (LLF) as follows [12, 60, 78]:

$$LLF\ (f_{error}, \tau_{error}) = \log\ p(r|f_{error}, \tau_{error}), \tag{4.29}$$

where $p(r|f_{error}, \tau_{error})$ denotes the probability density function of observing the received signal r, given a frequency error f_{error} and timing error τ_{error}. In Reference [78] it is shown that for $N_c + M$ samples

$$LLF(f_{error}, \tau_{error}) = |\gamma(\tau_{error})|\ \cos(2\pi f_{error} + \angle\gamma(\tau_{error})) - \rho\Phi(\tau_{error}), \tag{4.30}$$

$$\gamma(m) = \sum_{k=m}^{m+M-1} r(k)r^*(k+N_c),\ \Phi(m) = \sum_{k=m}^{m+M-1} |r(k)|^2 + |r(k+N_c)|^2, \tag{4.31}$$

where $\angle$ represents the argument of a complex number, ρ is a constant depending on the SNR, which represents the magnitude of the correlation between the sequences $r(k)$ and $r(k + N_c)$, and m is the start of the correlation function of the received sequence (in the case of no timing error one can start at $m = 0$). Note that the first term in Equation (4.30) is the weighted magnitude of $\gamma(\tau_{error})$, which is the sum of M consecutive correlations.

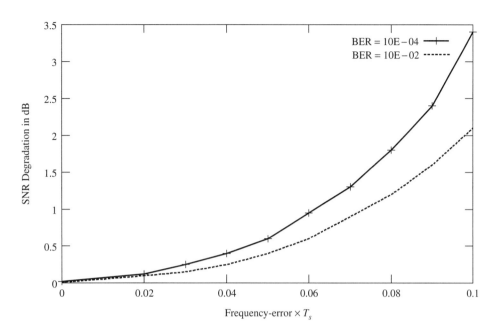

Figure 4-10 SNR degradation in dB versus the normalized frequency error $f_{error} T_s$; $N_c = 2048$ sub-carriers

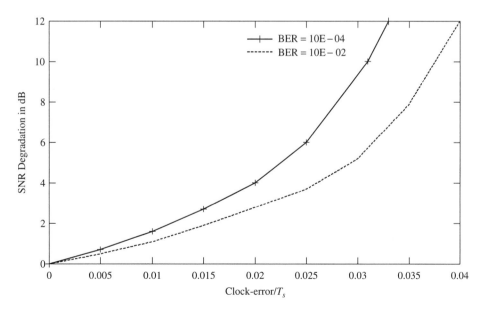

Figure 4-11 SNR degradation in dB versus the normalized timing error τ_{error} / T_s; $N_c = 2048$ sub-carriers

These sequences $r(k)$ could be known in the receiver by transmitting, for instance, two consecutive reference symbols $(M = N_c)$ as proposed by Moose [60], or one can exploit the presence of the guard time $(M = L_g)$ [78].

The maximization of the above *LLF* can be done in two steps:

- A first maximization can be performed to find the frequency error estimate $\hat{f}_{error}$,
- In a second step, the value of the given frequency error estimate is exploited for final maximization to find the timing error estimate $\hat{\tau}_{error}$.

The maximization of f_{error} is given by the partial derivation $\partial LLF(f_{error}, \tau_{error})/\partial f_{error} = 0$, which results in

$$\hat{f}_{error} = -\frac{1}{2\pi} \angle \gamma(\tau_{error}) + z = -\frac{1}{2\pi} \frac{\sum\limits_{k=m}^{m+M-1} \text{Im} \left[r(k)r^*(k + N_c) \right]}{\sum\limits_{k=m}^{m+M-1} \text{Re} \left[r(k)r^*(k + N_c) \right]} + z, \qquad (4.32)$$

where z is an integer value.

By inserting $\hat{f}_{error}$ in Equation (4.30), we obtain

$$LLF(\hat{f}_{error}, \tau_{error}) = |\gamma(\tau_{error})| - \rho \Phi(\tau_{error}), \qquad (4.33)$$

and maximizing Equation (4.33) gives us a joint estimate of $\hat{f}_{error}$ and $\hat{\tau}_{error}$:

$$\hat{f}_{error} = -\frac{1}{2\pi} \angle \gamma(\hat{\tau}_{error}),$$

$$\hat{\tau}_{error} = \arg(\max_{\tau_{error}} \{ |\gamma(\tau_{error})| - \rho \Phi(\tau_{error}) \}). \qquad (4.34)$$

Note that in case of $M = N_c$ (i.e. two reference symbols), $|\hat{f}_{error}| < 0.5$, $z = 0$, and no timing error $\tau_{error} = 0$ ($m = 0$), one obtains the same results as Moose [60] (see Figure 4-12).

The main drawback of the Moose maximum likelihood frequency detection is the small range of acquisition, which is only half of the sub-carrier spacing F_s. When $f_{error} \rightarrow 0.5F_s$, the estimate $\hat{f}_{error}$ may, due to noise and the discontinuity of arctangent, jump to -0.5. When this happens, the estimate is no longer unbiased and in practice it becomes useless. Thus, for frequency errors exceeding one-half of the sub-carrier spacing, an initial acquisition strategy, coarse frequency acquisition, should be applied. To enlarge the acquisition range of a maximum likelihood estimator, a modified version of this estimator was proposed in Reference [12]. The basic idea is to modify the shape of the LLF.

The joint estimation of frequency and timing error using the guard time may be sensitive in environments with several long echoes. In the following section, we will examine some approaches for time and frequency synchronization which are used in several implementations.

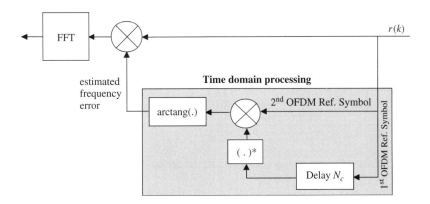

Figure 4-12   Moose maximum likelihood frequency estimator ($M = N_c$)

4.2.4   Time Synchronization

As we have explained before, the main objective of time synchronization for OFDM systems is to know when a received OFDM symbol starts. By using the guard time, the timing requirements can be relaxed. A time offset not exceeding the guard time gives rise to a phase rotation of the sub-carriers. This phase rotation is larger on the edge of the frequency band. If a timing error is small enough to keep the channel impulse response within the guard time, the orthogonality is maintained and a symbol timing delay can be viewed as a phase shift introduced by the channel. This phase shift can be estimated by the channel estimator (see Section 4.3) and corrected by the channel equalizer (see Section 4.5). However, if a time shift is larger than the guard time, ISI and ICI occur and signal orthogonality is lost.

Basically the task of the time synchronization is to estimate the two main functions: FFT window positioning (OFDM symbol/frame synchronization) and sampling rate estimation for A/D conversion control. The operation of time synchronization can be carried out in two steps: coarse and fine symbol timing.

4.2.4.1   Coarse Symbol Timing

Different methods, depending on the transmission signal characteristics, can be used for coarse timing estimation [24, 25, 78]. Basically, the power at baseband can be monitored prior to FFT processing and, for instance, the dips resulting from null symbols (see Figure 4-9) might be used to control a 'flywheel'-type state transition algorithm, as known from traditional frame synchronization [43].

Null symbol detection

A null symbol, containing no power, is transmitted, for instance, in DAB at the beginning of each OFDM frame (see Figure 4-13). By performing simple power detection at the receiver side before the FFT operation, the beginning of the frame can be detected; i.e. the receiver locates the null symbol by searching for a dip in the power of the received signal. This can be achieved, for instance, by using a flywheel algorithm to guard against

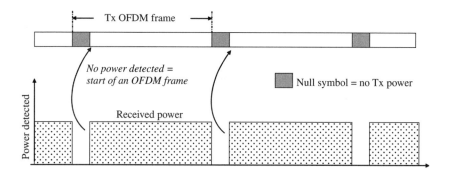

Figure 4-13   Coarse time synchronization based on null symbol detection

occasional failures to detect the null symbol once in lock [43]. The basic function of this algorithm is that, when the receiver is out-of-lock, it searches continuously for the null symbols, whereas when in-lock it searches for the symbol only at the expected null symbols. The null symbol detection gives only coarse timing information.

Two identical half reference symbols

In Reference [81] a timing synchronization is proposed that searches for a training symbol with two identical halves in the time domain, which can be sent at the beginning of an OFDM frame (see Figure 4-14). At the receiver side, these two identical time domain sequences may only be phase shifted $\phi = \pi T_s f_{error}$ due to the carrier frequency offset. The two halves of the training symbol are made identical by transmitting a PN sequence on the even frequencies, while zeros are used on the odd frequencies. Let there be M complex-valued samples in each half of the training symbol. The function for estimating

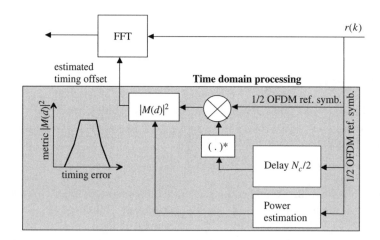

Figure 4-14   Time synchronization based on two identical half reference symbols

the timing error d is defined as

$$M(d) = \frac{\sum_{m=0}^{M-1} r^*_{d+m} r_{d+m+M}}{\sum_{m=0}^{M-1} |r_{d+m+M}|^2}. \tag{4.35}$$

Finally, the estimate of the timing error is derived by taking the maximum quadratic value of the above function, i.e. max $|M(d)|^2$. The main drawback of this metric is its 'plateau', which may lead to some uncertainties.

Guard time exploitation

Each OFDM symbol is extended by a cyclic repetition of the transmitted data (see Figure 4-15). As the guard interval is just a duplication of a useful part of the OFDM symbol, a correlation of the part containing the cyclic extension (guard interval) with the given OFDM symbol enables a fast time synchronization [78]. The sampling rate can also be estimated based on this correlation method. The presence of strong noise or long echoes may prevent accurate symbol timing. However, the noise effect can be reduced by integration (filtering) on several peaks obtained from subsequent estimates. As far as echoes are concerned, if the guard time is chosen long enough to absorb all echoes, this technique can still be reliable.

4.2.4.2 Fine Symbol Timing

For fine time synchronization, several methods based on transmitted reference symbols can be used [14]. One straightforward solution applies the estimation of the channel impulse response. The received signal without noise $r(t) = s(t) \otimes h(t)$ is the convolution of the transmit signal $s(t)$ and the channel impulse response $h(t)$. In the frequency domain after FFT processing we obtain $R(f) = S(f)H(f)$. By transmitting special reference symbols (e.g. CAZAC sequences [63]), $S(f)$ is *a priori* known by the receiver. Hence,

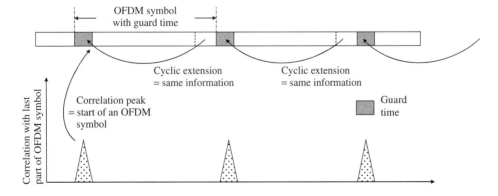

Figure 4-15 Time synchronization based on guard time correlation properties

after dividing $R(f)$ by $S(f)$ and IFFT processing, the channel impulse response $h(t)$ is obtained and an accurate timing information can be derived.

If the FFT window is not properly positioned, the received signal becomes

$$r(t) = s(t - t_0) \otimes h(t), \tag{4.36}$$

which turns into

$$R(f) = S(f)H(f)\, e^{-j2\pi f t_0} \tag{4.37}$$

after the FFT operation. After division of $R(f)$ by $S(f)$ and again performing an IFFT, the receiver obtains $h(t - t_0)$ and with that t_0. Finally, the fine time synchronization process consists of delaying the FFT window so that t_0 becomes quasi zero (see Figure 4-16).

In case of multi-path propagation, the channel impulse response is made up of multiple Dirac pulses. Let C_p be the power of each constructive echo path and I_p be the power of a destructive path. An optimal time synchronization process is to maximize the C/I, the ratio of the total constructive path power to the total destructive path power. However, for ease of implementation a sub-optimal algorithm might be considered, where the FFT window positioning signal uses the first significant echo, i.e. the first echo above a fixed threshold. The threshold can be chosen from experience, but a reasonable starting value can be derived from the minimum carrier-to-noise ratio required.

4.2.4.3 Sampling Clock Adjustment

As we have seen, the received analogue signal is first sampled at instants determined by the receiver clock before FFT operation. The effect of a clock frequency offset is twofold: the useful signal component is rotated and attenuated and, furthermore, ICI is introduced.

The sampling clock could be considered to be close to its theoretical value so it may have no effect on the result of the FFT. However, if the oscillator generating this clock is left free-running, the window opened for FFT may gently slide and will not match the useful interval of the symbols.

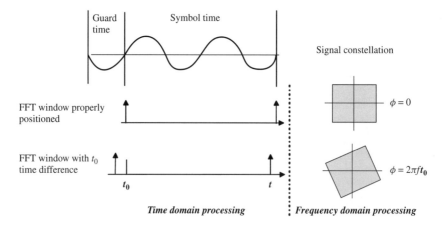

Figure 4-16 Fine time synchronization based on channel impulse response estimation

A first simple solution is to use the method described above to evaluate the proper position of the window and to readjust it dynamically. However, this method generates a phase discontinuity between symbols where a readjustment of the FFT window occurs. This phase discontinuity requires additional filtering or interpolation after FFT operation.

A second method, although using a similar strategy, is to evaluate the shift of the FFT window that is proportional to the frequency offset of the clock oscillator. The shift can be used to control the oscillator with better accuracy. This method allows a fine adjustment of the FFT window without the drawback of phase discontinuity from one symbol to the other.

4.2.5   Frequency Synchronization

Another fundamental function of an OFDM receiver is the carrier frequency synchronization. Frequency offsets are introduced by differences in oscillator frequencies in the transmitter and receiver, Doppler shifts, and phase noise. As we have seen earlier, the frequency offset leads to a reduction of the signal amplitude since the *sinc* functions are shifted and no longer sampled at the peak and to a loss of orthogonality between sub-carriers. This loss introduces ICI, which results in a degradation of the global system performance [60, 75, 76].

In the previous sections we have seen that in order to avoid severe SNR degradation, the frequency synchronization accuracy should be better than 2 % (see Section 4.2.2.3). Note that a multi-carrier system is much more sensitive to a frequency offset than a single-carrier system [67].

As shown in Figure 4-8, the frequency error in an OFDM system is often corrected by a tracking loop with a frequency detector to estimate the frequency offset. Depending on the characteristics of the transmitted signal (pilot-based or not) several algorithms for frequency detection and synchronization can be applied:

- algorithms based on the analysis of special synchronization symbols embedded in the OFDM frame [8, 55, 60, 62, 63, 81];
- algorithms based on the analysis of the received data at the output of the FFT (non-pilot-aided) [12]; and
- algorithms based on the analysis of guard time redundancy [13, 38, 78].

Like time synchronization, frequency synchronization can be performed in two steps: coarse and fine frequency synchronization.

4.2.5.1   Coarse Frequency Synchronization

We assume that the frequency offset is greater than half of the sub-carrier spacing, i.e.

$$f_{error} = \frac{2z}{T_s} + \frac{\phi}{\pi T_s}, \qquad (4.38)$$

where the first term of the above equation represents the frequency offset, which is a multiple of the sub-carrier spacing where z is an integer and the second term is the additional frequency offset being a fraction of the sub-carrier spacing, i.e. ϕ is smaller than π.

The aim of the coarse frequency estimation is mainly to estimate z. Depending on the transmitted OFDM signal, different approaches for coarse frequency synchronization can be used [12–14, 63, 78, 81].

CAZAC/M sequences

A general approach is to analyze the transmitted special reference symbols at the beginning of an OFDM frame, for instance, the CAZAC/M sequences [63] specified in the DVB-T standard [19]. As shown in Figure 4-17, CAZAC/M sequences are generated in the frequency domain and are embedded in I and R sequences. The CAZAC/M sequences are differentially modulated. The length of the M sequences is much larger than the length of the CAZAC sequences. The I and R sequences have the same length N_1, while in the I sequence (respectively R sequence) the imaginary (respectively real) components are 1 and the real (respectively imaginary) components are 0. The I and R sequences are used as start positions for the differential encoding/decoding of M sequences. A wide-range coarse synchronization is achieved by correlating with the transmitted known M sequence reference data, shifted over $\pm N_1$ sub-carriers (e.g. $N_1 = 10$ to 20) from the expected center point [24, 63]. The results from different sequences are averaged. The deviation of the correlation peak from the expected center point z with $-N_1 < z < +N_1$ is converted to an equivalent value used to correct the offset of the RF oscillator, or the baseband signal is corrected before the FFT operation. This process can be repeated until the deviation is less than $\pm N_2$ sub-carriers (e.g. $N_2 = 2$ to 5). For a fine-range estimation, in a similar manner the remaining CAZAC sequences can be applied that may reduce the frequency error to a few hertz.

The main advantage of this method is that it only uses one OFDM reference symbol. However, its drawback is the high amount of computation needed, which may not be adequate for burst transmission.

Schmidl and Cox

Similar to Moose [60], Schmidl and Cox [81] propose the use of two OFDM symbols for frequency synchronization (see Figure 4-18). However, these two OFDM symbols

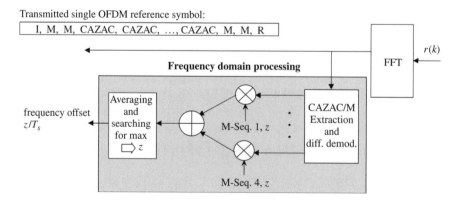

Figure 4-17 Coarse frequency offset estimation based on CAZAC/M sequences

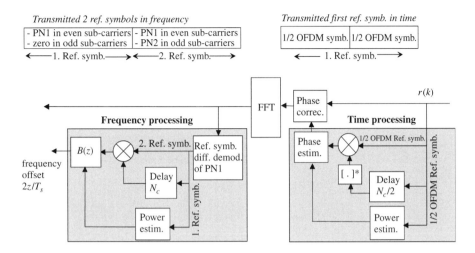

Figure 4-18   Schmidl and Cox frequency offset estimation using two OFDM symbols

have a special construction that allows a frequency offset estimation greater than several sub-carrier spacings. The first OFDM training symbol in the time domain consists of two identical symbols generated in the frequency domain by a PN sequence on the even sub-carriers and zeros on the odd sub-carriers. The second training symbol contains a differentially modulated PN sequence on the odd sub-carriers and another PN sequence on the even sub-carriers. Note that the selection of a particular PN sequence has little effect on the performance of the synchronization.

In Equation (4.38), the second term can be estimated in a similar way to the Moose approach [60] by employing the two halves of the first training symbols, $\hat{\phi} = angle[M(d)]$ (see Equation (4.35)). These two training symbols are frequency-corrected by $\hat{\phi}/(\pi T_s)$. Let their FFT be $x_{1,k}$ and $x_{2,k}$, the differentially modulated PN sequence on the even frequencies of the second training symbol be v_k, and X be the set of indices for the even sub-carriers. For the estimation of the integer sub-carrier offset given by z, the following metric is calculated:

$$
B(z) = \frac{\left| \displaystyle\sum_{k \in X} x^*_{1,k+2z}\, v^*_k x_{2,k+2z} \right|^2}{2 \left(\displaystyle\sum_{k \in X} |x_{2,k}|^2 \right)^2}.
\tag{4.39}
$$

The estimate of z is obtained by taking the maximum value of the above metric $B(z)$.

The main advantage of this method is its simplicity, which may be adequate for burst transmission. Furthermore, it allows a joint estimation of timing and frequency offset (see Section 4.2.4.1).

4.2.5.2   Fine Frequency Synchronization

Under the assumption that the frequency offset is less than half of the sub-carrier spacing, there is a one-to-one correspondence between the phase rotation and the frequency offset.

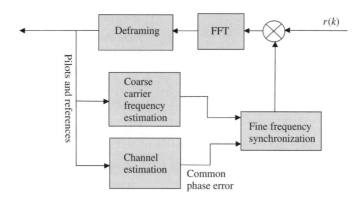

Figure 4-19   Frequency synchronization using reference symbols

The phase ambiguity limits the maximum frequency offset value. The phase offset can be estimated by using pilot/reference aided algorithms [81]. Furthermore, as explained in Section 4.2.5.1, for fine frequency synchronization some other reference data (i.e. CAZAC sequences) can be used. Here, the correlation process in the frequency domain can be done over a limited number of sub-carrier frequencies (e.g. $\pm N_2$ sub-carriers).

As shown in Figure 4-19, channel estimation (see Section 4.3) can additionally deliver a common phase error estimation (see Section 4.7.1.3), which can be exploited for fine frequency synchronization.

4.2.6   Automatic Gain Control (AGC)

In order to maximize the input signal dynamic and to avoid saturation, the variation of the received signal field strength before FFT operation or before A/D conversion can be adjusted by an AGC function [14, 81]. Two kinds of AGC can be implemented:

- *Controlling the time domain signal before A/D conversion.* First, in the digital domain, the average received power is computed by filtering. Then, the output signal is converted to analogue (e.g. by a sigma-delta modulator), which controls the signal attenuation before the A/D conversion.
- *Controlling the time domain signal before FFT.* In the frequency domain the output of the FFT signal is analyzed and the result is used to control the signal before the FFT.

4.3   Channel Estimation

When applying receivers with coherent detection in fading channels, instantaneous information about the channel state is required and has to be estimated by the receiver. The basic principle of pilot symbol aided channel estimation is to multiplex reference symbols, so-called pilot symbols, into the data stream. The receiver estimates the channel state information based on the received, known pilot symbols. The pilot symbols can be scattered in the time and/or frequency directions in OFDM frames (see Figure 4-9).

Special cases are either pilot tones, which are sequences of pilot symbols in the time direction on certain sub-carriers, or OFDM reference symbols, which are OFDM symbols consisting completely of pilot symbols.

4.3.1 *Two-Dimensional Channel Estimation*

4.3.1.1 **Two-Dimensional Filter**

Multi-carrier systems enable channel estimation in two dimensions by inserting pilot symbols on several sub-carriers in the frequency direction in addition to the time direction with the intention to estimate the channel transfer function $H(f, t)$ [35–37, 48]. By choosing the distances of the pilot symbols in the time and frequency directions sufficiently small with respect to the channel coherence bandwidth and time, estimates of the channel transfer function can be obtained by interpolation or filtering.

The described channel estimation operates on OFDM frames where $H(f, t)$ is estimated separately for each transmitted OFDM frame, allowing burst transmission based on OFDM frames. The discrete frequency and time representation $H_{n,i}$ of the channel transfer function introduced in Section 1.1.6 is used here. The values $n = 0, \ldots, N_c - 1$ and $i = 0, \ldots, N_s - 1$ are the frequency and time indices of the fading process, where N_c is the number of sub-carriers per OFDM symbol and N_s is the number of OFDM symbols per OFDM frame. The estimates of the discrete channel transfer function $H_{n,i}$ are denoted as $\hat{H}_{n,i}$. An OFDM frame consisting of 13 OFDM symbols, each with 11 sub-carriers, is shown as an example in Figure 4-20. The rectangular arrangement of the pilot symbols is referred to as a rectangular grid. The discrete distance in sub-carriers between two pilot symbols in the frequency direction is N_f and in OFDM symbols in the time direction is N_t. In the example given in Figure 4-20, N_f is equal to 5 and N_t is equal to 4.

The received symbols of an OFDM frame are given by

$$R_{n,i} = H_{n,i} S_{n,i} + N_{n,i}, \quad n = 0, \ldots, N_c - 1, \quad i = 0, \ldots, N_s - 1, \qquad (4.40)$$

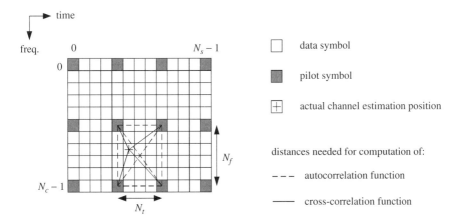

Figure 4-20 Pilot symbol grid for two-dimensional channel estimation

where $S_{n,i}$ and $N_{n,i}$ are the transmitted symbols and the noise components respectively. The pilot symbols are written as $S_{n',i'}$, where the frequency and time indices at locations of pilot symbols are marked as n' and i'. Thus, for equally spaced pilot symbols we obtain

$$n' = pN_f, \quad p = 0, \ldots, \lceil N_c/N_f \rceil - 1 \tag{4.41}$$

and

$$i' = qN_t, \quad q = 0, \ldots, \lceil N_s/N_t \rceil - 1, \tag{4.42}$$

assuming that the first pilot symbol in the rectangular grid is located at the first sub-carrier of the first OFDM symbol in an OFDM frame. The number of pilot symbols in an OFDM frame results in

$$N_{grid} = \left\lceil \frac{N_c}{N_f} \right\rceil \left\lceil \frac{N_s}{N_t} \right\rceil. \tag{4.43}$$

Pilot symbol aided channel estimation operates in two steps. In the first step, the initial estimate $\breve{H}_{n',i'}$ of the channel transfer function at positions where pilot symbols are located is obtained by dividing the received pilot symbol $R_{n',i'}$ by the originally transmitted pilot symbol $S_{n',i'}$, i.e.

$$\breve{H}_{n',i'} = \frac{R_{n',i'}}{S_{n',i'}} = H_{n',i'} + \frac{N_{n',i'}}{S_{n',i'}}. \tag{4.44}$$

In the second step, the final estimates of the complete channel transfer function belonging to the desired OFDM frame are obtained from the initial estimates $\breve{H}_{n',i'}$ by two-dimensional interpolation or filtering. The two-dimensional filtering is given by

$$\hat{H}_{n',i'} = \sum_{\{n',i'\} \in \Psi_{n,i}} \omega_{n',i',n,i} \breve{H}_{n',i'}, \tag{4.45}$$

where $\omega_{n',i',n,i}$ is the shift-variant two-dimensional impulse response of the filter. The subset $\Psi_{n,i}$ is the set of initial estimates $\breve{H}_{n',i'}$ that is actually used for estimation of $\hat{H}_{n,i}$.

The number of filter coefficients is

$$N_{tap} = \|\Psi_{n,i}\| \leqslant N_{grid}. \tag{4.46}$$

In the OFDM frame illustrated in Figure 4-20, N_{grid} is equal to 12 and N_{tap} is equal to 4.

Two-Dimensional Wiener Filter

The criterion for the evaluation of the channel estimator is the mean square value of the estimation error

$$\varepsilon_{n,i} = H_{n,i} - \hat{H}_{n,i}. \tag{4.47}$$

The mean square error is given by

$$J_{n,i} = E\{|\varepsilon_{n,i}|^2\}. \tag{4.48}$$

The optimal filter in the sense of minimizing $J_{n,i}$ with the minimum mean square error criterion is the two-dimensional Wiener filter. The filter coefficients of the two-dimensional Wiener filter are obtained by applying the orthogonality principle in linear mean square estimation,

$$E\{\varepsilon_{n,i} \check{H}^*_{n'',i''}\} = 0, \quad \forall\{n'', i''\} \in \Psi_{n,i}. \tag{4.49}$$

The orthogonality principle states that the mean square error $J_{n,i}$ is minimum if the filter coefficients $\omega_{n',i',n,i}, \forall\{n', i'\} \in \Psi_{n,i}$ are selected such that the error $\varepsilon_{n,i}$ is orthogonal to all initial estimates $\check{H}^*_{n'',i''}, \forall\{n'', i''\} \in \Psi_{n,i}$. The orthogonality principle leads to the Wiener–Hopf equation, which states that

$$E\{H_{n,i}\check{H}^*_{n'',i''}\} = \sum_{\{n',i'\}\in\Psi_{n,i}} \omega_{n',i',n,i} E\{\check{H}_{n',i'}\check{H}^*_{n'',i''}\}, \quad \forall\{n'', i''\} \in \Psi_{n,i}. \tag{4.50}$$

With Equation (4.44) and by assuming that $N_{n'',i''}$ has zero mean and is statistically independent from the pilot symbols $S_{n'',i''}$, the cross-correlation function $E\{H_{n,i}\check{H}^*_{n'',i''}\}$ is equal to the discrete time–frequency correlation function $E\{H_{n,i}\check{H}^*_{n'',i''}\}$, i.e. the cross-correlation function is given by

$$\theta_{n-n'',i-i''} = E\{H_{n,i}H^*_{n'',i''}\}. \tag{4.51}$$

The auto-correlation function in Equation (4.50) is given by

$$\phi_{n'-n'',i'-i''} = E\{\check{H}_{n',i'}\check{H}^*_{n'',i''}\}. \tag{4.52}$$

When assuming that the mean energy of all symbols $S_{n,i}$ including pilot symbols is equal, the auto-correlation function can be written in the form

$$\phi_{n'-n'',i'-i''} = \theta_{n'-n'',i'-i''} + \sigma^2 \delta_{n'-n'',i'-i''}. \tag{4.53}$$

The cross-correlation function depends on the distances between the actual channel estimation position n, i and all pilot positions n'', i'', whereas the auto-correlation function depends only on the distances between the pilot positions and, hence, is independent of the actual channel estimation position n, i. Both relations are illustrated in Figure 4-20. Inserting Equations (4.51) and (4.52) into Equation (4.53) yields, in vector notation,

$$\boldsymbol{\theta}^T_{n,i} = \boldsymbol{\omega}^T_{n,i}\boldsymbol{\Phi}, \tag{4.54}$$

where $\boldsymbol{\Phi}$ is the $N_{tap} \times N_{tap}$ auto-correlation matrix and $\boldsymbol{\theta}_{n,i}$ is the cross-correlation vector of length N_{tap}. The vector $\boldsymbol{\omega}_{n,i}$ of length N_{tap} represents the filter coefficients $\omega_{n',i',n,i}$

required to obtain the estimate $\hat{H}_{n,i}$. Hence, the filter coefficients of the optimum two-dimensional Wiener filter are

$$\omega_{n,i}^T = \theta_{n,i}^T \Phi^{-1}, \tag{4.55}$$

when assuming that the auto- and cross-correlation functions are perfectly known and $N_{tap} = N_{grid}$. Since in practice the auto-correlation function Φ and cross-correlation function $\theta_{n,i}$ are not perfectly known in the receiver, estimates or assumptions about these correlation functions are necessary in the receiver.

4.3.1.2 Two Cascaded One-Dimensional Filters

Two-dimensional filters tend to have a large computational complexity. The choice of two cascaded one-dimensional filters working sequentially can give a good tradeoff between performance and complexity. The principle of two cascaded one-dimensional filtering is depicted in Figure 4-21. Filtering in the frequency direction on OFDM symbols containing pilot symbols, followed by filtering in the time direction on all sub-carriers is shown. This ordering is chosen to enable filtering in the frequency direction directly after receiving a pilot symbol bearing OFDM symbol and, thus, to reduce the overall filtering delay. However, the opposite ordering would achieve the same performance due to the linearity of the filters.

The mean square error of the two cascaded one-dimensional filters working sequentially is obtained in two steps. Values and functions related to the first filtering are marked with the index [1] and values and functions related to the second filtering are marked with the index [2]. The estimates delivered by the first one-dimensional filter are

$$\hat{H}_{n,i'}^{[1]} = \sum_{\{n',i'\} \in \Psi_{n,i'}} \omega_{n',n}^{[1]} \check{H}_{n',i'}. \tag{4.56}$$

The filter coefficients $\omega_{n',n}^{[1]}$ only depend on the frequency index n. This operation is performed in all $\lceil N_s/N_t \rceil$ pilot symbol bearing OFDM symbols. The estimates delivered

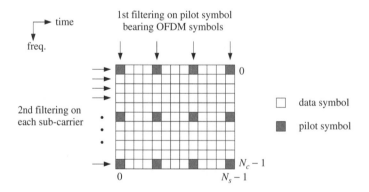

Figure 4-21   Two cascaded one-dimensional filter approach

by the second one-dimensional filter are

$$\hat{H}_{n,i} = \hat{H}_{n,i}^{[2]} = \sum_{\{n,i'\} \in \Psi_{n,i}} \omega_{i',i}^{[2]} \hat{H}_{n,i'}^{[1]}. \tag{4.57}$$

The filter coefficients $\omega_{i',i}^{[2]}$ only depend on the time index i. The estimates $\hat{H}_{n,i'}^{[1]}$ obtained from the first filtering are used as pilot symbols for the second filtering on sub-carrier n. The second filtering is performed on all N_c sub-carriers.

4.3.2 One-Dimensional Channel Estimation

One-dimensional channel estimation can be considered as special case of two-dimensional channel estimation, where the second dimension is omitted. These schemes require a higher overhead on pilot symbols, since the correlation of the fading in the second dimension is not exploited in the filtering process. The overhead on pilot symbols with one-dimensional channel estimation Λ_{1D} compared to two-dimensional channel estimation Λ_{2D} is

$$\Lambda_{1D} = N_f \Lambda_{2D} \tag{4.58}$$

with one-dimensional filtering in the time direction or

$$\Lambda_{1D} = N_t \Lambda_{2D} \tag{4.59}$$

with one-dimensional filtering in the frequency direction.

4.3.3 Filter Design

4.3.3.1 Adaptive Design

A filter is designed by determining the filter coefficients $\omega_{n,i}$. In the following, two-dimensional filtering is considered. The filter coefficients for one-dimensional filters are obtained from the two-dimensional filter coefficients by omitting one of the dimensions that is not required in the corresponding one-dimensional filter. The two-dimensional filter coefficients can be calculated, given the discrete time–frequency correlation function of the channel $\theta_{n-n'',i-i''}$ and the variance of the noise σ^2. In the mobile radio channel, it can be assumed that the delay power density spectrum $\rho(\tau)$ and the Doppler power density spectrum $S(f_D)$ are statistically independent. Thus, the time–frequency correlation function $\theta_{n-n'',i-i''}$ can be separated in the frequency correlation function $\theta_{n-n''}$ and the time correlation function $\theta_{i-i''}$. Hence, the optimum filter has to adapt the filter coefficients to the actual power density spectra $\rho(\tau)$ and $S(f_D)$ of the channel. The resulting channel estimation error can be minimized with this approach since the filter mismatch can be minimized. Investigations with adaptive filters show significant performance improvements with adaptive filters [61, 65]. Of importance for the adaptive filter scheme is that the actual power density spectra of the channel should be estimated with high accuracy, low delay, and reasonable effort.

4.3.3.2 Robust Nonadaptive Design

A low-complex selection of the filter coefficients is to choose a fixed set of filter coefficients which is designed such that a great variety of power density spectra with different shapes and maximum values is covered [37, 48]. No further adaptation to the time-variant channel statistics is performed during the estimation process. A reasonable approach is to adapt the filters to uniform power density spectra. By choosing the filter parameter τ_{filter} equal to the maximum expected delay of the channel τ_{max}, the normalized delay power density spectrum used for the filter design is given by

$$\rho_{filter}(\tau) = \begin{cases} \dfrac{1}{\tau_{filter}} & |\tau| < \dfrac{\tau_{filter}}{2} \\ 0 & \text{otherwise} \end{cases}. \qquad (4.60)$$

Furthermore, by choosing the filter parameter $f_{D,\,filter}$ equal to the maximum expected Doppler frequency of the channel $f_{D,\max}$, the normalized Doppler power density spectrum used for the filter design is

$$S_{filter}(f_D) = \begin{cases} \dfrac{1}{2f_{D,\,filter}} & |f_D| < f_{D,\,filter} \\ 0 & \text{otherwise} \end{cases}. \qquad (4.61)$$

With the selection of uniform power density spectra, the discrete frequency correlation functions results in

$$\theta_{n-n''} = \frac{\sin(\pi\,\tau_{filter}(n - n'')F_s)}{\pi\,\tau_{filter}(n - n'')F_s} = \text{sinc}(\pi\,\tau_{filter}(n - n'')F_s) \qquad (4.62)$$

and the discrete time correlation function yields

$$\theta_{i-i''} = \frac{\sin(2\pi f_{D,\,filter}(i - i'')T_s')}{2\pi f_{D,\,filter}(i - i'')T_s'} = \text{sinc}(2\pi f_{D,\,filter}(i - i'')T_s'). \qquad (4.63)$$

The auto-correlation function is obtained according to Equation (4.53).

4.3.4 Implementation Issues

4.3.4.1 Pilot Distances

In this section, the arrangement of the pilot symbols in an OFDM frame is shown by applying the two-dimensional sampling theorem. The choice of a rectangular grid is motivated by the results presented in Reference [37] where channel estimation performance with rectangular, diagonal, and random grids is investigated. Channel estimation either with a rectangular or diagonal grid shows similar performance but outperforms channel estimation with a random grid.

Given the normalized filter bandwidths $\tau_{filter}F_s$ and $f_{D,\,filter}T'_s$, the sampling theorem requires that the distance of the pilot symbols in the frequency direction is

$$N_f \leqslant \frac{1}{\tau_{filter}F_s}, \tag{4.64}$$

and in the time direction is

$$N_t \leqslant \frac{1}{2f_{D,\,filter}T'_s}. \tag{4.65}$$

An optimum sampling of the channel transfer function is given by a balanced design which guarantees that the channel is sampled in time and in frequency with the same sampling rate. A balanced design is defined as [36]

$$N_f\tau_{filter}F_s \approx 2N_t f_{D,\,filter}T'_s. \tag{4.66}$$

A practically proven value of the sampling rate is the selection of approximately two-times oversampling to achieve a reasonably low complexity with respect to the filter length and performance. A practical hint concerning the performance of the channel estimation is to design the pilot grid such that the first and the last OFDM symbol and sub-carrier respectively in an OFDM frame contain pilot symbols (see Figure 4-20). This avoids the channel estimation having to perform channel prediction, which is more unreliable than interpolation. In the special case that the downlink can be considered as a broadcasting scenario with continuous transmission, it is possible to continuously use pilot symbols in the time direction without requiring additional pilots at the end of an OFDM frame.

4.3.4.2 Overhead

Besides the mean square error of a channel estimation, a criterion for the efficiency of a channel estimation is the overhead and the loss in SNR due to pilot symbols. The overhead due to pilot symbols is given by

$$\Lambda = \frac{N_{\text{grid}}}{N_c N_s} \tag{4.67}$$

and the SNR loss in dB is defined as

$$V_{\text{pilot}} = 10\log_{10}\left(\frac{1}{1-\Lambda}\right). \tag{4.68}$$

4.3.4.3 Signal-to-Pilot Power Ratio

The pilot symbols $S_{n',i'}$ can be transmitted with higher average energy than the data-bearing symbols $S_{n,i}, n \neq n', i \neq i'$. Pilot symbols with increased energy are called *boosted pilot symbols* [19]. The boosting of pilot symbols is specified in the DVB-T, LTE, and IEEE802.16x standards and achieves better estimates of the channel but reduces the average SNR of the data symbols. The choice of an appropriate boosting level for the pilot symbols is investigated in Reference [37].

4.3.4.4 Complexity and Performance Aspects

A simple alternative to optimum Wiener filtering is a DFT-based channel estimator illustrated in Figure 4-22. The channel is first estimated in the frequency domain on the N_{pilot} sub-carriers where pilots have been transmitted. In the next step, this N_{pilot} estimates are transformed with a N_{pilot} point IDFT in the time domain and the resulting time sequence can be weighted before it is transformed back in the frequency domain with an N_c point DFT. As long as $N_{\text{pilot}} < N_c$, the DFT-based channel estimation performs interpolation. The DFT-based channel estimation can also be applied in two dimensions, while the time direction is processed in the same way as described above for the frequency direction. An appropriate weighting function between IDFT and DFT applies the minimum mean square error criterion [89].

The application of singular value decomposition for channel estimation is an approach to reduce the complexity of the channel estimator as long as an irreducible error floor can be tolerated in the system design [15].

The performance of pilot symbol based channel estimation concepts can be further improved by iterative channel estimation and decoding, where reliable decisions obtained from the decoding are exploited for channel estimation. A two-dimensional implementation is possible with estimation in the time and frequency directions [79].

When assumptions about the channel can be made in advance, the channel estimator performance can be improved. An approach is based on a parametric channel model where the channel frequency response is estimated by using an N_p path channel model [90]. The parametric channel modeling requires additional complexity to estimate the multi-path time delays.

4.3.5 Performance Analysis

In this section, the mean square error performance of pilot symbol aided channel estimation in multi-carrier systems is shown. To evaluate and optimize the channel estimation, a multi-carrier reference scenario typical for mobile radio systems is defined. The frequency band has a bandwidth of $B = 2\,\text{MHz}$ and is located at a carrier frequency of $f_c = 1.8\,\text{GHz}$. The OFDM operation and its inverse are achieved with an IFFT and FFT respectively of size 512. The considered multi-carrier transmission scheme processes one OFDM frame per estimation cycle. An OFDM frame consists of $N_s = 24$ OFDM symbols. The OFDM frame duration results in $T_{fr} = 6.6\,\text{ms}$. The filter parameters are chosen as $\tau_{filter} = 20\,\mu\text{s}$

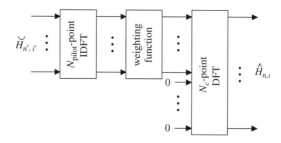

Figure 4-22 Low complexity DFT-based channel estimation

and $f_{D,filter} = 333.3\,\text{Hz}$, where $f_{D,filter}$ corresponds to a velocity of $v = 200\,\text{km/h}$. A balanced design with approximately two-times oversampling is given with a pilot symbol spacing of $N_f = 6$ in the frequency direction and of $N_t = 3$ in the time direction. The resulting system parameters are summarized in Table 4-1. The chosen OFDM frame structure for pilot symbol aided channel estimation in two dimensions is illustrated in Figure 4-23.

In the time direction, the last OFDM symbol with pilot symbols of the previous OFDM frame can be used for filtering. With a pilot spacing of 6 in the frequency direction, while starting and ending with a sub-carrier containing pilot symbols, a number of 511 used sub-carriers per OFDM symbol is obtained and is considered in the following. The resulting overhead Λ due to pilot symbols is 5.6 %. With pilot symbols and data symbols having the same average energy, the loss in SNR V_{pilot} due to pilot symbols is only 0.3 dB for the defined multi-carrier scenario.

The performance criterion for the evaluation of the channel estimation is the mean square error, which is averaged over an OFDM frame. Thus, edge effects are also taken into account. The information about the variance σ^2 required for the calculation of the autocorrelation matrix Φ is assumed to be known perfectly at the receiver unless otherwise stated. In practice, the variance σ^2 can be estimated by transmitting a null symbol without signal energy at the beginning of each OFDM frame (see Section 4.2.1). Alternatively, the auto-correlation function can be optimized for the highest σ^2 at which successful data transmission should be possible.

The mean square error of two-dimensional (2-D) filtering without model mismatch and two cascaded one-dimensional (2×1-D) filtering applied in a multi-carrier system is presented and compared in the following. The mobile radio channel used has a uniform delay power spectrum with $\tau_{max} = 20\,\mu\text{s}$ and a uniform Doppler power density spectrum with $f_{D,max} = 333.3\,\text{Hz}$.

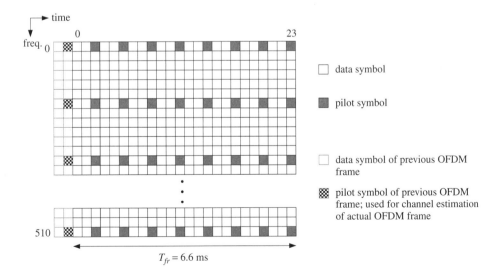

Figure 4-23 OFDM frame with pilot grid for channel estimation in two dimensions

Table 4-1 Parameters for pilot symbol aided channel estimation in two dimensions

Parameter	Value
Bandwidth	$B = 2\,\text{MHz}$
Carrier frequency	$f_c = 1.8\,\text{GHz}$
OFDM frame duration	$T_{fr} = 6.6\,\text{ms}$
OFDM symbols per OFDM frame	$N_s = 24$
FFT size	512
OFDM symbol duration	$T_s = 256\,\mu\text{s}$
Cyclic prefix duration	$T_g = 20\,\mu\text{s}$
Sub-carrier spacing	$F_s = 3.9\,\text{kHz}$
Number of used sub-carriers	511
Pilot symbol distance in frequency direction	$N_f = 6$
Pilot symbol distance in time direction	$N_t = 3$
Delay filter bandwidth	$\tau_{filter} = 20\,\mu\text{s}$
Doppler filter bandwidth	$f_{D,\,filter} = 333.3\,\text{Hz}$
Filter characteristic	Wiener filter with and without model mismatch

In Figure 4-24, the mean square error (MSE) versus the SNR for 2-D filtering without model mismatch with different numbers of filter taps is shown. The corresponding results for $2 \times 1-\text{D}$ filtering are presented in Figure 4-25. It can be observed in both figures that the mean square error decreases with increasing numbers of taps. The mean square error presented with 2-D filtering using 100 taps can be considered as a lower bound. In the case of 2-D filtering, 25 taps seems to be reasonable with respect to mean square error performance and complexity. In the case of $2 \times 1-\text{D}$ filtering, 2×5 taps is a reasonable choice. A further increase of the number of taps only reduces the mean square error slightly. Moreover, with 2×5 taps, the performance with $2 \times 1-\text{D}$ filtering is similar to the performance with 2-D filtering with 25 taps. Based on these results, $2 \times 1-\text{D}$ filtering with 2×5 taps is chosen for channel estimation in this section.

In the following, the focus is on the degradation due to model mismatch in the filter design. In the first step, the mean square error of the defined channel estimation in different COST 207 channel models versus the SNR is shown in Figure 4-26 for $2 \times 1-\text{D}$ filtering with 2×5 taps. The velocity of the terminal station is equal to 3 km/h. As a reference, the mean square error curve without model mismatch is given. The presented channel estimation provides a better performance in channels with a large delay spread than in channels with a low delay spread. The reason for this effect is that a channel with a delay power density spectrum that matches more closely with that chosen for the filter design also matches more closely to the performance without model mismatch and, thus, shows a better mean square error performance.

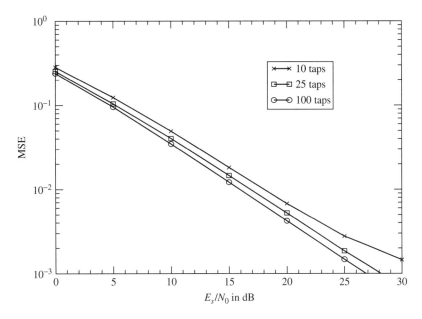

Figure 4-24 MSE for 2-D channel estimation with different numbers of filter taps; no model mismatch

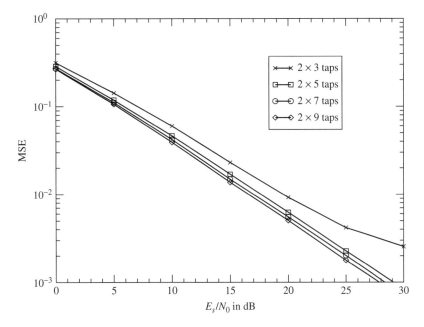

Figure 4-25 MSE for 2 × 1D channel estimation with different numbers of filter taps; no model mismatch

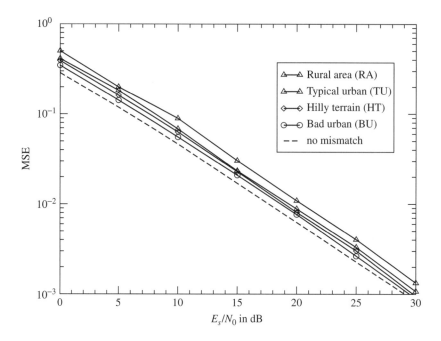

Figure 4-26   MSE for 2×1D channel estimation with model mismatch; $v = 3$ km/h

In the second step, the influence of different Doppler power density spectra is considered. Furthermore, the effect when the condition of two-times oversampling is not fulfilled in mobile radio channels with Doppler frequencies larger than 333.3 Hz is investigated. Figure 4-27 shows the mean square error of the presented channel estimation with 2×1–D filtering and 2×5 taps for different velocities v of the terminal station in the COST 207 channel model HT versus the SNR. As a reference, the mean square error curve without model mismatch is given. It can be observed that there is approximately no change in the mean square error with different velocities v of the mobile station as long as two-times oversampling of the fading process is guaranteed. For the defined multi-carrier system, two-times oversampling is given for a velocity of 200 km/h. As soon as the rule of two-times oversampling is not fulfilled, shown for velocity of $v = 250$ km/h, the mean square error increases considerably.

Figure 4-28 shows the mean square error obtained with the proposed channel estimation using 2×1–D filtering and 2×5 taps when the autocorrelation matrix Φ is optimized for an SNR of 10 dB. During the channel estimation no information about the actual variance σ^2 is used to adapt the filter coefficients optimally. It can be seen that accurate results are obtained in a span of about 10 dB with mean at 10 dB. For higher SNRs, the mean square error obtained with the simplified channel estimation flattens out. Thus, application of the simplified channel estimation that requires no estimation of σ^2 depends on whether the error floor is acceptable or not. Otherwise, it depends on the SNR dynamic range at the input of the receiver whether the simplified channel estimation is applicable or not.

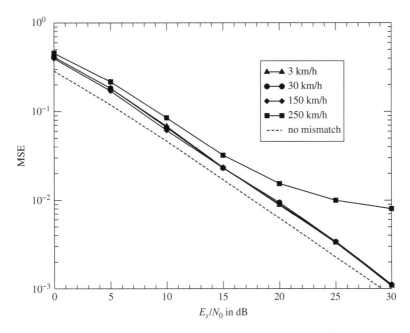

Figure 4-27   MSE for $2 \times 1D$ channel estimation with model mismatch; COST 207 channel model HT

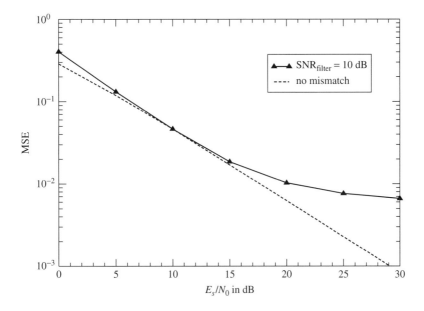

Figure 4-28   MSE for $2 \times 1D$ channel estimation with σ^2 mismatch

4.3.6 Time Domain Channel Estimation

Channel estimation for OFDM systems can be applied in the time domain, as in single-carrier systems. The basic principle is to insert PN sequences periodically in the time domain between OFDM symbols where the channel impulse response is obtained by correlation in the time domain [91, 92]. In OFDM systems, the signal used for time domain channel estimation can be generated in the transmitter in the frequency domain. The insertion of equal-spaced pilot symbols in the frequency domain results in an OFDM symbol that can be considered as split into several time slots, each with a pilot. Assuming that the impulse response of each time slot is identical, the channel response in the time domain can be obtained by averaging the impulse responses of the time slots [58, 91]. The cost of time domain channel estimation compared to frequency domain channel estimation is its higher complexity. Time domain channel estimation can be combined with time domain channel equalization [9].

4.3.7 Decision Directed Channel Estimation

Under the assumption that the channel is quasi-stationary over two OFDM symbols, the decisions from the previous OFDM symbol can be used for data detection in the current OFDM symbol. These schemes can outperform differentially modulated schemes and are efficient if FEC coding is included in the reliable reconstruction of the transmitted sequence [56]. The principle of decision directed channel estimation including FEC decoding is illustrated in Figure 4-29.

In the first estimation step (start), reference symbols have to be used for the initial channel estimation. These reference symbols are the symbols $S_{n,0}, n = 0, \ldots, N_c - 1$, of the initial OFDM symbol. The initial estimation of the channel transfer function is given by

$$\hat{H}_{n,0} = \frac{R_{n,0}}{S_{n,0}} = H_{n,0} + \frac{N_{n,0}}{S_{n,0}}. \tag{4.69}$$

The channel can be oversampled in the frequency and time direction with respect to filtering, since all re-encoded data symbols can be used as pilot symbols. Thus, the channel transfer function can be low-pass filtered in the frequency and/or time direction

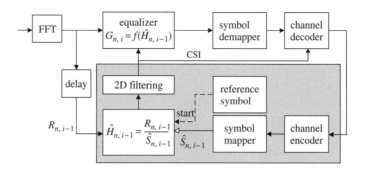

Figure 4-29   Decision directed channel estimation

in order to reduce the noise. Without filtering in either the frequency or time domain, the BER performance of differential modulation is obtained. Based on the obtained channel estimate $\hat{H}_{n,i-1}$, the received symbols $R_{n,i}$ are equalized according to

$$U_{n,i} = G_{n,i} R_{n,i}, \tag{4.70}$$

where

$$G_{n,i} = f(\hat{H}_{n,i-1}) \tag{4.71}$$

is an equalization coefficient depending on the previous channel estimate $\hat{H}_{n,i-1}$. In order to achieve a reliable estimation of the transmitted sequence, the error correction capability of the channel coding is used. The bit sequence at the output of the channel decoder is re-encoded and re-modulated to obtain reliable estimates $\hat{S}_{n,i-1}$ of the transmitted sequence $S_{n,i-1}$. These estimates are fed back to obtain a channel estimate $\hat{H}_{n,i-1}$, which is used for the detection of the following symbol $S_{n,i}$. With decision directed channel estimation the channel estimates are updated symbol-by-symbol.

Decision directed channel estimation can reduce the amount of reference data required to only an initial OFDM reference symbol. This can significantly reduce the overhead due to pilot symbols at the expense of additional complexity [28]. The achievable estimation accuracy with decision directed channel estimation including FEC decoding is comparable to pilot symbol aided channel estimation and with respect to BER outperforms classical differential demodulation schemes [56]. Classical differential modulation schemes can also benefit from decision directed channel estimation where the correlations in time and frequency of previously received symbols are filtered and used for estimation of the actual $\hat{H}_{n,i}$.

4.3.8 Blind and Semi-Blind Channel Estimation

The cyclic extension of an OFDM symbol can be used as an inherent reference signal within the data, enabling channel estimation based on the cyclic extension [34, 64, 88]. Since no additional pilot symbols are required with this method in OFDM schemes, this can be considered as blind channel estimation. The advantage of blind channel estimation based on the cyclic prefix is that this channel estimation concept is standard-compliant and can be applied to all commonly used OFDM systems that use a cyclic prefix.

A further approach of blind detection without the necessity of pilot symbols for coherent detection is possible when joint equalization and detection is applied. This is possible by trellis decoding of differentially encoded PSK signals [52] where the trellis decoding can efficiently be achieved by applying the Viterbi algorithm. The differential encoding can be performed in the frequency or time direction, while the detector exploits correlations between adjacent sub-carriers and/or OFDM symbols. These blind detection schemes require a low number of pilot symbols and outperform classical differential detection schemes. The complexity of a blind scheme with joint equalization and detection is higher than that of differential or coherent receivers due to the additional implementation of the Viterbi algorithm.

Statistical methods for blind channel estimations have also been proposed [34] which, however, require several OFDM symbols to be able to estimate the channel and might fail in mobile radio channels with fast fading. Blind channel estimation concepts exploiting the feature that the transmitted data are confined to a finite alphabet set can perform channel estimation from a single OFDM symbol [93]. Modifications of the finite alphabet approach attempt to reduce the enormous computational effort of these schemes [44].

Semi-blind channel estimation takes advantage of pilot symbols that are included in the data stream. For example, most existing OFDM systems have pilot symbols multiplexed in the data streams such that these symbols in combination with blind algorithms result in semi-blind schemes with improved estimation accuracy.

4.3.9 Channel Estimation in MC-SS Systems

4.3.9.1 Downlink

MC-CDMA

The synchronous downlink of MC-CDMA systems is a broadcast scenario where all K users can exploit the same pilot symbols within an OFDM frame [42]. The power of the common pilot symbols has to be adjusted such that the terminal station with the most critical channel conditions is able to estimate the channel. It is therefore possible to realize a pilot symbol scheme with adaptive power adjustment.

The performance of an MC-CDMA mobile radio system with two-dimensional channel estimation in the downlink is shown in Figure 4-30. The transmission bandwidth is 2 MHz and the carrier frequency is located at 1.8 GHz. The number of sub-carriers is 512. MMSE equalization is chosen as a low complex detection technique in the terminal station and the system is fully loaded. The system parameters are given in Table 4-1 and the pilot symbol grid corresponds to the structure shown in Figure 4-23. The COST 207 channel models have been chosen as propagation models. It can be observed that two-dimensional channel estimation can handle different propagation scenarios with high velocities up to 250 km/h. For the chosen scenario, the SNR degradation compared to perfect channel estimation is in the order of 2 dB.

4.3.9.2 Uplink

Channel estimation in the uplink requires separate sets of pilot symbols for each user for the estimation of the user specific channel state information. This leads to a much higher overhead in pilot symbols in the uplink compared to the downlink. The increase of the overhead in the uplink is proportional to the number of users K. Alternatives to this approach are systems with pre-equalization in the uplink (see Section 2.1.6).

MC-CDMA

In synchronous uplinks of MC-CDMA systems, the channel impulse responses of the different uplink channels can be estimated by assigning each user an exclusive set of pilot symbol positions that the user can exploit for channel estimation. This can either be a one- or two-dimensional channel estimation per user. The channel estimation concept per user is identical to the concept in the downlink used for all users. In mobile radio channels with high time and frequency selectivity, the possible number of active users can be quite small, due to the high number of required pilot symbols.

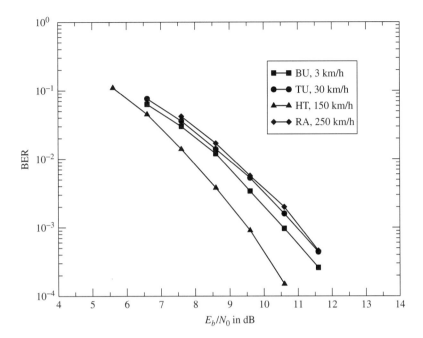

Figure 4-30   BER versus SNR of MC-CDMA with pilot symbol aided channel estimation in the downlink

Alternatively, a pilot symbol design in the time domain is possible by using for each user a different fraction of an OFDM symbol [83]. These methods have restrictions on the maximum length of the channel impulse response τ_{max}. When fulfilling the condition

$$\tau_{max} \leqslant \frac{T_s}{K}, \tag{4.72}$$

the system is able to estimate K different channel impulse responses within one OFDM symbol. Since the OFDM system design typically results in an OFDM symbol duration T_s much larger than τ_{max}, several users can be estimated in one OFDM symbol. The principle of time domain channel estimation in the uplink of an MC-CDMA system with eight users is illustratively shown in Figure 4-31, where $h^{(k)}$ is the channel impulse response $h^{(k)}(\tau, t)$ of user k.

The structure for channel estimation shown in Figure 4-31 is obtained by inserting pilot symbols in an OFDM symbol and rotating the phase of the complex pilot symbols according to the user index. This results in a delayed version of the time reference signal for each user. Since whole OFDM symbols are used for this type of channel estimation, the overhead in pilot symbols is comparable to that of scattering pilots in the frequency domain, as described at the beginning of this section.

Additionally, filtering in the time direction can be introduced, which reduces the overhead on pilot symbols. The distance between OFDM symbols with reference information is N_t OFDM symbols. The filter in the time direction can be designed as described in

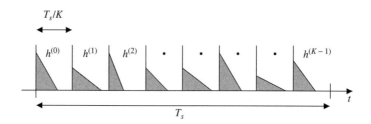

Figure 4-31   MC-CDMA uplink channel estimation

Section 4.3.3. Alternatively, in a synchronous uplink, a single-carrier modulated midamble can be inserted between OFDM symbols of an OFDM frame, such that classical single-carrier multi-user channel estimation concepts can be applied [4].

Another approach is to use differential modulation in an MC-CDMA uplink to overcome the problem with channel estimation. Using chip- and symbol-level differential encoding for the uplink has been investigated in Reference [85]. The results show that the capacity is limited to moderate system loads and that the performance of this approach strongly depends on the propagation environment.

SS-MC-MA

In the following, the performance of an SS-MC-MA mobile radio system with pilot symbol aided channel estimation in the uplink is presented. The SS-MC-MA system transmits on 256 sub-carriers. The transmission bandwidth is 2 MHz and the carrier frequency is located at 1.8 GHz. The channel estimation in the SS-MC-MA system is based on filtering in the time direction. An OFDM frame consists of 31 OFDM symbols as illustrated in Figure 4-32.

Each user exclusively transmits on a subset of eight sub-carriers. The spreading is performed with Walsh–Hadamard codes of length $L = 8$. The receiver applies maximum likelihood detection. Convolutional codes of rate 1/2 and memory 6 are used for channel coding.

The BER versus the SNR for an SS-MC-MA system in the uplink is shown in Figure 4-33. The COST 207 channel models, each with a different velocity of the terminal station, are considered. It should be noted that the performance of the SS-MC-MA system is independent of the number of active users due to the avoidance of multiple access interference; i.e. the performance presented is valid for any system load, assuming that ICI can be neglected. It can be observed that one-dimensional channel estimation can handle different propagation scenarios with high velocities of 250 km/h. One-dimensional uplink channel estimation requires additional overhead compared to two-dimensional channel estimation for the downlink, which results in higher SNR degradation (of about 1 dB) due to the required power for additional pilot symbols.

MC-DS-CDMA

Due to the close relationship of MC-DS-CDMA and DS-CDMA, a channel estimation technique design for DS-CDMA can be applied in the same way to MC-DS-CDMA.

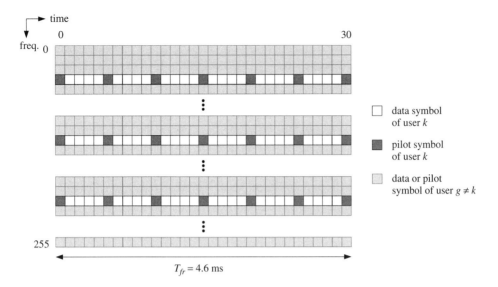

Figure 4-32 Exemplary OFDM frame of an SS-MC-MA system in the uplink

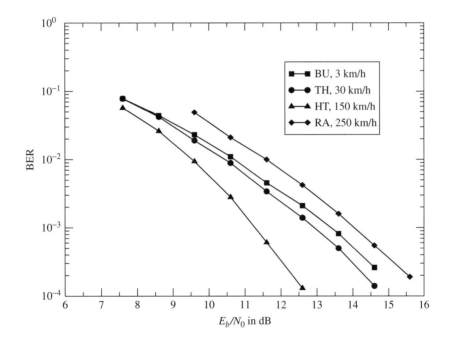

Figure 4-33 BER versus SNR of SS-MC-MA with pilot symbol aided channel estimation

A common approach is that each user transmits a known user-specific reference sequence at a certain period of time. These user-specific sequences are used to estimate the channel impulse response $h^{(k)}(\tau, t)$ by evaluating the correlation function of the known reference sequences with the received sequences.

In the case of a synchronous uplink, a similar approach as described above for the MC-CDMA uplink can be applied to MC-DS-CDMA.

4.3.10   Channel Estimation in MIMO-OFDM Systems

The basic concept of applying OFDM in *multiple input multiple output* (MIMO) systems, i.e. employing multiple transmit and receive antennas, is described in Chapter 6.

Regarding channel estimation, pilot-based or decision directed channel estimation with multiple antennas can simultaneously estimate multiple channel transfer functions if the channel has moderate delay spread, such that the correlations between adjacent sub-carriers are high. Based on the assumption that fading on adjacent sub-carriers is equal, it is possible to decouple the channel transfer functions corresponding to the different transmit antennas as long as transmitted data on the different antennas are known at the receiver by either pilot symbols, or the decoded data can be used to generate the reference data [3, 10, 47, 49, 59]. Making use of the fact that the channel delay profiles of the various channels in a MIMO scheme should have similar delay profiles, this knowledge allows to improve the accuracy of the channel estimator further [50].

Blind channel estimation in MIMO-OFDM systems can be achieved using statistical methods and separating signals from the different transmit antennas by performing a periodic pre-coding of the individual data streams prior to transmission [6]. Each transmit antenna is assigned a different pre-coding sequence.

4.4   Channel Coding and Decoding

Channel coding is an inherent part of any multi-carrier system. By using channel state information (CSI) in a maximum likelihood type FEC decoding process a high diversity and, hence, high coding gain can be achieved, especially in fading channels [1, 24]. Therefore, it is crucial to choose the encoder in such a way that it enables the exploitation of soft information for decoding. Furthermore, flexibility on the coding scheme to derive different code rates (e.g. for unequal error protection) from the same mother code is always preferred. This flexibility may allow one to adapt the transmission scheme to different transmission conditions.

The following channel coding schemes have been proposed for multi-carrier transmission [1, 16, 17, 19, 20, 24]:

- punctured convolutional coding;
- concatenated coding (e.g. inner convolutional and outer block code, i.e. Reed Solomon code);
- Turbo coding (block or convolutional); and recently the
- low density parity check code (LDPC).

4.4.1 Punctured Convolutional Coding

A punctured convolutional code that provides from the mother code rate 1/2, memory v (e.g. $v = 6$ resulting in 64 states), a wide range of higher inner code rates R (e.g. $R = 2/3, 3/4, 5/6,$ and 7/8) is usually applied, for instance, with a generator polynomial $G_1 = 171_{oct}, G_2 = 133_{oct}$.

The puncturing patterns of a convolutional code with 64 states for different inner code rates R are given in Table 4-2. In this table '0' means that the coded bit is not transmitted (i.e. punctured or masked) and '1' means that the coded bit is transmitted. It should be noticed that each matrix has two rows and several columns, where the puncturing vector for each row corresponds to the outputs of the encoder X and Y respectively (see Figures 4-34 and 4-35). For decoding the received data a soft input maximum likelihood sequence estimator efficiently realized with the Viterbi algorithm can be employed [70]. Deriving the soft values by taking the channel state information gives a high diversity for decoding, resulting in high performance. The number of bits that could be used for soft values is typically 3–4 bits.

Table 4-3 shows the performance of punctured convolutional coding (CC) for different modulation schemes in AWGN, Rayleigh, and Ricean (10 dB Rice factor) fading channels with perfect channel estimation.

4.4.2 Concatenated Convolutional and Reed Solomon Coding

Compared to a single code, the main advantage of concatenated coding schemes is to obtain much higher coding gains at low BERs with reduced complexity. For concatenated coding, usually as the outer code a shortened Reed Solomon code and as the inner code punctured convolutional codes are used (see Figures 4-36 and 4-37). An optional interleaving between these codes can be inserted. The role of this byte interleaving is to scatter the bursty errors at the output of the inner decoder, i.e the Viterbi decoder [70].

Table 4-2 Puncturing patterns of a 64-state convolutional code (X_1 is sent first)

Inner code rate R	Puncturing patterns	Minimum distance d_{min}	Transmitted sequences (after P/S conversion)
1/2	X: 1 Y: 1	10	$X_1 Y_1$
2/3	X: 1 0 Y: 1 1	6	$X_1 Y_1 Y_2$
3/4	X: 1 0 1 Y: 1 1 0	5	$X_1 Y_1 Y_2 X_3$
5/6	X: 1 0 1 0 1 Y: 1 1 0 1 0	4	$X_1 Y_1 Y_2 X_3 Y_4 X_5$
7/8	X: 1 0 0 0 1 0 1 Y: 1 1 1 1 0 1 0	3	$X_1 Y_1 Y_2 Y_3 Y_4 X_5 Y_6 X_7$

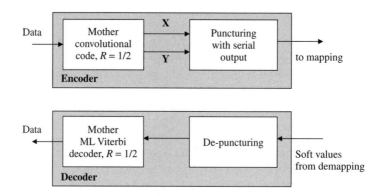

Figure 4-34 Punctured convolutional coding

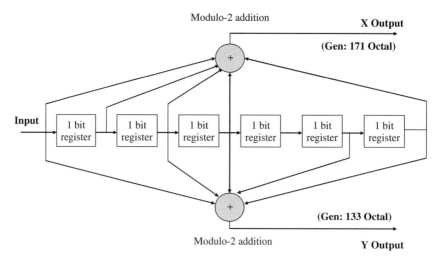

Figure 4-35 Example of inner mother convolutional code of rate 1/2 with memory 6

This type of coding scheme is quite flexible, providing different outer and inner code rates.

The outer shortened Reed Solomon RS($K + 2t$, K, t) code can transmit up to K bytes. This code can be derived, for instance, from the original systematic Reed Solomon RS(255, 239, $t = 8$) code, able to correct up to $t = 8$ byte errors. As the field generator polynomial for Reed Solomon codes over Galois field GF(256), the generator polynomial $P(x) = x^8 + x^4 + x^3 + x^2 + 1$ can be employed. As the inner code, the same convolutional coding as described in Section 4.4.1 can be used.

Shortened RS codes together with convolutional coding can easily be adapted for the downlink and uplink packet transmission:

- One or several MAC packets can be mapped to the information part (K bytes of the outer code).

Table 4-3 E_s/N_0 for punctured convolutional codes with perfect CSI for M-QAM modulation in AWGN, independent Rayleigh, and Ricean fading channels for BER $= 2 \times 10^{-4}$

Modulation	CC rate R	AWGN	Ricean fading (Rice factor 10 dB)	Rayleigh fading
QPSK	1/2	3.1 dB	3.6 dB	5.4 dB
	2/3	4.9 dB	5.7 dB	8.4 dB
	3/4	5.9 dB	6.8 dB	10.7 dB
	5/6	6.9 dB	8.0 dB	13.1 dB
	7/8	7.7 dB	8.7 dB	16.3 dB
16-QAM	1/2	8.8 dB	9.6 dB	11.2 dB
	2/3	11.1 dB	11.6 dB	14.2 dB
	3/4	12.5 dB	13.0 dB	16.7 dB
	5/6	13.5 dB	14.4 dB	19.3 dB
	7/8	13.9 dB	15.0 dB	22.8 dB
64-QAM	1/2	14.4 dB	14.7 dB	16.0 dB
	2/3	16.5 dB	17.1 dB	19.3 dB
	3/4	18.0 dB	18.6 dB	21.7 dB
	5/6	19.3 dB	20.0 dB	25.3 dB
	7/8	20.1 dB	21.0 dB	27.9 dB

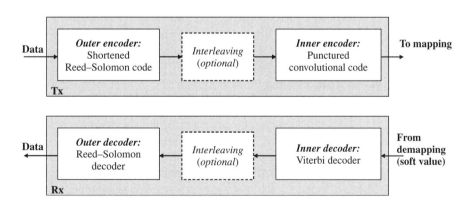

Figure 4-36 Concatenated Reed Solomon and convolutional code

– If the total number of K bytes is smaller than 239 bytes, the remaining $239 - K$ bytes are filled by zero bytes. Systematic RS (255, 239, $t = 8$) coding is applied. After RS coding, the systematic structure of the RS code allows one to shorten the code, i.e. remove the filled $239 - K$ zero bytes before transmission. Then, each coded packet of length $K + 16$ bytes will be serial bit converted.

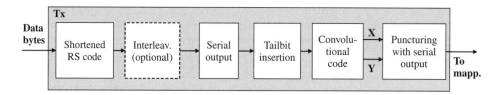

Figure 4-37   Coding procedure for packet transmission

- At the end of the each packet, tailbits (e.g. 6 bits for memory 6) can be inserted for inner code trellis termination purposes.
- A block consisting of $[(K + 16) \times 8 + 6]$ bits is encoded by the inner convolutional mother binary code of rate 1/2. After convolutional coding, the puncturing operation is applied following the used inner code rate R for the given packet. This results in a total of $[(K + 16) \times 8 + 6]/R$ bits. Finally, the punctured bits are serial-to-parallel converted and submitted to the symbol mapper.

If the BER before RS decoding is guaranteed to be about 2×10^{-4}, then with sufficient interleaving (e.g. eight RS code words) for the same SNR values given in Table 4-3, a quasi error-free (i.e. BER $< 10^{-12}$) transmission after RS decoding is guaranteed. However, if no interleaving is employed, depending on the inner coding rate, a loss of about 1.5–2.5 dB has to be considered to achieve a quasi error-free transmission [22].

4.4.3   Turbo Coding

Recently, interest has focused on iterative decoding of parallel or serial concatenated codes using *soft-in/soft-out* (SISO) decoders with simple code components in an interleaved scheme [5, 31–33, 71]. These codes, after several iterations, provide near-Shannon performance [32, 33]. We will consider two classes of codes with iterative decoding: convolutional and block Turbo codes. These codes have already been adopted in several standards.

4.4.3.1   Convolutional Turbo Coding

By applying systematic recursive convolutional codes in an iterative scheme and by introducing an interleaver between the two parallel encoders, promising results can be obtained with so-called convolutional Turbo codes [5]. Convolutional Turbo codes are of great interest because of their good performance at low SNRs.

Figure 4-38 shows the block diagram of a convolutional Turbo encoder. The code structure consists of two parallel recursive systematic punctured convolutional codes. A block of encoded bits consists of three parts, the two parity bit parts and the systematic part. The systematic part is the same in both code bit streams and, hence, has to be transmitted only once. The code bit sequence at the output of the Turbo encoder is given by the vector $\mathbf{b}^{(k)}$.

In the receiver, the decoding is performed iteratively. Figure 4-39 shows the block diagram of the convolutional Turbo decoder. The component decoders are soft output decoders providing log-likelihood ratios (LLRs) of the decoded bits (see Section 2.1.8). The basic idea of iterative decoding is to feed-forward/backward the soft decoder output

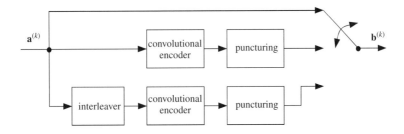

Figure 4-38   Convolutional Turbo encoder

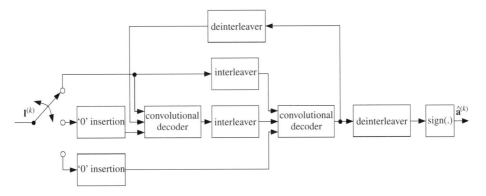

Figure 4-39   Convolutional Turbo decoder

in the form of LLRs, improving the next decoding step. In the initial stage, the non-interleaved part of the coded bits $\mathbf{b}^{(k)}$ is decoded. Only the LLRs given by the vector $\mathbf{l}^{(k)}$ at the input of the Turbo decoder are used. In the second stage, the interleaved part is decoded. In addition to the LLRs given by $\mathbf{l}^{(k)}$, the decoder uses the output of the first decoding step as *a priori* information about the coded bits. This is possible due to the separation of the two codes by the interleaver. In the next iteration cycle, this procedure is repeated, but now the non-interleaved part can be decoded using the *a priori* information delivered by the last decoding step. Hence, this decoding run has a better performance than the first one and the decoding improves. Since in each individual decoding step the decoder combines soft information from different sources, the representation of the soft information is crucial.

It is shown in References [32] and [33] that the soft value at the decoder input should be an LLR to guarantee that after combining the soft information at the input of the decoder LLRs are available again. The size of the Turbo code interleaver and the number of iterations essentially determine the performance of the Turbo coding scheme.

The performance of Turbo codes as channel codes in different multi-carrier multiple access schemes is analyzed for the following Turbo coding scheme. The component codes of the Turbo code are recursive systematic punctured convolutional codes, each of rate 2/3, resulting in an overall Turbo code rate of $R = 1/2$. Since the performance with Turbo codes in fading channels cannot be improved with a memory greater than 2 for a BER of 10^{-3} [33], we consider a convolutional Turbo code with memory 2 in order to minimize the computational complexity. The component decoders exploit the soft output

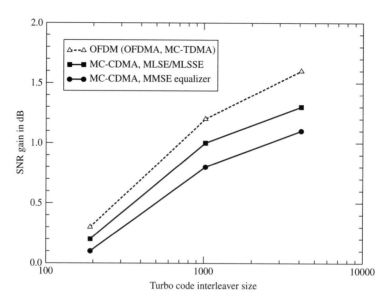

Figure 4-40 SNR gain with Turbo codes relative to convolutional codes versus Turbo code interleaver size I_{TC}

Viterbi algorithm (SOVA) [31]. The Turbo code interleaver is implemented as a random interleaver. Iterative Turbo decoding in the channel decoder uses 10 iterations. The SNR gain with Turbo codes relative to convolutional codes with $R = 1/2$ and memory 6 versus the Turbo code interleaver size I_{TC} is given in Figure 4-40 for the BER of 10^{-3}.

The results show that OFDMA and MC-TDMA systems benefit more from the application of Turbo codes than MC-CDMA systems. It can be observed that the improvements with Turbo codes at interleaver sizes smaller than 1000 are small. Due to the large interleaver sizes required for convolutional Turbo codes, they are of special interest for non-real-time applications.

4.4.3.2 Block Turbo Coding

The idea of product block or block Turbo coding is to use the well-known product codes with block codes as components for two-dimensional coding (or three dimensions) [71]. The two-dimensional code is depicted in Figure 4-41. The k_r information bits in the rows are encoded into n_r bits by using a binary block code $C_r(n_r, k_r)$. The redundancy of the code is $r_r = n_r - k_r$ and d_r the minimum distance. After encoding the rows, the columns are encoded using another block code $C_c(n_c, k_c)$, where the check bits of the first code are also encoded.

The two-dimensional code has the following characteristics:

- overall block size $n = n_r \cdot n_c$;
- number of information bits $k_r \cdot k_c$;
- code rate $R = R_r \cdot R_c$, where $R_i = k_i/n_i$, $i = c, r$; and
- minimum distance $d_{\min} = d_r \cdot d_c$.

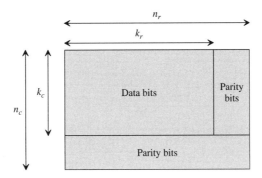

Figure 4-41   Two-dimensional product code matrix

Table 4-4   Generator polynomials of Hamming codes as block Turbo code components

n_i	k_i	Generator
7	4	$x^3 + x + 1$
15	11	$x^4 + x + 1$
31	26	$x^5 + x^2 + 1$
63	57	$x^6 + x + 1$
127	120	$x^7 + x^3 + 1$

The binary block codes employed for rows and columns can be systematic BCH (Bose–Chaudhuri–Hocquenghem) or Hamming codes [51]. Furthermore, the constituent codes of rows or columns can be extended with an extra parity bit to obtain *extended* BCH or Hamming codes. Table 4-4 gives the generator polynomials of the Hamming codes used in block Turbo codes.

The main advantage of block Turbo codes is in their application for packet transmission, where an interleaver, as it is used in convolutional Turbo coding, is not necessary. Furthermore, as typical for block codes, block Turbo codes are efficient at high code rates.

To match packet sizes, a product code can be shortened by removing symbols. In the two-dimensional case, either rows or columns can be removed until the appropriate size is reached. Unlike one-dimensional codes (such as Reed Solomon codes), parity bits are removed as part of the shortening process, helping to keep the code rate high.

The decoding of block Turbo codes is done in an iterative way [71], as in the case of convolutional Turbo codes. First, all of the horizontal received blocks are decoded and then all of the vertical received blocks are decoded (or vice versa). The decoding procedure is iterated several times to maximize the decoder performance. The core of the decoding process is the soft-in/soft-out (SISO) constituent code decoder. High performance iterative decoding requires the constituent code decoders to not only determine a transmitted sequence but also to yield a soft decision metric (i.e. LLR), which is a measure of the likelihood or confidence of each bit in that sequence. Since most algebraic block

Table 4-5 Performance of block Turbo codes in an AWGN channel after three iterations for BPSK

BTC constituent codes	Coded packet size	Code rate	E_b/N_0 at BER = 10^{-9}
(23,17)(31,25)	53 bytes	0.596	4.5 dB
(16,15)(64,57)	106 bytes	0.834	6.2 dB
(56,49)(32,26)	159 bytes	0.711	3.8 dB
(49,42)(32,31)	159 bytes	0.827	6.5 dB
(43,42)(32,31)	159 bytes	0.945	8.5 dB

decoders do not operate with soft inputs or generate soft outputs, such block decoders have primarily been applied using the soft output Viterbi algorithm (SOVA) [31] or a soft output variant of the modified Chase algorithm(s) [7]. However, other SISO block decoding algorithms can also be used for deriving the LLR.

The decoding structure of block Turbo codes is similar to that of Figure 4-39, where instead of convolutional decoders, the row and column decoders are applied. Note that here the interleaving is simply a read/write mechanism of rows and columns of the code matrix. The performance of block Turbo codes with three iterations for different packet sizes in an AWGN channel is given in Table 4-5.

4.4.4 Low Density Parity Check (LDPC) Codes

An LDPC code $C(n, k)$ is a special class of linear block codes whose parity check matrix $\mathbf{H}[n - k, n]$ has mainly '0's and only a small number of '1's, i.e. is sparse, where each block of k information bits is encoded to a codeword of size n.

LDPC codes were originally introduced by Gallager in 1962 [29]. However, they did not get much attention for decades due to the absence of an efficient and feasible implementation technology. Motivated by the success of iterative soft Turbo decoding, MacKay and Neal re-invented LDPC codes in 1995 [53, 54]; later on these codes received a great deal of interest from the information theory community, where it was shown that their performance is similar to that of Turbo codes; i.e. they achieve near-Shannon performances. Today, LDPC codes are part of several standards (e.g. DVB-S2, WiMAX- IEEE 802.16e, and IEEE 802.11n).

To understand them better let us take the following LDPC code of size $n = 8$, information bits $k = 4$, and parity bits $m = n - k = 4$. This code has rate 1/2 and can be specified by the following parity check matrix $\mathbf{H}$:

$$H = \begin{matrix} & n_1 & n_2 & n_3 & n_4 & .. & & .. & n_8 & \\ & \begin{bmatrix} 1 & 0 & 0 & 1 & 1 & 0 & 0 & 1 \\ 0 & 1 & 1 & 0 & 1 & 0 & 1 & 0 \\ 1 & 0 & 1 & 0 & 0 & 1 & 0 & 1 \\ 0 & 1 & 0 & 1 & 0 & 1 & 1 & 0 \end{bmatrix} & \begin{matrix} m_1 \\ m_2 \\ m_3 \\ m_4 \end{matrix} \end{matrix} \qquad (4.73)$$

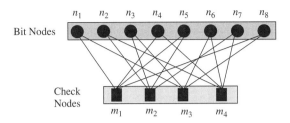

Figure 4-42   The Tanner graph representation of the LDPC code

Note that the number of '1's in each column of this matrix $\mathbf{H}$ is only 2, which makes this matrix sparse. Due to this sparse property, the same code can be equivalently represented by a bipartite graph, called a 'Tanner' graph, which is illustrated in Figure 4-42. This graph connects each check equation (check node) to its participating bits (bit nodes). A connection between a bit node n_i and a check node m_j is established if there is a '1' in the column of the matrix $\mathbf{H}$.

Parity check equations imply that for a valid codeword, the modulo-2 sum of adjacent bits of every check node has to be zero. In other words, the vector $\mathbf{x}$ is a part of the codeword $\mathbf{C}$ if it satisfies the following condition:

$$\mathbf{H}\mathbf{x}^T = 0, \quad \forall \ \mathbf{x} \in \mathbf{C}. \tag{4.74}$$

The 'Tanner graph' code representation enables the LDPC codes to have a parallelizable decoding implementation, which consists of simple operations such as addition, comparison, and table look-up. The degree of parallelism is tunable, which makes it easy to find a tradeoff between throughput, decoding delay, and the overall complexity.

The decoding process is based on the so-called 'belief-propagation' i.e. trying to determine in an iterative manner the best possible codeword from this graph. In each iteration process, the bit nodes and check nodes communicate with each other. The decoding process starts first by assigning the received channel $LLR(x_i)$ value of every bit to all the outgoing edges, i.e. from the corresponding bit node n_i to its adjacent check nodes m_j. By receiving this information, each check node updates the bit node information by referring to the parity check equations, i.e. looking at the connections within the Tanner graph, and sends it back. In each bit node a soft majority vote among the information from its adjacent check nodes will be taken. At this stage, if the hard decisions $\mathbf{x}$ on these bits satisfy all of the parity check equations, i.e. $\mathbf{H}\mathbf{x}^T = 0$, this means that a valid codeword has been found and the process stops. Otherwise, the bit nodes send the result of their soft majority votes to the check nodes again and hence the process will be repeated until the right codeword is found. In the following, we briefly describe this decoding algorithm.

4.4.4.1   Initialization

We will denote $LLR(x_i)$ as the *a priori* log-likelihood ratio of the transmitted bit x_i. Note that the sign of this LLR indicates the hard decision on the received bit, whereas the magnitude of LLR gives the reliability of the decision. The decoding is initiated by

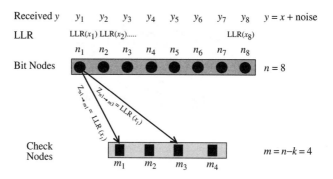

Figure 4-43   Tanner graph; decoding initialization

assigning the channel $LLR(x_i)$ to all of the outgoing edges of every bit node as follows (see Figure 4-43):

$$z_{n_i \to m_j} = LLR(x_i), \quad i = 1, \ldots, N, \quad j = 1, \ldots, \deg(\text{bit node } n_i), \tag{4.75}$$

where the number of edges corresponding to a node is called the '*deg*' of that node.

4.4.4.2   Tanner Graph's Check Node Refresh

The outgoing values from the check node m_j back to adjacent bit nodes n_i is computed as follows (see Figure 4-44):

$$w_{m_j \to n_i} = g(z_{n_1 \to m_j}, z_{n_2 \to m_j}, \ldots, z_{n_{\deg} \to m_j}), \tag{4.76}$$

where

$$g(a, b) = \text{sign}(a) \times \text{sign}(b) \times \min(|a|, |b|) + \log(1 + e^{-|a+b|}) - \log(1 + e^{-|a-b|}). \tag{4.77}$$

Note that if the log terms in the above equation are neglected, the expression of $g(a, b)$ results in the well known simple 'sum-min' algorithm used, for instance, by the soft-output-Viterbi decoder to add two soft values.

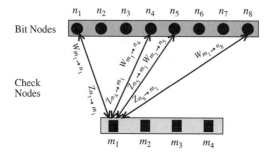

Figure 4-44   Tanner graph; check node refresh

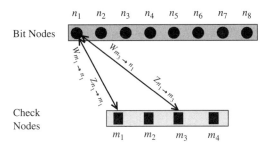

Figure 4-45 Tanner graph; bit node refresh

4.4.4.3 Tanner Graph's Bit Nodes Refresh

The outgoing values from the bit nodes n_i back to the check node m_j are computed as follows (see Figure 4-45):

$$z_{n_i \to m_j} = LLR(x_i) + \sum_{j \neq i} w_{m_j \to n_i}. \tag{4.78}$$

Note that this is a soft majority vote on the value of the bit x_i using all information except $w_{n_i \to m_i}$.

After the bit nodes updates, the hard decision x_i can be made for each bit by looking at the sign of $x_i = \text{sign}(v_{n_i \to m_i} + w_{n_i \to m_i})$. If the hard decisions satisfy the parity check equations, a valid codeword is found and the decoding is stopped. Otherwise another iteration will be started and a check node/bit node update is performed. If no convergence is achieved after a predetermined number of iterations, the current output is given out and a decoding failure is declared.

4.4.4.4 Performance and Complexity

For very large specific codes ($> 10\,000$ bits, similar to random codes) and a high number of iterations (e.g. 50), near-Shannon performance can be achieved. For instance, in the second version of the DVB-S2 specifications [18] it is shown that with an LDPC code of $n = 64\,800$, more than $2–2.5\,\text{dB}$ of extra coding gain compared to the classical concatenated RS + convolutional code can be achieved. However, the length n for wireless communication applications will be small due to the delay constraints. For instance, in IEEE 802.16e, the chosen maximum LDPC code length is 2304 bits.

In general, like finding a good Turbo interleaver, finding a good LDPC code by simulation is difficult, since some good LDPC codes at BER $= 10^{-6}$ might not be good at BER $= 10^{-9}$. This is due to the so-called 'error-step' behavior of LDPC codes, which is difficult to predict by simulations. Therefore, a programmable hardware platform (e.g. DSP or FPGA) is usually needed parallel to simulations to test in real time the performance of a given code.

The decoding complexity depends strongly on the code length n, the number of iterations, and the soft value computation. A so-called unified architecture based on two-phase

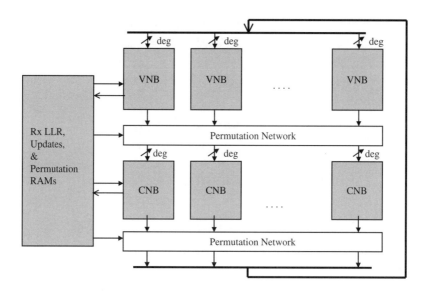

Figure 4-46   Two-step phased LDPC decoding processing

data path permutation networks is usually used for decoding (see Figure 4-46). In a given iteration, in the first phase the variable node blocks (VNB) (see Equation (4.76)) and in the second phase the check node blocks (CNB) (see Equation (4.78)) are calculated and then updated. The number of processing blocks depends strongly on the characteristics of the sub-matrices of the matrix **H**.

4.4.5   OFDM with Code Division Multiplexing: OFDM-CDM

OFDM-CDM is a multiplexing scheme that is able to exploit diversity better than conventional OFDM systems. Each data symbol is spread over several sub-carriers and/or several OFDM symbols, exploiting additional frequency and/or time diversity [39, 40]. By using orthogonal spreading codes, self-interference between data symbols can be minimized. Nevertheless, self-interference occurs in fading channels due to a loss of orthogonality between the spreading codes. To reduce this degradation, an efficient data detection and decoding technique is required. The principle of OFDM-CDM is shown in Figure 4-47.

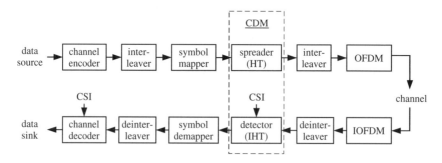

Figure 4-47   OFDM-CDM transmitter and receiver

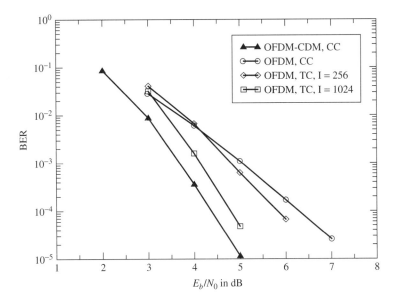

Figure 4-48 Performance of OFDM-CDM with classical convolutional codes versus OFDM with Turbo codes and interleaver size 256 and 1024 respectively

MC-CDMA and SS-MC-MA can be considered as special cases of OFDM-CDM. In MC-CDMA, CDM is applied for user separation and in SS-MC-MA different CDM blocks of spread symbols are assigned to different users [41].

The OFDM-CDM receiver applies single-symbol detection or more complex multi-symbol detection techniques which correspond to single-user or multi-user detection techniques respectively in the case of MC-CDMA. The reader is referred to Section 2.1.5 for a description of the different detection techniques.

In Figure 4-48, the performance of OFDM-CDM using classical convolutional codes (CC) is compared with the performance of OFDM using Turbo codes. The BER versus the SNR for code rate 1/2 and QPSK symbol mapping is shown. Results are given for OFDM-CDM with soft IC after the first iteration and for OFDM using Turbo codes with interleaver sizes $I = 256$ and $I = 1024$ and iterative decoding with 10 iterations. As a reference, the performance of OFDM with classical convolutional codes is given. It can be observed that OFDM-CDM with soft IC and classical convolutional codes can outperform OFDM with Turbo codes.

4.5 Signal Constellation, Mapping, De-Mapping, and Equalization

4.5.1 Signal Constellation and Mapping

The modulation employed in multi-carrier systems is usually based on quadrature amplitude modulation (QAM) with 2^m constellation points, where m is the number of bits transmitted per modulated symbol and $M = 2^m$ is the number of constellation points. The

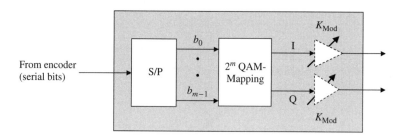

Figure 4-49   Signal mapping block diagram

general principle of modulation schemes is illustrated in Figure 4-49, which is valid for both uplink and downlink.

For the downlink, high-order modulation can be used such as 4-QAM ($m = 2$) up to 64-QAM ($m = 6$). For the uplink, more robust 4-QAM and 16-QAM are preferred. Typically, constellation mappings are based on Gray mapping, where adjacent constellation points differ by only one bit. Table 4-6 defines the constellation for 4-QAM modulation. In this table, $b_l, l = 0, \ldots, m - 1$, denotes the modulation bit order after serial-to-parallel conversion.

The complex modulated symbol takes the value $I + jQ$ from the 2^m point constellation (see Figure 4-50). In the case of transmission of mixed constellations in the downlink frame, i.e. adaptive modulation (from 4-QAM up to 64-QAM), a constant transmit power should be guaranteed. Unlike the uplink transmission, this would provide the advantage that the downlink interference from all base stations has a quasi-constant behavior. Therefore, the output complex values are formed by multiplying the resulting $I + jQ$ value by a normalization factor K_{MOD}, as shown in Figure 4-49. The normalization K_{MOD} depends on the modulation as prescribed in Table 4-7.

Symbol mapping can also be performed differentially as with D-QPSK applied in the DAB standard [16]. Differential modulation avoids the necessity of estimating the carrier phase. Instead, the received signal is compared to the phase of the preceding symbol [70]. However, since one wrong decision results in two decision errors, differential modulation performs worse than non-differential modulation with accurate knowledge of the channel in the receiver. Differential demodulation can be improved by applying a two-dimensional demodulation, where the correlations of the channel in the time and frequency directions are taken into account in the demodulation [30].

Table 4-6   Bit mapping with 4-QAM

b_0	b_1	I	Q
0	0	1	1
0	1	1	-1
1	0	-1	1
1	1	-1	-1

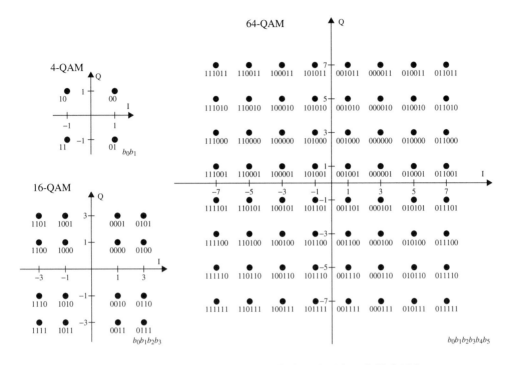

Figure 4-50 M-QAM signal constellation (4-, 16- and 64-QAM)

Table 4-7   Modulation-dependent normalization factor K_{MOD}

Modulation	K_{MOD}
4-QAM	1
16-QAM	$1/\sqrt{5}$
64-QAM	$1/\sqrt{21}$

4.5.2   Equalization and De-Mapping

The channel estimation unit in the receiver provides for each sub-carrier n an estimate of the channel transfer function $H_n = a_n \, e^{j\varphi_n}$. In mobile communications, each sub-carrier is attenuated (or amplified) by a Rayleigh or Ricean distributed variable $a_n = |H_n|$ and is phase distorted by φ_n. Therefore, after FFT operation a correction of the amplitude and the phase of each sub-carrier is required. This can be done by a simple channel inversion, i.e. multiplying each sub-carrier by $1/H_n$. Since each sub-carrier suffers also from noise, this channel correction leads to a noise amplification for small values of a_n. To counteract this effect, the SNR value $\gamma_n = 4|H_n|^2/\sigma_n^2$ of each sub-carrier (where σ_n^2 is the noise variance at sub-carrier n) should be considered for soft metric estimation. Moreover, in order to provide soft information for the channel decoder, i.e. Viterbi decoder, the received channel corrected data (after FFT and equalization with CSI coefficients) should

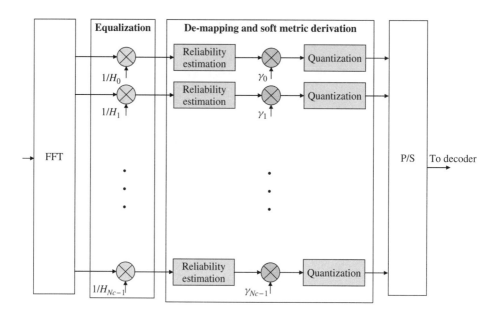

Figure 4-51   Channel equalization and soft metric derivation

be optimally converted to soft metric information. Thus, the channel-corrected data have to be combined with the reliability information exploiting channel state information for each sub-carrier so that each encoded bit has an associated soft metric value and a hard decision, which are provided to the Viterbi decoder.

As shown in Figure 4-51, after channel correction, i.e. equalization, a reliability information is provided for each mapped bit of the constellation. This reliability information corresponds to the minimum distance from the nearest decision boundary that affects the decision of the current bit. This metric corresponds to LLR values (see Section 2.1.8) [70] after it is multiplied with the corresponding value of the SNR of each sub-carrier $\gamma_n = 4|H_n|^2/\sigma_n^2$. Finally, after quantization (typically 3–4 bits for amplitude and 1 bit for the sign), these soft values are submitted to the channel decoder.

4.6   Adaptive Techniques in Multi-Carrier Transmission

As shown in Chapter 1, the signal suffers especially from time and frequency selectivity of the radio channel. Co-channel and adjacent channel interference (CCI and ACI) are further impairments that are present in cellular environments due to the high frequency re-use. Each terminal station may have different channel conditions. For instance, the terminal stations located near the base station receive the highest power which results in a high carrier-to-noise and -interference power ratio $C/(N + I)$. However, the terminal station at the cell border has a lower $C/(N + I)$.

In order to exploit the channel characteristics and to use the spectrum in an efficient way, several adaptive techniques can be applied, namely adaptive FEC coding, adaptive modulation, and adaptive power leveling. Note that the criteria for these adaptive techniques can be based on the measured $C/(N + I)$ or the received average power per symbol or per sub-carrier. These measured data have to be communicated to the transmitter via a return channel, which may be seen as a disadvantage for adaptive techniques.

In TDD systems this disadvantage can be reduced, since the channel coefficients are typically highly correlated between successive uplink and downlink slots and, thus, are also available at the transmitter. Only if significant interference occurs at the receiver, this has to be communicated to the transmitter via a return channel.

4.6.1 Nulling of Weak Sub-Carriers

The most straightforward solution for reducing the effect of noise amplification during equalization is the technique of nulling weak sub-carriers, which can be applied in an adaptive way. Sub-carriers with the weakest received power are discarded at the transmission side. However, by using strong channel coding or long spreading codes, the gain obtained by nulling weak sub-carriers is reduced.

4.6.2 Adaptive Channel Coding and Modulation

Adaptive coding and modulation in conjunction with multi-carrier transmission can be applied in several ways. The most commonly used method is to adapt channel coding and modulation during each transmit OFDM frame/burst assigned to a given terminal station [19, 20]. The most efficient coding and modulation will be used for the terminal station having the highest $C/(N + I)$, where the most robust one will be applied for the terminal station having the worst $C/(N + I)$ (see Figure 4-52). The spectral efficiency in a cellular environment is doubled using this adaptive technique [22].

An alternative technique that can be used in multi-carrier transmission is to apply the most efficient modulation for sub-carriers with the highest received power, where the most robust modulation is applied for sub-carriers suffering from multipath fading (see Figure 4-53). Furthermore, this technique can be applied in combination with power control to reduce *out-of-band* emission, where for sub-carriers located at the channel bandwidth border low order modulation with low transmit power and for sub-carriers in the middle of the bandwidth higher order modulation with higher power can be used.

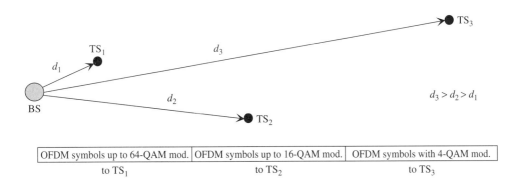

Figure 4-52   Adaptive channel coding and modulation per OFDM symbol

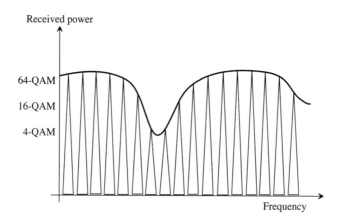

Figure 4-53 Adaptive modulation per sub-carrier

4.6.3 Adaptive Power Control

Besides the adaptation of coding and modulation, the transmit power of each OFDM symbol or each sub-carrier can be adjusted to counteract, for instance, the near–far problem or shadowing. A combination of adaptive coding and modulation (the first approach) with power adjustment per OFDM symbol is usually adopted [20].

4.7 RF Issues

A simplified OFDM transmitter front-end is illustrated in Figure 4-54. The transmitter comprises an I/Q generator with a local oscillator with carrier frequency f_c, low-pass filters, a mixer, channel pass-band filters, and a power amplifier. After power amplification and filtering, the RF analogue signal is submitted to the transmit antenna. The receiver front-end comprises similar components.

Especially in cellular environments, due to employing low gain antennas, i.e. non-directive antennas, high power amplifiers are needed to guarantee a given coverage and

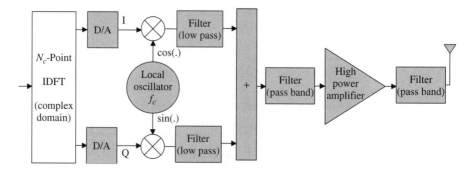

Figure 4-54 Simplified OFDM transmitter front-end

hence reduce, for instance, infrastructure costs by installing fewer base stations. Unfortunately, high power amplifiers are non-linear devices, where the maximum efficiency is achieved at the saturation point. Furthermore, at high carrier frequencies (e.g. IEEE 802.11h at 5 GHz) low cost RF transmit and receive oscillators can be applied at the expense of higher phase noise.

The main objective of this section is to analyze the performance of multi-carrier and multi-carrier CDMA transmission with a high number of sub-carriers in the presence of low cost oscillators with phase noise and HPAs with both AM/AM and AM/PM non-linear conversions. First, a commonly accepted *phase noise model* is described. After analyzing its effects in multi-carrier transmission with high order modulation, measures in the digital domain based on *common phase error* (CPE) correction are discussed. The effects of two classes of non-linear power amplifiers are presented, namely traveling wave tube amplifiers (TWTAs) and solid state power amplifiers (SSPAs). Two techniques to reduce the effects of non-linear HPAs based on pre-distortion and spreading code selection are discussed. Finally, in order to estimate the required transmit RF power for a given coverage area, a link budget analysis is carried out.

4.7.1 Phase Noise

The performance of multi-carrier synchronization tracking loops depends strongly on the RF oscillator phase noise characteristics. Phase noise instabilities can be expressed and measured in the time and/or frequency domain.

4.7.1.1 Phase Noise Modeling

Various phase noise models exist for the analysis of phase noise effects. Two often-used phase noise models that assume instability of the phase only are described in the following.

Lorenzian power density spectrum
The phase noise generated by the oscillators can be modeled by a Wiener–Lèvy process [68], i.e.

$$\theta(t) = 2\pi \int_0^t \mu(\tau) \, d\tau, \qquad (4.79)$$

where $\mu(\tau)$ represents white Gaussian frequency noise with power spectral density N_0. The resulting power density spectrum is Lorenzian, i.e.

$$H(f) = \frac{2}{\left(\pi\beta + \left(\dfrac{2\pi f}{\beta}\right)^2\right)}. \qquad (4.80)$$

The two-sided 3 dB bandwidth of the Lorenzian power density spectrum is given by β, also referred to as the line-width of the oscillator.

Measurement-Based power density spectrum
An approach used within the standardization of DVB-T is the application of the power density spectrum defined by [74]

$$H(f) = 10^{-c} + \begin{cases} 10^{-a} & |f| \leqslant f_1 \\ 10^{b(f_1-f)/(f_2-f_1)-a} & f > f_1 \\ 10^{b(f_1+f)/(f_2-f_1)-a} & f < -f_1 \end{cases} . \tag{4.81}$$

The parameters a and f_1 characterize the phase lock loop (PLL) and the parameter c the noise floor. The steepness of the linear slope is given by b and the frequency f_2 indicates where the noise floor becomes dominant. A plot of the power density spectrum with typical parameters ($a = 6.5, b = 4, c = 10.5, f_1 = 1\,\text{kHz}$, and $f_2 = 10\,\text{kHz}$) is shown in Figure 4-55.

This phase noise process can be modeled using two white Gaussian noise processes, as shown in Figure 4-56. The first noise term is filtered by an analogue filter with a transfer function as shown in Figure 4-55, while the second term gives a phase noise floor that depends on the tuner technology. A digital model of the phase noise process can be obtained by sampling the above analogue model at frequency f_{samp}.

Further phase noise models can be found in Reference [2].

4.7.1.2 Effects of Phase Noise in Multi-Carrier Transmission

For reliable demodulation in OFDM systems, orthogonality of the sub-carriers is essential, which is threatened in the receiver by phase noise caused by local oscillator inaccuracies. The local oscillators are applied in receivers for converting the RF signal to a baseband signal. The effects of local oscillator inaccuracies are severe for low cost mobile receivers.

The complex envelope of an OFDM signal is given by

$$s(t) = \frac{1}{N_c} \sum_{n=0}^{N_c-1} S_n \, e^{j2\pi f_n t}. \tag{4.82}$$

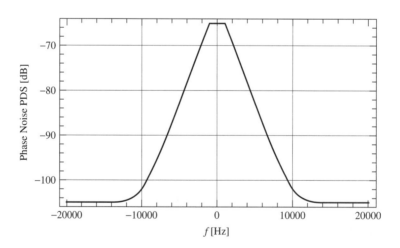

Figure 4-55 Phase noise power density spectrum

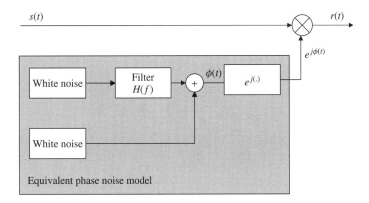

Figure 4-56 Phase noise model

For brevity but without loss of generality, it is assumed that the transmitted signal is only affected by phase noise $\varphi_N(t)$. The oscillator output can be written as

$$r(t) = s(t)\, e^{j\varphi_N(t)}. \tag{4.83}$$

The signal at the FFT output corresponding to sub-carrier n is given by

$$R_n = S_n I_0 + \sum_{\substack{m=0 \\ m \neq n}}^{N_c - 1} S_m I_{n-m}, \tag{4.84}$$

where

$$I_n = \frac{1}{T_s} \int_0^{T_s} e^{j 2\pi f_n t}\, e^{j\varphi_N(t)}\, \mathrm{d}t. \tag{4.85}$$

According to Equation (4.84), the effects of phase noise can be separated into two parts [74]. The component I_0 in the first term in Equation (4.84) represents a common phase error due to phase noise that is independent of the sub-carrier index and is common to all sub-carriers. The sum of the contributions from the $N_c - 1$ sub-carriers given by the second term in Equation (4.84) represents the ICI caused by phase noise. The ICI depends on the data and channel coefficients of all different $N_c - 1$ sub-channels such that the ICI can be considered as Gaussian noise for large N_c.

The effects of the common phase error and ICI are shown in Figure 4-57, where the mixed time/frequency representation of the total phase error Ψ_E of the sub-carriers per OFDM symbol is shown for an OFDM system with an FFT size of 2048. The mixed time/frequency representation of the total phase error shows in the frequency direction the phase error over all sub-carriers within one OFDM symbol. The time direction is included by illustrating this for 30 subsequent OFDM symbols. It can be shown that each OFDM symbol is affected by a common phase error and noise like ICI. The auto-correlation function (ACF) of the common phase error between adjacent OFDM symbols is shown in Figure 4-58.

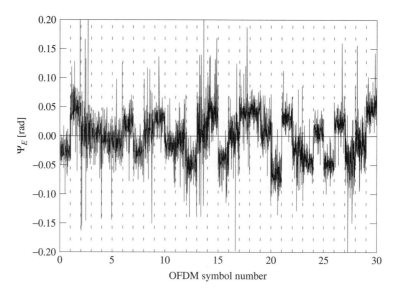

Figure 4-57   Mixed time/frequency representation of the total phase error caused by phase noise

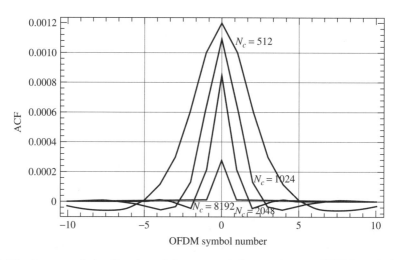

Figure 4-58   Auto-correlation function of the common phase error over OFDM symbols for different numbers of sub-carriers N_c

The phase noise is modeled as described in Section 4.7.1.1 for the measurement-based power density spectrum. It can be observed that with an increasing number of sub-carriers the correlation of the common phase error between adjacent OFDM symbols decreases.

4.7.1.3   Common Phase Error Correction

The block diagram of a common phase error correction proposed in Reference [74] is shown in Figure 4-59.

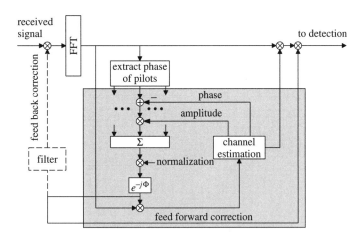

Figure 4-59 Common phase error correction

The common phase error is evaluated and corrected before detection and further processing. The phase of the pilot symbols is extracted after the FFT and the channel phase obtained from the previous OFDM symbol is subtracted on each pilot carrier. After weighting each pilot phase with the amplitude of the previous estimate, all remaining phase errors are added and normalized, which results in the common phase error estimate. This estimate is used in the data stream for common phase error correction as well as to correct the pilots used for channel estimation. Since with an increasing number of sub-carriers N_c the common phase error between adjacent OFDM symbols becomes more uncorrelated, it might be necessary to estimate and correct the common phase error within each OFDM symbol, i.e. pilot symbols have to be transmitted in each OFDM symbol.

4.7.1.4 Analysis of the Effects of Phase Noise

OFDM systems are more sensitive to phase noise than single-carrier systems due to the N_c times longer OFDM symbol duration and ICI [67, 68]. In OFDM systems, differential modulation schemes are more robust against phase noise than non-differential modulated schemes. Moreover, the OFDM sensitivity to phase noise increases as high order modulation is applied [11, 84].

In Figure 4-60, the degradation due to phase noise is shown for a coded DVB-T transmission since this is a system with high numbers of sub-carriers [74]. The chosen modulation scheme is 64-QAM. OFDM is performed with 2k FFT and the channel fading is Ricean. The performance for a receiver with and without common phase error correction is presented, where it can be observed that common phase error correction can significantly improve the system performance. Results of the effects of phase noise on MC-CDMA systems can be found in References [82] and [86].

4.7.2 *Non-linearities*

Multi-carrier modulated systems using OFDM are more sensitive to high power amplifier (HPA) non-linearities than single-carrier modulated systems [80]. The OFDM signal

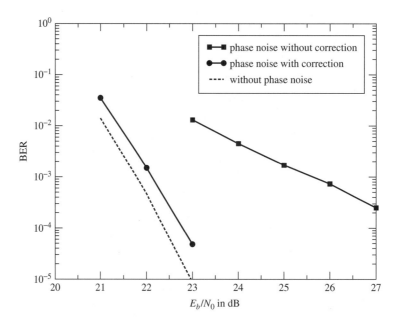

Figure 4-60 Effects of phase noise with and without common phase error correction for a DVB-T transmission with 64-QAM in a Ricean fading channel

requires higher output back-off values to keep an acceptable performance in the presence of non-linear amplifiers. This is due to the presence of a high peak-to-average power ratio (PAPR) in an OFDM signal, leading to severe clipping effects. The PAPR of an OFDM signal is defined as

$$PAPR_{OFDM} = \frac{\max |z_i|^2}{\frac{1}{N_c} \sum_{i=0}^{N_c-1} |z_i|^2}, \quad i = 0, \ldots, N_c - 1, \qquad (4.86)$$

where z_i are the transmitted time samples of an OFDM symbol.

The transmitted samples z_i result from the IFFT operation. Following the central limit theorem, the samples z_i have a complex Gaussian distribution for large numbers of sub-carriers N_c, where their amplitudes $|z_i|$ are Rayleigh distributed. At the output of the HPA, all OFDM signal points with amplitudes higher than the saturation amplitude A_{sat} will be mapped to a point in a circle with radius A_{sat}, which causes high interference, i.e. high degradation of the transmitted signal.

The aim of this section is to analyze the influence of the effects of the non-linearity due to a traveling waves tube amplifier (TWTA) and a solid state power amplifier (SSPA) in an MC-CDMA system. First, we present the influence of non-linear distortions on the downlink and uplink transmissions and then we examine some techniques to reduce the effects of non-linear distortions. These techniques are based on pre-distortion or the appropriate selection of the spreading codes to reduce the PAPR for an MC-CDMA transmission system.

4.7.2.1 Effects of Non-linear Distortions in DS-CDMA and MC-CDMA

The fact that multi-carrier CDMA transmission is more sensitive to HPA non-linearities than single-carrier CDMA transmission is valid in the single-user case, i.e. in the uplink transmitter. However, for the downlink the situation is different. The downlink transmitted signal is the sum of all active user signals, where the spread signal for both transmission schemes (single-carrier or multi-carrier) will have a high PAPR [23]. Therefore, for the downlink both systems may have quite similar behavior. To justify this, we additionally consider a single-carrier DS-CDMA system, where the transmitter consists of a spreader, transmit filter, and the HPA. The receiver is made out of the receive filter and the de-spreader, i.e. detector. The transmit and receive filters are chosen such that the channel is free of inter-chip interference when the HPA is linear. For instance, raised cosine filters with the roll-off factor α, equally split between the transmitter and receiver side, can be considered.

For both systems, conventional correlative detection is used. As disturbance, we consider only the effects of HPAs in the presence of additive white Gaussian noise. In order to compensate for the effects of the HPA non-linearities, an automatic gain control (AGC) with phase compensation is used in both systems. This is equivalent to a complex-valued one tape equalizer. Both schemes use BPSK modulation. The processing gain for MC-CDMA is $P_{G,MC} = 64$ and for DS-CDMA $P_{G,DS} = 63$. The total number of users is K.

The non-linear HPA can be modeled as a memory-less device [77]. Let $x(t) = r(t) \, e^{-j\varphi(t)}$ be the HPA complex-valued input signal with amplitude $r(t)$ and phase $\varphi(t)$. The corresponding output signal can be written as

$$y(t) = R(t) \, e^{-j\phi(t)}, \qquad (4.87)$$

where $R(t) = f(r(t))$ describes the AM/AM conversion representing the non-linear function between the input and the output amplitudes. The AM/PM distortion $\Phi(t) = g(r(t), \varphi(t))$ produces additional phase modulation. In the following, we consider two types of HPAs, namely a TWTA and a SSPA, which are commonly used in the literature [77].

The non-linear distortions of HPAs depend strongly on the output back-off (OBO):

$$OBO = \frac{P_{sat}}{P_{out}}, \qquad (4.88)$$

where $P_{sat} = A_{sat}^2$ represents the saturation power and $P_{out} = E\{|y(t)|^2\}$ is the mean power of the transmitted signal $y(t)$. Small values of the OBO cause the amplifier operation point to be near saturation. In this case a good HPA efficiency is achieved, but as a consequence the HPA output signal is highly distorted.

HPA models
Traveling Wave Tube Amplifier (TWTA)
For this type of amplifier the AM/AM and AM/PM conversions are [77]

$$R_n(t) = \frac{2r_n(t)}{1 + r_n^2(t)},$$

$$\phi(t) = \varphi(t) + \pi/3 \frac{r_n^2(t)}{1 + r_n^2(t)}, \qquad (4.89)$$

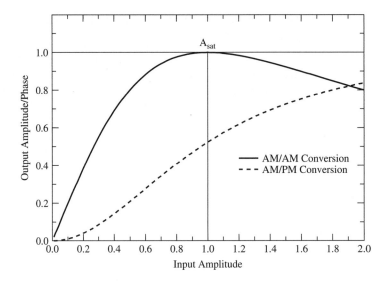

Figure 4-61 TWTA characteristics

where in the above expressions the input and the output amplitudes are normalized by the saturation amplitude A_{sat}. This kind of amplifier has the most critical characteristics due to no one-to-one AM/AM conversion (no bijection) and the AM/PM conversion. In Figure 4-61, the normalized characteristics of a TWTA are illustrated. The AM/PM conversion versus the normalized input amplitude is given in radians.

Solid State Power Amplifier (SSPA)
For this type of amplifier the AM/AM and AM/PM conversions are given with some specific parameters as follows [72, 80]:

$$R_n(t) = \frac{r_n(t)}{(1 + r_n^{10}(t))^{1/10}}$$

$$\phi(t) = \varphi(t) \tag{4.90}$$

It can be seen that the SSPA adds no phase distortion.

Influence of HPA Non-linearity
DS-CDMA
After spreading with Gold codes [26] and BPSK modulation, the overall transmitted signal $x(t)$ is obtained. For low OBO values, the signal $y(t)$ at the output of the amplifier is highly disturbed, which, after the receive filter $h(t)$, leads to a non-linear channel with memory characterized by wrapped output chips containing clusters, resulting in inter-chip interference.

Figure 4-62 shows the amplitude distribution (derived by simulations) of the filtered signal before the HPA for the uplink with spreading code length $L = 63$ and energy per

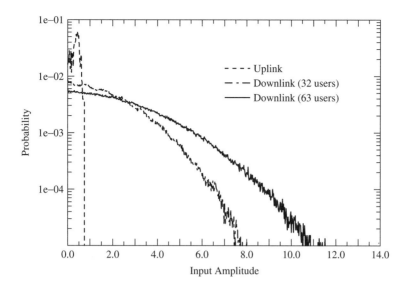

Figure 4-62   Signal amplitude distribution with single-carrier DS-CDMA

chip $E_c = -8$ dB. It should be noticed that the distribution is presented in logarithmic scaling. It can be observed that this signal has low peak amplitudes.

However, for high numbers of users in the downlink, the amplitude distribution of the transmitted signal at the base station in the case of BPSK modulation can be approximated by a Gaussian distribution with zero mean and variance ψ^2. In Figure 4-62 the amplitude distributions (derived by simulations) for the downlink case for $K = 32$ (respectively $K = 63$) active users with $L = 63$ and average power $\Omega = 10 \log(K) + E_c = 7$ dB (respectively 10 dB) are presented. It can be seen that high numbers of active users result in high peak amplitude values. The presence of high peak amplitudes causes a high degradation in the transmitted signal for low OBO values.

MC-CDMA

The BPSK modulated information bits of each user $k = 0, \ldots, K - 1$, are spread by the corresponding spreading code, where Walsh–Hadamard codes are applied. For the uplink, the spread data symbol of user k is mapped on to N_c sub-carriers of an OFDM symbol. For the downlink, the spread data of all active users are added synchronously and mapped on to the N_c sub-carriers.

Assuming perfect frequency interleaving, the input signal of the OFDM operation is statistically independent. By using a high number of sub-carriers, the complex-valued OFDM signal can be approximated by a complex-valued Gaussian distribution with zero mean and variance ψ^2. Hence, the amplitude of the OFDM signal is Rayleigh distributed and the phase is uniformly distributed.

Figure 4-63 shows the amplitude distribution of an MC-CDMA signal for the uplink case with a chip energy $E_c = -8$ dB. This curve is derived by simulations with the following parameters: $L = 64$, $N_c = 512$ [27]. The MC-CDMA signal is Rayleigh

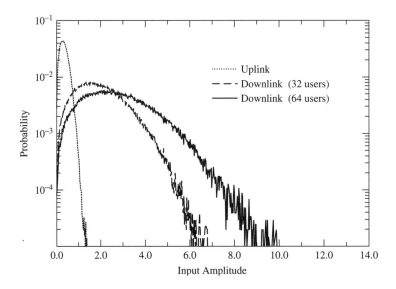

Figure 4-63   Signal amplitude distribution with MC-CDMA

distributed. Compared to the uplink of a DS-CDMA system, the MC-CDMA signal has a higher peak amplitude. This high signal amplitude leads to a higher degradation for small OBO values and results in severe clipping effects.

In the same figure, the signal amplitude distribution (derived by simulations) for the downlink case with $L = 64$, $N_c = 512$, and $K = 32$ (respectively $K = 64$) active users is presented. This signal is also Rayleigh distributed with average power $\Omega = 10 \log(K) + E_c = 7$ dB (respectively 10 dB). Comparing this distribution to that of the downlink of a DS-CDMA system, one can see that the MC-CDMA system has a lower peak amplitude. Therefore, in this example the MC-CDMA signal is more resistant to non-linearities than the DS-CDMA signal in the downlink.

For low OBO values, the signal $y(t)$ at the output of the amplifier is highly disturbed, which leads to a non-linear channel with memory characterized by wrapped output chips containing clusters, which results in inter-chip/inter-carrier interference.

4.7.2.2   Reducing the Influence of HPA Non-linearities

In this section two techniques to reduce the effects of non-linear amplifiers are discussed. The first one is to adapt the transmitted signal to the HPA characteristics by pre-correction, i.e. pre-distortion methods. The second method concentrates on the choice of the spreading code to reduce the transmitted signal PAPR.

Pre-distortion techniques
Through the above analysis, we have seen that a high OBO might be needed for the downlink to guarantee a given bit error rate. To make better use of the available HPA power, compensation techniques can be used at the transmitter side. Several methods

of data pre-distortion for non-linearity compensation have been previously introduced for single-carrier systems [45, 72]. To highlight the gain obtained by pre-distortion techniques for MC-CDMA, we consider a simple method based on the analytical inversion of the HPA characteristics. The presence of extra memory using pre-distortion techniques with memory that may reduce the effects of interference [45] is not considered in this section due to its higher complexity.

Let $z(t) = |z(t)|\, e^{-j\Psi(t)}$ be the signal that has to be amplified and $y(t) = |y(t)|\, e^{j\phi(t)}$ be the amplified signal (see Figure 4-64). To limit the distortions of an HPA, a device with output signal $x(t) = |x(t)|\, e^{-j\varphi(t)} = r(t)\, e^{-j\varphi(t)}$ can be inserted in the baseband before the HPA so that the HPA output $y(t)$ becomes as close as possible to the original signal $z(t)$. Hence the pre-distortion function will be chosen such that the global function between $z(t)$ and $y(t)$ will be equivalent to an idealized amplifier

$$
y(t) = \begin{cases} z(t) & \text{if } |z(t)| < A_{sat} \\[2mm] A_{sat}\,\dfrac{z(t)}{|z(t)|} & \text{if } |z(t)| \geqslant A_{sat} \end{cases}
\tag{4.91}
$$

TWTA

The inversion of the TWTA leads to the normalized function

$$
r_n(t) = \begin{cases} \dfrac{1}{z_n(t)}\left[1 - \sqrt{1 - z_n^2(t)}\right] & \text{if } |z_n(t)| < 1 \\[3mm] \dfrac{z_n(t)}{|z_n(t)|} & \text{if } |z_n(t)| \geqslant 1 \end{cases}
\tag{4.92}
$$

and

$$
\varphi(t) = \begin{cases} \psi(t) - \dfrac{\pi}{3}\,\dfrac{z_n^2(t)}{(1 + z_n^2(t))} & \text{if } |z_n(t)| < 1 \\[3mm] \psi(t) - \dfrac{\pi}{6} & \text{if } |z_n(t)| \geqslant 1 \end{cases}
\tag{4.93}
$$

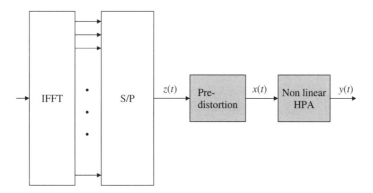

Figure 4-64   Analytical pre-distortion technique for multi-carrier transmission

SSPA

The inversion of the SSPA with the above specified parameters leads to the normalized function

$$r_n(t) = \frac{z_n(t)}{(1 - z_n(t)^{10})^{1/10}},$$

$$\varphi(t) = \psi(t). \tag{4.94}$$

Performance of analytical pre-distortion

For performance evaluation, the total degradation (TD) for a given BER is considered as the main criterion. It is given as follows:

$$TD = SNR - SNR_{AWGN} + OBO, \tag{4.95}$$

where SNR is the signal-to-noise ratio in the presence of non-linear distortions and SNR_{AWGN} is the signal-to-noise ratio in the case of a linear channel, i.e. only Gaussian noise.

One can see that as the OBO decreases, the HPA is more efficient, but, on the other hand, the non-linear distortion effects increase and a higher SNR is needed to compensate for this effect compared to a linear channel. For high OBO values the HPA works in its linear zone and there are no distortions. However, the loss of HPA efficiency through the high OBO value is taken into account in the total degradation expression. Since the total degradation changes from a decreasing to an increasing function, we can expect an optimal value for the OBO that minimizes the total degradation. In order to be independent of the channel coding, we consider an uncoded BER of 10^{-2} and 10^{-3} for our analysis.

The performance of analytical pre-distortion for the MC-CDMA system is presented in Figure 4-65. As Table 4-8 shows, for the uplink one can achieve about 2.1 dB gain with

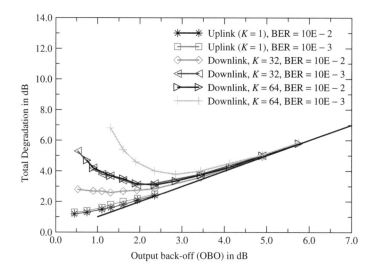

Figure 4-65 Performance of pre-distortion technique for MC-CDMA, TWTA

Table 4-8 Minimum total degradation for different transmission schemes, BER $= 10^{-2}$

Transmission scheme	Downlink, $K = 32$		Downlink, $K = 64$		Uplink	
	SSPA	TWTA	SSPA	TWTA	SSPA	TWTA
MC-CDMA without pre-distortion	2.6 dB	4.0 dB	3.2 dB	4.7 dB	1.25 dB	3.0 dB
DS-CDMA without pre-distortion	5.0 dB	5.5 dB	5.3 dB	5.8 dB	0.9 dB	1.1 dB
MC-CDMA with pre-distortion	–	2.5 dB	–	3.1 dB	–	0.9 dB

respect to a non-pre-distorted scheme using the TWTA. Similar gains for the downlink case for different BER cases have been obtained.

In Table 4-8 the minimum total degradations TD_{min} at a BER of 10^{-2} for MC-CDMA and DS-CDMA systems are summarized. The lower degradation of the MC-CDMA system for the downlink is due to its lower peak signal amplitude compared to the DS-CDMA signal and to the presence of frequency interleaving. In the case of TWTAs, the gain provided by pre-distortion for MC-CDMA is more than 1.5 dB for both the uplink and the downlink. From this table we conclude that with respect to non-linearities MC-CDMA is a good choice for the downlink and MC-DS-CDMA with a small number of sub-carriers could be adequate for the uplink.

Appropriate selection of spreading codes

As we have seen before, MC-CDMA is sensitive to amplifier non-linearity effects. We have used Walsh–Hadamard codes rather than Gold codes for the downlink in the case of multi-carrier transmission. In this section we highlight the effects of the spreading code selection and detail the appropriate choice for the uplink and the downlink [66, 69].

In Section 2.1.4.2, upper bounds for the PAPR of different spreading codes have been presented for MC-CDMA systems. Both uplink and downlink have been analyzed and it is shown that different codes are optimum with respect to the PAPR for up- and downlinks. Exemplary values for these upper bounds are given in Table 4-9 for different spreading code lengths in the uplink. As shown in Table 4-9, Zadoff–Chu codes have the lowest PAPR for the uplink.

Table 4-9 PAPR upper bounds for different spreading codes in the uplink

Spreading code	$L = 16$	$L = 64$
Walsh–Hadamard	32	128
Gold	~ 15.5	~ 31.5
Golay	4	4
Zadoff–Chu	2	2

For the downlink, the simulation results given in Reference [66] show that for different numbers of active users and a given spreading factor the PAPR in the case of Walsh–Hadamard codes decreases as the number of users increases, while for other codes (e.g. Golay codes) it increases. For instance, for a spreading factor of $L = 16$, in the case of Walsh–Hadamard codes the PAPR is equal to 7.5 and for Golay codes it is about 26. The simulation results given in Reference [66] confirm that for the downlink the best choice, also for the minimization of multiple access interference, are Walsh–Hadamard codes.

4.7.3 Narrowband Interference Rejection in MC-CDMA

The spread spectrum technique in combination with rake receivers is an interesting approach to remove the effects of interference and multi-path propagation [87]. In order to combat strong narrowband interference different techniques of notch filtering in the time domain (based on the LMS algorithm) and in the transform domain (based on the FFT) have been analyzed that provide promising results for spread spectrum systems [57]. However, in a frequency- and time-selective fading channel with coherent detection, its high performance is no longer guaranteed without perfect knowledge about the impulse response of the channel and without performing an optimum detection.

Another interesting approach is based on the MC-CDMA technique [21]. The interference can be considered as narrowband multitone *sine* interference. A method for evaluating both interference and the fading process based on frequency domain analysis is studied in the following. The estimated interference and fading process is used for weighting each received chip before de-spreading.

The narrowband interference can be modeled as consisting of a number of Q continuous *sine* wave tones [46]

$$Int(t) = \sum_{m=0}^{Q-1} A_m \cos(2\pi f_m t + \phi_m), \tag{4.96}$$

where A_m is the amplitude of the interferer at frequency f_m and ϕ_m is a random phase.

4.7.3.1   Interference Estimation

The interference estimation in multi-carrier systems can be based on the transmission of a null symbol, i.e. one OFDM symbol with a non-modulated signal (see Section 4.2.1). At the receiver side this null symbol contains only interference and noise. The power of the interference Int_l in each sub-carrier l can be estimated by performing an FFT operation on this null symbol. The estimated interference power is used for weighting the received sub-carriers.

However, interference can also be detected without the use of a null symbol. It can be done, for instance, by performing an envelope detection of the received signal after the FFT operation. Of course, this method is not accurate, since it suffers from the presence of fading and ICI.

Detection strategy

In the following, we consider the single-user case. In the presence of fading, interference and noise, the received signal vector $\mathbf{r} = (R_0, R_1, \ldots, R_{L-1})^T$ can be written as

$$\mathbf{r} = \mathbf{Hs} + \mathbf{Int} + \mathbf{n}, \tag{4.97}$$

where $\mathbf{H}$ is a diagonal matrix of dimension $L \times L$ with elements $H_{l,l} = a_l\, e^{j\varphi_l}$ corresponding to the fading and phase rotation disturbance, $\mathbf{s} = (S_0, S_1, \ldots, S_{L-1})^T$ is the transmitted signal, $\mathbf{Int} = (Int_0, Int_1, \ldots, Int_{L-1})^T$ is the narrowband interference, and $\mathbf{n} = (N_0, N_1, \ldots, N_{L-1})^T$ represents the Gaussian noise. The optimal detection consists of choosing the best transmitted sequence by minimizing the distance between the received sequence and all possible transmitted sequences by using the values a_l, φ_l, Int_l as channel state information (CSI). Let us denote the possible transmitted sequences as $(v_0^{(i)}, v_1^{(i)}, \ldots, v_{L-1}^{(i)})$, $i = 1, 2$. The information bit is detected if we maximize the following expression [21]:

$$\Delta_i^2 = \sum_{l=0}^{L-1} \frac{H_{l,l}^*}{\sqrt{1 + Int_l^2}} R_l v_l^{(i)}, \quad i = 1, 2. \tag{4.98}$$

This corresponds to a conventional correlative detection using the CSI, where the received chips are weighted by $H_{l,l}^*/\sqrt{1 + Int_l^2}$. There are many advantages of such a weighting. The first one is that it is equivalent to *soft erasure* or *soft switching-off* of the sub-carriers containing interference. The second advantage is that no pre-defined threshold for the decision of switching-off the interferers is needed, unlike in the classical transform domain notch-filtering method [57]. Finally, the system does not need extra complexity in terms of hardware, since the FFT operation is already part of the system.

If the number of interference tones is Q, the decision process is equivalent to considering the case where Q chips are erased at the transmitter as we perform erasing of the interferers. Therefore, the above expression can be approximated as

$$\Delta_i^2 \approx \sum_{l=0}^{L-Q-1} \frac{H_{l,l}^*}{\sqrt{1 + Int_l^2}} \overline{R}_l v_l^{(i)}, \quad i = 1, 2, \tag{4.99}$$

where $\overline{R}_l$ corresponds to the received chips with low interference.

Performance evaluation

In Figure 4-66 the performance of an MC-CDMA system in the presence of narrowband interference in the case of an uncorrelated Rayleigh fading channel is plotted. OFDM is realized by an inverse FFT with $N_c = 64$ points. The sub-carriers are spaced 20 kHz apart. The useful bandwidth is 1.28 MHz. In all simulations the spreading factor is $L = 64$ where the average received power per chip is normalized to one. Note that the interference rejection technique, i.e. soft erasing, performs well. The analytical curves are also plotted and match with the simulation results.

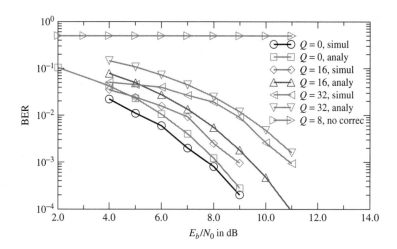

Figure 4-66   Performance of narrowband interference rejection with MC-CDMA

These results show that the combination of spread spectrum with OFDM in the presence of multi-tone narrowband interference in a frequency- and time-selective fading channel is a promising approach.

4.7.4   Link Budget Evaluation

In order to estimate the transmit power for a given coverage area a link budget analysis is always necessary, especially for a cellular mobile communications system. This analysis allows one to estimate the minimum transmit power that a terminal station should select to guarantee a given link availability. Especially for mobile terminal stations, it is important to save power in order to enable a longer battery life cycle. For any link budget evaluation, especially for the base station, the transmit power must be checked to find out whether it is compliant with the maximum allowed transmission power specified by the regulatory bodies.

The transmit power P_{TX} can be calculated as follows:

$$P_{TX} = Path\ loss + P_{Noise} - G_{Antenna} + \text{Offset} + \text{Rx}_{loss} + C/N, \qquad (4.100)$$

where

$$Path\ loss = 10 \log_{10} \left(\frac{4\pi f_c d^{\eta/2}}{c} \right)^2 \qquad (4.101)$$

is the propagation loss in dB. In the case of mobile communications η can be estimated to be in the order of 3 to 5 (see Section 1.1.1). The variable d represents the distance between the transmitter and the receiver, f_c is the carrier frequency, and c is the speed of light. In the presence of line of sight (LOS), as in case of free space, $\eta = 2$ and the above formula becomes

$$Path\ loss_{LOS} = 32.4 + 20\log_{10}\left(\frac{d}{km}\right) + 20\log_{10}\left(\frac{f_c}{MHz}\right). \qquad (4.102)$$

The noise power at the receiver input is

$$P_{Noise} = FN_{Thermal} = FKTB \qquad (4.103)$$

where F is the receiver noise factor, K is the Boltzmann constant ($K = 1.38 \times 10^{-23}$ J/K), T is the temperature in Kelvin, and B is the total occupied Nyquist bandwidth. The noise power is expressed in dBm. $G_{Antenna}$ is the sum of the transmit and receive antenna gains expressed in dBi. *Offset* is the margin for all uncertainties such as interference, shadowing, and fading margin reserve. Rx_{loss} is the margin for all implementation uncertainties. C/N is the carrier-to-noise power ratio to guarantee a given BER. The receiver sensitivity threshold is given by

$$Rx_{th} = P_{Noise} + Rx_{loss} + C/N. \qquad (4.104)$$

In Table 4-10, an example of a link budget evaluation for using MC-TDMA in mobile indoor and outdoor communications is given. We consider for the base station transmit antenna a gain of about 16 dBi. However, for the terminal side, we consider an omni-directional antenna with 0 dBi gain. This table shows that about 25–27 dBm transmit

Table 4-10 Example of link budget evaluation for mobile and indoor communications

Parameters	Mobile outdoor communications	WLAN indoor communications
Carrier frequency	2 GHz	5 GHz
Coverage	Macro cell (10 km)	Pico cell (30 m)
Path loss power, n	2.2	3
Path loss	130 dB	112 dB
Indoor penetration	N/A	10 dB
Noise factor	6 dB	6 dB
Bandwidth	2 MHz	20 MHz
Base station antenna gain (90°)	∼ 16 dBi	∼ 16 dBi
Terminal station antenna gain	0 dBi	0 dBi
Offset	∼ 5 dB	∼ 5 dB
C/N for coded QPSK (rate 1/2)	6 dB	6 dB
Receiver sensitivity	−94 dBm	−84 dBm
Needed Tx power	25 dBm	27 dBm

power would be required to achieve a coverage of 10 km with QPSK for outdoor mobile applications at 2 GHz carrier frequency and a coverage of 30 m for WLAN indoor reception at 5 GHz carrier frequency.

References

[1] Alard M. and Lassalle R., "Principles of modulation and channel coding for digital broadcasting for mobile receivers," *European Broadcast Union Review*, no. 224, pp. 168–190, Aug. 1987.

[2] Armada A. G., "Understanding the effects of phase noise in orthogonal frequency division multiplexing (OFDM)," *IEEE Transactions on Broadcasting*, vol. 47, pp. 153–159, June 2001.

[3] Auer G. and Cosovic I., "On pilot grid design for an OFDM air interface," in *Proc. IEEE Wireless Communications and Networking Conference (WCNC 2007)*, Hong Kong, China, March 2007.

[4] Berens F., Jung P., Plechinger J., and Baier P. W., "Multi-carrier joint detection CDMA mobile communications," in *Proc. IEEE Vehicular Technology Conference (VTC '97)*, Phoenix, USA, pp. 1897–1901, May 1997.

[5] Berrou C., Glavieux A., and Thitimajshima P., "Near Shannon limit error-correcting coding and decoding: Turbo-codes (1)," in *Proc. IEEE International Conference on Communications (ICC '93)*, Geneva, Switzerland, pp. 1064–1070, May 1993.

[6] Bölcskei H., Heath R. W., and Paulraj A. J., "Blind channel identification and equalization in OFDM-based multi-antenna systems," *IEEE Transactions on Signal Processing*, vol. 50, pp. 96–109, Jan. 2002.

[7] Chase D., "A class of algorithms for decoding block codes with channel measurement information," *IEEE Transactions on Information Theory*, vol. 18, pp. 170–182, Jan. 1972.

[8] Classen F. and Meyer H., "Frequency synchronization for OFDM systems suitable for communication over frequency selective fading channels," in *Proc. IEEE Vehicular Technology Conference (VTC '94)*, Stockholm, Sweden, pp. 1655–1659, June 1994.

[9] Choi Y.-S., Voltz P. J., and Cassara F. A., "On channel estimation and detection for multicarrier signals in fast and selective Rayleigh fading channels," *IEEE Transactions on Communications*, vol. 49, pp. 1375–1387, Aug. 2001.

[10] Cosovic I. and Auer G. "Capacity analysis of MIMO-OFDM including pilot overhead and channel estimation errors," in *Proc. IEEE International Conference on Communications (ICC 2007)*, Glasgow, UK, June 2007.

[11] Costa E. and Pupolin S., "*M*-QAM-OFDM system performance in the presence of a non-linear amplifier and phase noise," *IEEE Transactions on Communications*, vol. 50, pp. 462–472, March 2002.

[12] Daffara F. and Chouly A., "Maximum likelihood frequency detectors for orthogonal multi-carrier systems," in *Proc. IEEE International Conference on Communications (ICC '93)*, Geneva, Switzerland, pp. 766–771, May 1993.

[13] Daffara F. and Adami O., "A new frequency detector for orthogonal multi-carrier transmission techniques," in *Proc. IEEE Vehicular Technology Conference*, Chicago, USA, pp. 804–809, July 1995.

[14] De Bot P., Le Floch B., Mignone V., and Schütte H.-D., "An overview of the modulation and channel coding schemes developed for digital terrestrial TV broadcasting within the dTTb project," in *Proc. International Broadcasting Convention (IBC)*, Amsterdam, The Netherlands, pp. 569–576, Sept. 1994.

[15] Edfors O., Sandell M., van de Beek J.-J., Wilson S. K., and Börjesson P. O., "OFDM channel estimation by singular value decomposition," *IEEE Transactions on Communications*, vol. 46, pp. 931–939, July 1998.

[16] ETSI DAB (EN 300 401), "Radio broadcasting systems; digital audio broadcasting (DAB) to mobile, portable and fixed receivers," Sophia Antipolis, France, April 2000.

[17] ETSI DVB-RCT (EN 301 958), "Interaction channel for digital terrestrial television (RCT) incorporating multiple access OFDM," Sophia Antipolis, France, March 2001.

[18] ETSI DVB-S2 (EN 302 307), "DVB-S2; 2nd generation framing structure, channel coding and modulation systems for broadcasting, interactive services," Sophia Antipolis, France, June 2006.

[19] ETSI DVB-T (EN 300 744), "Digital video broadcasting (DVB); framing structure, channel coding and modulation for digital terrestrial television," Sophia Antipolis, France, July 1999.

[20] ETSI HIPERMAN (TS 102 177), "High performance metropolitan area network, Part A1: physical layer," Sophia Antipolis, France, Feb. 2003.

[21] Fazel K., "Narrow-band interference rejection in orthogonal multi-carrier spread spectrum communications," in *Proc. IEEE International Conference on Universal Personal Communications (ICUPC '94)*, San Diego, USA, pp. 46–50, Oct. 1994.

[22] Fazel K., Decanis C., Klein J., Licitra G., Lindh L., and Lebert Y. Y., "An overview of the ETSI-BRAN HA physical layer air interface specification," in *Proc. IEEE International Symposium on Personal, Indoor and Mobile Radio Communications (PIMRC 2002)*, Lisbon, Portugal, pp. 102–106, Sept. 2002.

[23] Fazel K. and Kaiser S., "Analysis of non-linear distortions on MC-CDMA", in *Proc. IEEE International Conference on Communications (ICC '98)*, Atlanta, USA, pp. 1028–1034, June 1998.

[24] Fazel K., Kaiser S., and Robertson P., "OFDM: A key component for terrestrial broadcasting and cellular mobile radio," in *Proc. International Conference on Telecommunication (ICT '96)*, Istanbul, Turkey, pp. 576–583, April 1996.

[25] Fazel K., Kaiser S., Robertson P., and Ruf M. J., "A concept of digital terrestrial television broadcasting," *Wireless Personal Communications*, vol. 2, nos. 1 and 2, pp. 9–27, 1995.

[26] Fazel K., Kaiser S., and Schnell M., "A flexible and high-performance cellular mobile communications system based on multi-carrier SSMA," *Wireless Personal Communications*, vol. 2, nos. 1 and 2, pp. 121–144, 1995.

[27] Fazel K. and Papke L., "On the performance of convolutionally-coded CDMA/OFDM for mobile communications system," in *Proc. IEEE International Symposium on Personal, Indoor and Mobile Radio Communications (PIMRC '93)*, Yokohama, Japan, pp. 468–472, Sept. 93.

[28] Frenger P. K. and Svensson N. A. B., "Decision-directed coherent detection in multi-carrier systems on Rayleigh fading channels," *IEEE Transactions on Vehicular Technology*, vol. 48, pp. 490–498, March 1999.

[29] Gallager R. G., "Low density parity check codes," *IRE Transactions on Information Theory*, IT-8, pp. 21–28, 1962.

[30] Haas E. and Kaiser S., "Two-dimensional differential demodulation for OFDM," *IEEE Transactions on Communications*, vol. 51, pp. 580–586, April 2003.

[31] Hagenauer J. and Höher P., "A Viterbi algorithm with soft-decision outputs and its applications," in *Proc. IEEE Global Telecommunications Conference (GLOBECOM '89)*, Dallas, USA, pp. 1680–1686, Nov. 1989.

[32] Hagenauer J., Offer E., and Papke L., "Iterative decoding of binary block and convolutional codes," *IEEE Transactions on Information Theory*, vol. 42, pp. 429–445, Mar. 1996.

[33] Hagenauer J., Robertson R., and Papke L., "Iterative ('turbo') decoding of systematic convolutional codes with the MAP and SOVA algorithms," in *Proc. ITG Conference on Source and Channel Coding*, Munich, Germany, pp. 21–29, Oct. 1994.

[34] Heath R. W. and Giannakis G. B., "Exploiting input cyclostationarity for blind channel identification in OFDM systems," *IEEE Transactions on Signal Processing*, vol. 47, pp. 848–856, Mar. 1999.

[35] Höher P., "TCM on frequency-selective land-mobile fading channels," in *Proc. 5th Tirrenia International Workshop on Digital Communications*, Tirrenia, Italy, pp. 317–328, Sept. 1991.

[36] Höher P., Kaiser S., and Robertson P., "Two-dimensional pilot-symbol-aided channel estimation by Wiener filtering," in *Proc. IEEE International Conference on Acoustics, Speech and Signal Processing (ICASSP '97)*, Munich, Germany, pp. 1845–1848, April 1997.

[37] Höher P., Kaiser S., and Robertson P., "Pilot-symbol-aided channel estimation in time and frequency," in *Proc. IEEE Global Telecommunications Conference (GLOBECOM '97), Communication Theory Mini Conference*, Phoenix, USA, pp. 90–96, Nov. 1997.

[38] Hsieh M. H. and Wei C. H., "A low complexity frame synchronization and frequency offset compensation scheme for OFDM systems over fading channels," *IEEE Transactions on Communications*, vol. 48, pp. 1596–1609, Sept. 1999.

[39] Kaiser S., "Trade-off between channel coding and spreading in multi-carrier CDMA systems," in *Proc. IEEE International Symposium on Spread Spectrum Techniques and Applications (ISSSTA '96)*, Mainz, Germany, pp. 1366–1370, Sept. 1996.

[40] Kaiser S., "OFDM code division multiplexing in fading channels," *IEEE Transactions on Communications*, vol. 50, pp. 1266–1273, Aug. 2002.

[41] Kaiser S. and Fazel K., "A flexible spread spectrum multi-carrier multiple-access system for multimedia applications," in *Proc. IEEE International Symposium on Personal, Indoor and Mobile Radio Communications (PIMRC '97)*, Helsinki, Finland, pp. 100–104, Sept. 1997.

[42] Kaiser S. and Höher P., "Performance of multi-carrier CDMA systems with channel estimation in two dimensions," in *Proc. IEEE International Symposium on Personal, Indoor and Mobile Radio Communications (PIMRC '97)*, Helsinki, Finland, pp. 115–119, Sept. 1997.

[43] Kamal S. S. and Lyons R. G., "Unique-word detection in TDMA: acquisition and retention," *IEEE Transactions on Communications*, vol. 32, pp. 804–817, 1984.

[44] Kammeyer K.-D., Petermann T., and Vogeler S., "Iterative blind channel estimation for OFDM receivers," in *Proc. International Workshop on Multi-Carrier Spread Spectrum and Related Topics (MC-SS 2001)*, Oberpfaffenhofen, Germany, pp. 283–292, Sept. 2001.

[45] Karam G., *Analysis and Compensation of Non-linear Distortions in Digital Microwaves Systems*, ENST-Paris, 1989, PhD thesis.

[46] Ketchum J. W. and Proakis J. G., "Adaptive algorithms for estimating and suppressing narrow band interference in PN spread spectrum systems," *IEEE Transactions on Communications*, vol. 30, pp. 913–924, May 1982.

[47] Li Y., "Simplified channel estimation for OFDM systems with multiple transmit antennas," *IEEE Transactions on Wireless Communications*, vol. 1, pp. 67–75, Jan. 2002.

[48] Li Y., Cimini L. J., and Sollenberger N. R., "Robust channel estimation for OFDM systems with rapid dispersive fading channels," *IEEE Transactions on Communications*, vol. 46, pp. 902–915, July 1998.

[49] Li Y., Seshadri N., and Ariyavisitakul S., "Channel estimation for OFDM systems with transmitter diversity in mobile wireless channels," *IEEE Journal on Selected Areas in Communications*, vol. 17, pp. 461–471, March 1999.

[50] Li Y., Winters J. H., and Sollenberger N. R., "MIMO-OFDM for wireless communications: Signal detection with enhanced channel estimation," *IEEE Transactions on Communications*, vol. 50, pp. 1471–1477, Sept. 2002.

[51] Lin S. and Costello D, *Error Control Coding: Fundamentals and Applications*, Englewood Cliffs, New Jersey: Prentice Hall, 1983.

[52] Luise M., Reggiannini R., and Vitetta G. M., "Blind equalization/detection for OFDM signals over frequency-selective channels," *IEEE Journal on Selected Areas in Communications*, vol. 16, pp. 1568–1578, Oct. 1998.

[53] MacKay D. J. and Neal R. M., "Good codes based on very sparse matrices," in *Proc. 5th The Institute of Mathematics and its Applications (IMA) Conference 1995*, pp. 100–111, 1995.

[54] MacKay D. J. and Neal R. M., "Near Shannon limit performance of low density parity check codes," *Electronics Letters*, vol. 33, pp. 457–458, Mar. 1997.

[55] Massey J., "Optimum frame synchronization," *IEEE Transactions on Communications*, vol. 20, pp. 115–119, April 1997.

[56] Mignone V. and Morello A., "CD3-OFDM: a novel demodulation scheme for fixed and mobile receivers," *IEEE Transactions on Communications*, vol. 44, pp. 1144–1151, Sept. 1996.

[57] Milstein L. B., "Interference rejection techniques in spread spectrum communications," *Proceedings of the IEEE*, vol. 76, pp. 657–671, June 1988.

[58] Minn H. and Bhargava V. K., "An investigation into time-domain approach for OFDM channel estimation," *IEEE Transactions on Broadcasting*, vol. 46, pp. 240–248, Dec. 2000.

[59] Minn H., Kim D. I., and Bhargava V. K., "A reduced complexity channel estimation for OFDM systems with transmit diversity in mobile wireless channels," *IEEE Transactions on Communications*, vol. 50, pp. 799–807, May 2002.

[60] Moose P. H., "A technique for orthogonal frequency division multiplexing frequency offset correction", *IEEE Transactions on Communications*, vol. 42, pp. 2908–2914, Oct. 1994.

[61] Morelli M. and Mengali U., "A comparison of pilot-aided channel estimation methods for OFDM systems," *IEEE Transactions on Signal Processing*, vol. 49, pp. 3065–3073, Dec. 2001.

[62] Müller A., "Schätzung der Frequenzabweichung von OFDM-Signalen," in *Proc. ITG-Fachtagung Mobile Kommunikation, ITG-Fachberichte 124*, Neu-Ulm, Germany, pp. 89–101, Sept. 1993.

[63] Müller A., European Patent 0529421A2-1993, Priority date 29/08/1991.

[64] Muquet B., de Courville M., and Duhamel P., "Subspace-based blind and semi-blind channel estimation for OFDM systems," *IEEE Transactions on Signal Processing*, vol. 50, pp. 1699–1712, July 2002.

[65] Necker M, Sanzi F., and Speidel J., "An adaptive Wiener filter for improved channel estimation in mobile OFDM-systems," in *Proc. IEEE International Symposium on Signal Processing and Information Technology*, Cairo, Egypt, pp. 213–216, Dec. 2001.

[66] Nobilet S., Helard J.-F., and Mottier D., "Spreading sequences for uplink and downlink MC-CDMA systems: PAPR and MAI minimization," *European Transactions on Telecommunications (ETT)*, vol. 13, pp. 465–474, Sept. 2002.

[67] Pollet T., Moeneclaey M., Jeanclaude I., and Sari H., "Effect of carrier phase jitter on single-carrier and multi-carrier QAM systems," in *Proc. IEEE International Conference on Communications (ICC '95)*, Seattle, USA, pp. 1046–1050, June 1995.

[68] Pollet T., van Bladel M., and Moeneclaey M., "BER sensitivity of OFDM systems to carrier frequency offset and Wiener phase noise," *IEEE Transactions on Communications*, vol. 43, pp. 191–193, Feb./Mar./Apr. 1995.

[69] Popovic B. M., "Spreading sequences for multi-carrier CDMA systems," *IEEE Transactions on Communications*, vol. 47, pp. 918–926, June 1999.

[70] Proakis J. G., *Digital Communications*, New York: McGraw-Hill, 1995.

[71] Pyndiah R., "Near-optimum decoding of product codes: block Turbo codes," *IEEE Transactions on Communications*, vol. 46, pp. 1003–1010, Aug. 1999.

[72] Rapp C., *Analyse der nichtlinearen Verzerrungen modulierter Digitalsignale–Vergleich codierter und uncodierterter Modulationsverfahren und Methoden der Kompensation durch Vorverzerrung*, Düsseldorf: VDI Verlag, Fortschritt-Berichte VDI, series 10, no. 195, 1991, PhD thesis.

[73] Robertson P., "Effects of synchronization errors on multi-carrier digital transmission systems," *DLR Internal Report*, April 1994.

[74] Robertson P. and Kaiser S., "Analysis of the effects of phase-noise in orthogonal frequency division multiplex (OFDM) systems," in *Proc. IEEE International Conference on Communications (ICC '95)*, Seattle, USA, pp. 1652–1657, June 1995.

[75] Robertson P. and Kaiser S., "Analysis of the loss of orthogonality through Doppler spread in OFDM systems," in *Proc. IEEE Global Telecommunications Conference (GLOBECOM '99)*, Rio de Janeiro, Brazil, pp. 701–706, Dec. 1999.

[76] Robertson P. and Kaiser S., "Analysis of Doppler spread perturbations in OFDM(A) systems," *European Transactions on Telecommunications (ETT)*, vol. 11, pp. 585–592, Nov./Dec. 2000.

[77] Saleh A. M., "Frequency-independent and frequency-dependent non-linear models of TWTA," *IEEE Transactions on Communications*, vol. 29, pp. 1715–1720, Nov. 1981.

[78] Sandell M., *Design and Analysis of Estimators for Multi-Carrier Modulations and Ultrasonic Imaging*, Lulea University, Sweden, Sept. 1996, PhD thesis.

[79] Sanzi F. and ten Brink S., "Iterative channel estimation and detection with product codes in multi-carrier systems," in *Proc. IEEE Vehicular Technology Conference (VTC 2000-Fall)*, Boston, USA, Sept. 2000.

[80] Schilpp M., Sauer-Greff W., Rupprecht W., and Bogenfeld E., "Influence of oscillator phase noise and clipping on OFDM for terrestrial broadcasting of digital HDTV", in *Proc. IEEE International Conference on Communications (ICC '95)*, Seattle, USA, pp. 1678–1682, June 1995.

[81] Schmidl T. M. and Cox D. C., "Robust frequency and timing synchronization for OFDM," *IEEE Transactions on Communications*, vol. 45, pp. 1613–1621, Dec. 1997.

[82] Steendam H. and Moeneclaey M., "The effect of carrier phase jitter on MC-CDMA performance," *IEEE Transactions on Communications*, vol. 47, pp. 195–198, Feb. 1999.

[83] Steiner B., "Time domain uplink channel estimation in multi-carrier-CDMA mobile radio system concepts," in *Proc. International Workshop on Multi-Carrier Spread Spectrum (MC-SS '97)*, Oberpfaffenhofen, Germany, pp. 153–160, April 1997.

[84] Tomba L., "On the effect of Wiener phase noise in OFDM systems," *IEEE Transactions on Communications*, vol. 46, pp. 580–583, May 1998.

[85] Tomba L. and Krzymien W. A., "On the use of chip-level differential encoding for the uplink of MC-CDMA systems," in *Proc. IEEE Vehicular Technology Conference (VTC '98)*, Ottawa, Canada, pp. 958–962, May 1998.

[86] Tomba L. and Krzymien W. A., "Sensitivity of the MC-CDMA access scheme to carrier phase noise and frequency offset," *IEEE Transactions on Vehicular Technology*, vol. 48, pp. 1657–1665, Sept. 1999.

[87] Turin G. L., "Introduction to spread spectrum anti-multi-path techniques and their application to urban digital radio," *Proceedings of the IEEE*, vol. 68, pp. 328–353, March 1980.

[88] Wang X. and Liu K. J. R., "Adaptive channel estimation using cyclic prefix in multi-carrier modulation system," *IEEE Communications Letters*, vol. 3, pp. 291–293, Oct. 1999.

[89] Yang B., Cao Z., and Letaief K. B., "Analysis of low-complexity windowed DFT-based MMSE channel estimator for OFDM systems," *IEEE Transactions on Communications*, vol. 49, pp. 1977–1987, Nov. 2001.

[90] Yang B., Letaief K. B., Cheng R. S., and Cao Z., "Channel estimation for OFDM transmission in multipath fading channels based on parametric channel modeling," *IEEE Transactions on Communications*, vol. 49, pp. 467–479, March 2001.

[91] Yeh C.-S. and Lin Y., "Channel estimation using pilot tones in OFDM systems," *IEEE Transactions on Broadcasting*, vol. 45, pp. 400–409, Dec. 1999.

[92] Yeh C.-S., Lin Y., and Wu Y., "OFDM system channel estimation using time-domain training sequence for mobile reception of digital terrestrial broadcasting," *IEEE Transactions on Broadcasting*, vol. 46, pp. 215–220, Sept. 2000.

[93] Zhou S. and Giannakis G. B., "Finite-alphabet based channel estimation for OFDM and related multicarrier systems," *IEEE Transactions on Communications*, vol. 49, pp. 1402–1414, Aug. 2001.

5

Applications

5.1 Introduction

The deregulation of the telecommunications industry, creating pressure on new operators to innovate in service provision in order to compete with existing traditional telephone service providers, is and will be an important factor for an efficient use of spectrum. Most of the total information communicated over future digital networks will be *data* rather than purely *voice*. Hence, the demand for high rate packet-oriented services, such as mixed data, voice, and video services, that exceed the bandwidth of the conventional systems will increase.

Multi-media applications and computer communications are often bursty in nature. A typical user expects to have an instantaneous high bandwidth available delivered by his or her access mechanism when needed. This means that the average bandwidth required to deliver a given service will be low, even though the instantaneous bandwidth required is high.

Properly designed broadband systems allocate capacity to specific users instantly and, given a sufficiently large number of users, take advantage of statistical multiplexing to serve each user adequately with a fraction of the bandwidth needed to handle the peak data rate. The emergence of internet protocol (IP) networks exemplifies this trend.

As the examples given in Table 5-1 show, the user rate varies for different multi-media services. Generally, the peak data rate for a single user is required only for short periods of time. Therefore, the data rate that will be supported by future systems will be variable on demand up to a peak of at least 50 Mbit/s in the uplink direction and 100 Mbit/s in the downlink direction [5]. It may be useful in some systems to allow only lower data rates to be supported, thereby decreasing the overall traffic requirement, which could reduce costs and lead to longer ranges.

The user's demand for high bandwidth packet-oriented services with current delivery over low-bandwidth wireline copper loops (e.g. PSTN, ISDN, xDSL) might be adequate today but certainly not in the future.

Wireless technologies are currently limited to medium data rate services, but by offering *high mobility* the wireless technologies will offer new alternatives. In Figure 5-1 the data rate versus mobility for current and future standards (4G) is plotted. The current 2G systems like GSM provide high mobility but a low data rate. The 3G and 3G Evolution

Multi-Carrier and Spread Spectrum Systems Second Edition K. Fazel and S. Kaiser
© 2008 John Wiley & Sons, Ltd

Table 5-1   Examples of data rates and bit error rates versus services [25]

Service	Data rate	Bit error rate
Telephony and messaging	8–64 kbit/s	$10^{-3} - 10^{-6}$
Short control messages/signaling	8–64 kbit/s	10^{-9}
Lightweight browsing	64–512 kbit/s	10^{-6}
Video telephony/video conferencing	64 kbit/s – 5 Mbit/s	$10^{-3} - 10^{-6}$
Real time gaming	1–20 Mbit/s	$10^{-6} - 10^{-9}$
Video streaming	5–30 Mbit/s	$10^{-6} - 10^{-9}$
File exchange	Up to 50 Mbit/s	10^{-6}
LAN access	Up to 50 Mbit/s	10^{-6}

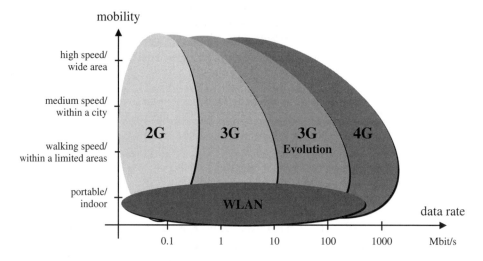

Figure 5-1   Data rate versus mobility in wireless standards

systems WCDMA/UMTS, HSPA, and LTE provide similar mobility as GSM but can deliver much higher data rates [14]. The supported data rates range from several Mbit/s with HSPA up to more than 300 Mbit/s provided with LTE [4]. The IEEE 802.11a and IEEE 802.11n WLAN standards have been designed for high rate data services with low mobility and low coverage (indoor environments). On the other hand, the ETSI BRAN HIPERMAN and IEEE 802.16x standards are specified to provide high data rates for broadband wireless access (BWA) with high coverage. HIPERMAN and IEEE 802.16x can provide high peak data rates of up to 60 Mbit/s.

On the broadcast side, DAB offers similar mobility as GSM, but with a much higher broadcast data rate. Although the DVB-T standard was originality designed for fixed or portable receivers, the results of several field trials have demonstrated its robustness at high speed as well [13]. Moreover, the DVB-H standard has been specified, which can be considered as an extension of DVB-T for hand-held devices with high mobility.

The common feature of the current wireless standards that offer high data rates is the use of *multi-carrier transmission*, i.e. OFDM [8, 15–19, 21–24]. In addition, these standards employ adaptive technologies by using several transmission modes, i.e. allowing different combinations of channel coding and modulation together with power control.

A simple adaptive strategy was introduced in DAB using multi-carrier differential QPSK modulation (and also in GSM using single-carrier GMSK modulation) with several punctured convolutional code rates. By applying a simple combination of source and channel coding, the primary goal was to protect the most important audio/speech message part with the most robust FEC scheme and to transmit the less important source coded data even without FEC. This technique allows one to receive the highest quality sound/speech in most reception conditions and an acceptable quality in the worst reception area, where it should be noticed that in the case of analogue transmission no signal would be received.

DVB-T and DVB-H employ different concatenated FEC coding rates with high order modulation up to 64-QAM and different numbers of sub-carriers and guard times. Here the objective was also to provide different video quality versus distance and different cell-planning flexibility, i.e. a countrywide single-frequency network or a regional network, for instance, using the so-called 'taboo' channels (free channels that cannot be used for analogue transmission due to the high amount of co-channel interference).

In WCDMA/UMTS, besides using different FEC coding rates, a variable spreading factor (VSF) with adaptive power control is introduced. As in GSM, the combination of FEC with source coding is also exploited. The variable spreading code allows one to have a good tradeoff between coverage, single-cell/multi-cell environments, and mobility. For high coverage areas with a high delay spread, large spreading factors can be applied and for low coverage areas with a low delay spread the smallest spreading factor can be used.

LTE provides an adaptive resource allocation by flexible coding and modulation (up to 64-QAM) with channel-dependent scheduling. HARQ with incremental redundancy is applied which triggers the rate matching procedure.

In IEEE 802.11a, HIPERMAN, and IEEE 802.16x standards/WiMAX, a solution based on the combination of multi-carrier transmission with high order modulation (up to 64-QAM) with adaptive FEC (variable rate coding) and adaptive power control is adopted. For each user, according to its required data rate and channel conditions the best combination of FEC, modulation scheme, and the number of time slots is allocated. The main objective is to offer the best tradeoff between data rate and coverage, where the mobility is not of big importance at the first stages. These standards also allow different guard times adapted to different cell coverage.

Offering a tradeoff between coverage, data rate, and mobility with a generic air interface architecture is the primary goal of the next-generation wireless systems. Users having no mobility and the lowest coverage distance (pico cells) with an ideal channel condition will be able to receive the highest data rate, while, on the other hand, the subscriber with the highest mobility conditions and highest coverage area (macro cells) will be able to receive the necessary data rate in order to establish the required communication link.

The aim of this chapter is to examine in detail the different application fields of multi-carrier transmission for multi-user environments. This chapter gives an overview with important technical parameters and highlights the strategy behind the choices.

First, concrete examples of the application of OFDM/OFDMA for B3G cellular mobile systems, i.e. LTE and WiMAX, are given. Then, future mobile radio concepts and field trials based on OFDM and MC-CDMA are described. The OFDM-based IEEE 802.11a standard is studied thereafter. Finally, the DVB-T return channel (DVB-RCT) specification will be presented.

5.2 3GPP Long Term Evolution (LTE)

5.2.1 Introduction

The *Third Generation Partnership Project* (3GPP) is the forum where WCDMA and its evolutions, *high speed packed access* (HSPA) and *3GPP Long Term Evolution* (LTE), have been specified. The standardization bodies ETSI, ARIB, TTC, TTA, CCSA, and ATIS are the organizational partners of 3GPP. The evolution of 3G standards under 3GPP is illustrated in Figure 5-2 and briefly described in the following.

WCDMA

A cellular mobile radio is characterized by providing mobile services with high geographical coverage. The introduction of WCDMA in Europe, referred to as UMTS [3] (see Section 1.3.3.2), was a move from mainly mobile voice communications towards mobile multimedia applications. The first version of WCDMA is Release 99 (R99), which is operated in the 3G carrier frequency bands at around 2 GHz using channels with 5 MHz bandwidth. Theoretical peak data rates of 2 Mbit/s are possible with R99 while in practice 384 kbit/s are implemented.

HSPA

The increasing demand for higher data rates resulted in the WCDMA evolutions *high speed downlink packet access* (HSDPA) [1] and *high speed uplink packet access* (HSUPA) [2]. Both extensions are based on the R99 single-carrier WCDMA. They increase the data rate up to 14.4 Mbit/s with HSDPA (Release 5, R5) and 5.7 Mbit/s with HSUPA (Release 6, R6) respectively. These improvements are obtained by introducing channel-dependent scheduling and *hybrid ARQ* (HARQ) with soft combining together with multiple code allocation. The downlink additionally supports adaptive coding and modulation. The set of HSDPA and HSUPA is termed *high speed packed access* (HSPA).

LTE

3GPP LTE is also referred to as *evolved universal terrestrial radio access* (E-UTRA) or *Super 3G* (S3G) and is introduced by 3GPP as Release 8 (R8) [8]. The specifications define a new physical air interface in order to further increase the data rate of the cellular mobile radio compared to HSPA. The key differentiation compared to WCDMA and

Figure 5-2   3G evolution within 3GPP

HSPA is the OFDM downlink and single-carrier FDMA uplink. LTE is targeting data rates of 100 Mbit/s in the downlink and 50 Mbit/s in the uplink. The improvements in data rate are due to enhanced channel-dependent scheduling and rate adaptation, also in the frequency domain, spatial multiplexing with MIMO, and larger channel bandwidths of up to 20 MHz.

4G (IMT-Advanced)

The *fourth generation* (4G) of the cellular mobile radio is referred to as IMT-Advanced and encompasses new radio technologies as well as existing technologies. The standardization of new physical air interfaces is targeting downlink peak data rates of 1 Gbit/s at low mobility and 100 Mbit/s at high mobility.

5.2.2 Requirements on LTE

The LTE radio access technology should be optimized for packet switched traffic with high data rate and low latency. The requirements that have to be fulfilled by LTE are defined in Reference [5] and summarized in Table 5-2.

With the figures from Table 5-2, the expected achievable spectrum efficiency in the downlink results in 5 bit/s/Hz with 100 Mbit/s in the 20 MHz bandwidth and correspondingly in 2.5 bit/s/Hz with 50 Mbit/s in the 20 MHz bandwidth in the uplink. The LTE performance presented in Section 5.2.11 shows that the LTE specification R8 exceeds these required spectrum efficiencies.

Table 5-2 Requirements for LTE [5]

Parameter	Target figures
Peak data rates	100 Mbit/s for downlink 50 Mbit/s for uplink
Average user throughput per MHz compared to HSPA Release 6	3–4 times higher for downlink 2–3 times higher for uplink
Spectrum efficiency in bit/s/Hz/cell compared to HSPA Release 6	3–4 times higher for downlink 2–3 times higher for uplink
Mobility	0–15 km/h (optimized for this range) 15–120 km/h (high performance guaranteed) 120–350 km/h (connection maintained)
Supported bandwidths	1.25–20 MHz
Spectrum allocation	Operation in paired spectrum (FDD) and unpaired spectrum (TDD) should be supported
Latency	5 ms user-plane latency at IP layer, for one-way 100 ms control-plane latency from idle to active state
Number of users per cell	At least 200 at 5 MHz bandwidth At least 400 at bandwidth higher than 5 MHz

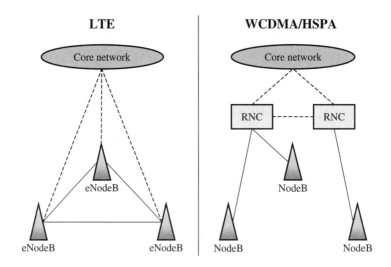

Figure 5-3   Comparison of RAN architecture between LTE and WCDMA/HSPA

5.2.3   Radio Access Network (RAN) Architecture

The architecture of the radio access network (RAN) of LTE is shown in Figure 5-3. For comparison, the RAN architecture of WCDMA/HSPA is also shown in order to highlight the differences. A significant difference between WCDMA and LTE is that LTE does not support macro diversity, which is the case for WCDMA/HSPA. A radio network controller (RNC) is omitted with LTE, which reduces the latency in the RAN. This leads LTE to shift more complexity into the eNodeB, which in LTE terminology refers to the base station.

Besides the physical layer processing, the LTE eNodeB has the tasks of mobility management and radio resource management, which both belong in the WCDMA/HSPA RAN to the tasks of the RNC. The eNodeBs in the LTE RAN are directly connected to each other and the handover decisions are taken by the eNodeB. In WCDMA/HSPA this task belongs to the RNC.

5.2.4   Radio Protocol Architecture

The LTE protocol stack is split into the user plane and the control plane. The user plane transports all user information from voice to data while the control plane is used only for control signalling. Figure 5-4 illustrates the LTE protocol stack.

All protocols shown here are located in the base station (eNodeB) and mobile terminal station (UE) respectively. The functions of the different protocol layers are summarized in the following.

Radio Resource Control (RRC)
The RRC is part of the control plane and is responsible for configuring the layer 1 and layer 2 protocols PDCP, RLC, MAC, and PHY. The main services and functions of the

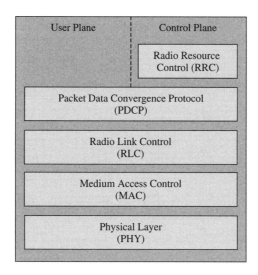

Figure 5-4   User and control plane protocol stack

RRC are admission control, handover management, QoS management, terminal station measurement reporting and control, MBMS control, and paging.

Packet Data Convergence Protocol (PDCP)

The main functions of the PDCP for the user plane are IP header compression, transfer of user data, and ciphering. At the control plane the corresponding functions are transfer of control data and ciphering.

Radio Link Control (RLC)

Segmentation and reassembly of packets from higher layers as well as error correction through ARQ are main functions of the RLC. Additionally, flow control between the eNodeB and the mobile terminal is handled by the RLC.

Medium Access Control (MAC)

The MAC is responsible for the scheduling in the up- and downlink, error correction through hybrid ARQ (HARQ), adaptive modulation, resource and power assignment, and antenna mapping.

Physical Layer (PHY)

Main functions of the physical layer are coding, modulation, and multiple antenna transmission (MIMO). The LTE physical layer is explained in detail in the following sections.

5.2.5   Downlink Transmission Scheme

The transmission scheme of the LTE downlink is OFDM with a cyclic prefix. It has been shown in Chapter 1 to 4 that besides the robustness of OFDM in frequency selective

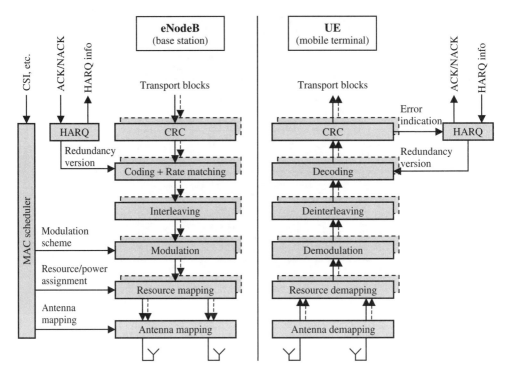

Figure 5-5   Block diagram of the LTE downlink

channels, OFDM guarantees high flexibility in resource allocation and scheduling in the frequency domain. The block diagram of the LTE downlink is shown in Figure 5-5.

Data packets from the higher layers, referred to as transport blocks, are delivered to the physical layer at the base station (eNodeB). Up to two transport blocks can be processed in parallel in the physical layer. The functions of the individual blocks illustrated in Figure 5-5 are explained in the following sub-sections for the base station. The mobile terminal, referred to as user equipment (UE), includes the respective counterparts.

5.2.5.1   Coding and Modulation

Cyclic Redundancy Check (CRC)
The CRC is inserted into each transport block to detect at the receiver side (UE) the presence of residual transmission errors.

Channel Coding, Rate Matching, and HARQ
Turbo coding of rate 1/3 is used for channel coding. It uses the same rate 1/2, memory 3, Turbo codes as WCDMA/HSPA as constituent codes to generate the overall code rate 1/3. Figure 5-6 illustrates the LTE Turbo encoder with the two constituent encoders.

Compared to WCDMA/HSPA, the interleaver between both constituent codes has been changed in order to better support parallelization in the decoding without contention. Rate

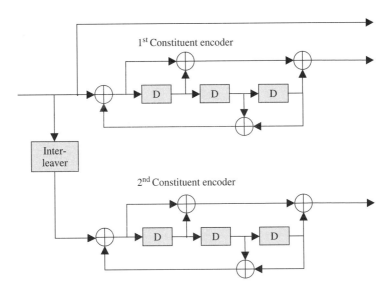

Figure 5-6 Rate 1/3 Turbo encoder

matching is applied by puncturing code bits to increase the effective code rate to more than 1/3 or by repeating code bits to reduce the effective code rate to less than 1/3 [7]. The selected rate depends on the actual available resources. The HARQ block applies incremental redundancy and triggers the rate matching procedure.

Interleaving
The interleaver applies cell-specific scrambling to increase the robustness against inter-cell interference.

Modulation
The modulation schemes applied in LTE are QPSK, 16-QAM, and 64-QAM. The MAC scheduler determines the actual modulation scheme.

5.2.5.2   Resource Mapping

Triggered by the MAC scheduler, the resource mapping assigns data symbols to the resource blocks by exploiting, for example, channel quality information at the transmitter. The downlink and uplink are structured in radio frames. The downlink frame structure is the same for FDD and TDD and is shown in Figure 5-7.

One radio frame has the duration of 10 ms and consists of 10 sub-frames. In TDD, one sub-frame is either allocated to the downlink or uplink. One sub-frame results in the duration of 1 ms and is built of two slots. A slot has a fixed length of 0.5 ms and includes either seven OFDM symbols with a normal cyclic prefix length or six OFDM symbols with an extended cyclic prefix length. The OFDM symbols with a normal cyclic prefix have a cyclic prefix length of $T_{cp} = 5.2\,\mu s$ for the first OFDM symbol and of $T_{cp} = 4.7\,\mu s$

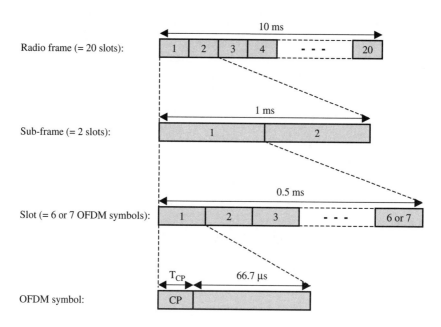

Figure 5-7 LTE downlink frame structure

for the remaining OFDM symbols. In the case of the extended cyclic prefix the length is $T_{cp} = 17.7\,\mu s$. The extended cyclic prefix is used in scenarios with a long delay spread and for multicast and broadcast services in the single-frequency network mode. The OFDM symbol duration without a guard interval is always $66.7\,\mu s$. A resource block has the duration of one slot and consists of 12 adjacent sub-carriers, as illustrated in Figure 5-8.

Thus, for the normal cyclic prefix length, 84 (12×7) resource elements are transmitted in one resource block while in the case of the extended cyclic prefix, 72 (12×6) resource elements are transmitted. A resource element corresponds to a symbol transmitted on one sub-carrier in one OFDM symbol. The sub-carrier distance in the downlink is $15\,kHz$ so that a resource block has a total bandwidth of $180\,kHz$.

Pilot symbols are inserted in each resource block for channel estimation at the receiver, which is required for coherent data detection. Figure 5-9 shows the pilot grid of a resource block with normal cyclic prefix length for different numbers of transmit antenna.

The basic pilot grid for the single antenna scheme has pilots in the first and third last OFDM symbols. The pilot symbols have a distance of six sub-carriers in the frequency direction and are shifted by three sub-carriers between the first and third last OFDM symbol. As described in Section 4.3, the channel estimator has to interpolate/filter the channel in the time and frequency directions. To improve the channel estimation accuracy, the channel estimator can, especially in the downlink, interpolate/filter over several resource blocks in time and frequency.

The receiver must be able to estimate the channel from each antenna in the case of multiple transmit antennas. To avoid interference between the pilot symbols from different transmit antennas, orthogonal sets of pilot symbols are assigned to the transmit antennas. As shown in Figure 5-9 for the case with two transmit antennas, the first antenna uses the

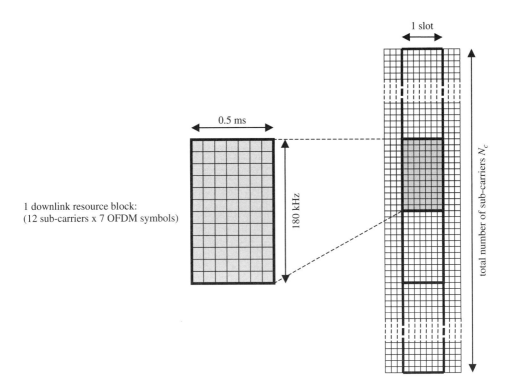

Figure 5-8 Resource block with normal cyclic prefix length in the downlink

original set of pilots as in the case with a single transmit antenna. The second transmit antenna places its pilots also in the first and third last OFDM symbols, but shifts the pilot symbol positions by three sub-carriers. In the case of four transmit antennas, the pilot symbols for the first and second transmit antennas are the same as for the two transmit antenna case. For the third transmit antenna, two pilot symbols are inserted in the second OFDM symbol with a distance of six sub-carriers. For the fourth transmit antenna, the pilot symbols are also inserted in the second OFDM symbols with a distance of six sub-carriers. However, relative to the pilots for the third transmit antenna, the pilots for the fourth transmit antenna are shifted by three sub-carriers. For all pilot grids hold that when one antenna transmits a pilot symbol in one resource element, the other antennas do not transmit in this resource element in order to avoid interference. Moreover, as the pilot density in time direction is reduced for the third and fourth transmit antennas, the case with four transmit antennas, applying for example spatial multiplexing, is mainly targeted for low mobility scenarios.

5.2.5.3 MAC Scheduler

The MAC scheduler in Figure 5-5 is responsible for scheduling and rate matching by controlling the HARQ and selecting the modulation scheme, resource and power assignment, and antenna mapping.

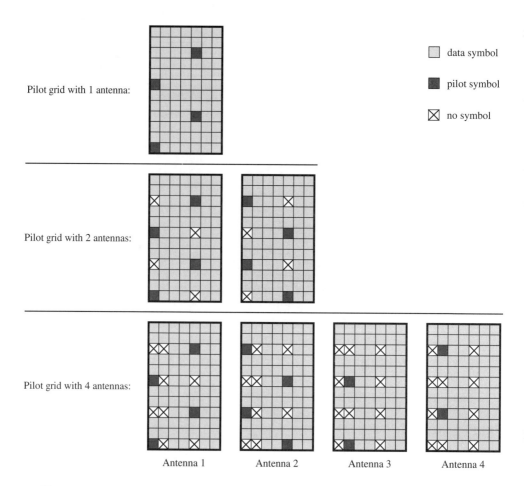

Antenna 1   Antenna 2   Antenna 3   Antenna 4

Figure 5-9   Pilot grid within one resource block for one, two, and four transmit antennas

5.2.5.4   Multi-Antenna Transmission

Multi-antenna transmission (antenna mapping) with up to four transmit antennas is specified in LTE. A maximum of two code words, i.e. transport blocks, can be transmitted in parallel by applying spatial multiplexing. The multi-antenna transmission is defined in a two-stage process indicated in Figure 5-10 [6].

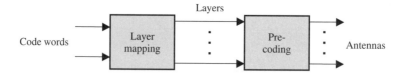

Figure 5-10   Antenna mapping in the downlink

The first stage is the layer mapping, where the transport blocks are mapped on to one or several layers. In the second stage, the layers are pre-coded according to the selected multi-antenna transmission scheme. This can be spatial multiplexing, beamforming, or transmit diversity. The specified multi-antenna transmission schemes are detailed in the following.

Spatial Multiplexing

Spatial multiplexing is used to transmit in parallel two transmission blocks using the same time–frequency resource. If the two transmission blocks belong to one user, the scheme is referred to as a single-user MIMO (SU-MIMO). Alternatively, the two transport blocks can belong to different users. This scheme is referred to as multi-user MIMO (MU-MIMO). The spatial multiplexing is based on a so-called codebook-based pre-coding. The codebook specifies a set of spatial pre-coding matrices that can be used at the transmitter. The actual pre-coding matrix is chosen based on channel quality information obtained by channel measurements per antenna in the downlink.

Spatial Multiplexing with Cyclic Delay Diversity (CDD)

To improve the performance of spatial multiplexing further, the combination of codebook pre-coding with CDD (see Section 6.3.1.3) is specified. Two types of CDD are distinguished. The first type is CDD with large cyclic shifts in order to achieve frequency diversity within one sub-band. The second type is CDD with small cyclic shifts where the channel remains flat within one sub-band but a frequency scheduling gain can be obtained by increased frequency selective fading over the total bandwidth.

Beamforming

Codebook pre-coding with only one layer, i.e. one transport block per time–frequency resource, results in beamforming.

Beamforming with Cyclic Delay Diversity (CDD)

CDD can be combined with beamforming in the same way as already described for spatial multiplexing.

Transmit Diversity

Space frequency block codes (SFBCs, see Section 6.3.4) are applied as transmit diversity scheme. Mapping schemes are defined for two and four transmit antennas. A single codeword is mapped to the layers, which are SFBC pre-coded before transmission via the antennas.

5.2.6 *Uplink Transmission Scheme*

The transmission scheme of the LTE uplink is localized DFT-spread OFDM (see Section 3.2.4). The block diagram for DFT-spread OFDM is shown in Figure 5-11.

DFT-spread OFDM is a single-carrier FDMA scheme that guarantees a low PAPR. This is essential for the power efficiency of the mobile terminal transmitter. A certain amount of flexibility in resource allocation and scheduling is achieved with DFT-spread OFDM in the uplink. The block diagram of the LTE uplink is shown in Figure 5-12.

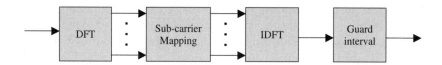

Figure 5-11   DFT-spread OFDM transmission scheme

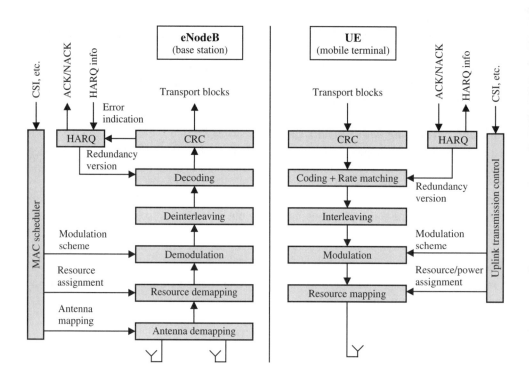

Figure 5-12   Block diagram of the LTE uplink

Transport blocks from the higher layers are delivered to the physical layer at the mobile terminal. In contrast to the downlink, only one transport block at a time can be processed in the uplink physical layer. The functions of the individual blocks illustrated in Figure 5-12 are explained in the following sub-sections for the mobile terminal. The base station includes the respective counterparts.

5.2.6.1   Coding and Modulation

Cyclic Redundancy Check (CRC)
The CRC is inserted into each transport block to detect at the receiver side (eNodeB) the presence of residual transmission errors.

Channel Coding, Rate Matching, and HARQ

Turbo coding of rate 1/3 is used for channel coding. The same code construction as for the downlink is used. The selected rate for rate matching depends on the actual available resources. The HARQ block applies incremental redundancy and triggers the rate matching procedure.

Interleaving

The interleaver applies cell-specific scrambling to increase the robustness against inter-cell interference.

Modulation

The modulation schemes applied in LTE are QPSK, 16-QAM, and 64-QAM. The MAC scheduler determines the actual modulation scheme.

5.2.6.2 Resource Mapping

Triggered by the MAC scheduler via the uplink transmission control, the resource mapping assigns data symbols to the resource blocks by exploiting, for example, channel quality information at the transmitter. The uplink frame structure is the same for FDD and TDD and is shown in Figure 5-13.

One radio frame has the duration of 10 ms and consists of 10 sub-frames. In TDD, one sub-frame is either allocated to the downlink or uplink. One sub-frame has the duration of 1 ms and is build of two slots. A slot has a fixed length of 0.5 ms and includes either seven DFT-spread OFDM (DFTS-OFDM) symbols with a normal cyclic prefix length or

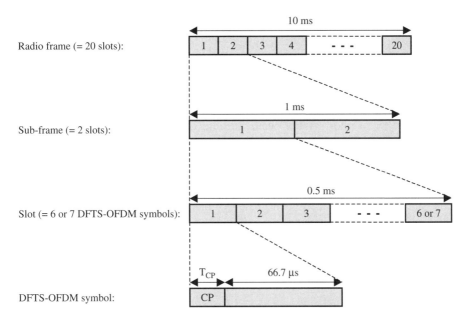

Figure 5-13 LTE uplink frame structure

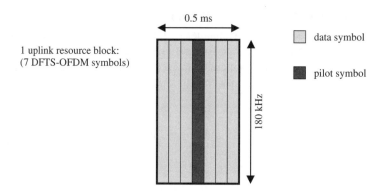

0.5 ms

1 uplink resource block:
(7 DFTS-OFDM symbols)

180 kHz

data symbol

pilot symbol

Figure 5-14 Uplink pilot allocation within one resource block

six DFT-spread OFDM symbols with an extended cyclic prefix length. The DFT-spread OFDM symbols with normal cyclic prefix have a cyclic prefix length of $T_{cp} = 5.2\,\mu s$ for the first DFT-spread OFDM symbol and of $T_{cp} = 4.7\,\mu s$ for the other DFT-spread OFDM symbols. In the case of the extended cyclic prefix the length is $T_{cp} = 17.7\,\mu s$. The extended cyclic prefix if used in scenarios with a long delay spread and for multi-cast and broadcast services in the single-frequency network mode. The DFT-spread OFDM symbol duration without a guard interval is always 66.7 μs.

A resource block is defined for one slot and consists of 12 adjacent sub-carriers. The sub-carrier distance in the uplink is 15 kHz so that a resource block has a bandwidth of 180 kHz. In contrast to the downlink, in the uplink resource blocks assigned to one mobile terminal in one OFDM symbol have to be adjacent to each other in the frequency domain.

One pilot symbol is inserted in each resource block for channel estimation at the receiver, which is required for coherent data detection. Figure 5-14 shows the pilot allocation within a resource block with seven DFT-spread OFDM symbols.

With DFT-spread OFDM, the pilot symbols are not scattered in frequency but are time multiplexed with the data symbols. Every fourth DFT-spread OFDM symbol within a resource block is a pilot symbol occupying the whole transmission bandwidth of the resource block. The pilot symbols are chosen such that they have nearly constant amplitude and good auto- and cross-correlation properties. This is achieved with Zadoff–Chu sequences (see Section 2.1.4.1). The correlation properties are important for separating the pilots of the resource block at the same positions between neighboring cells. Different Zadoff–Chu sequences are assigned to resource blocks of different cells.

Additionally, broadband channel sounding is required to enable channel-dependent scheduling in the uplink. This requires channel sounding over a much larger bandwidth than that of one resource block per user. While for the downlink pilot symbols are available over a large bandwidth for channel estimation, this is not always the case in the uplink. The uplink channel sounding is achieved by reserving blocks within one sub-frame for channel sounding. Data transmission is not possible in these blocks. The blocks with sounding signals occupy a bandwidth much larger than that of one resource block. These blocks are shared between different users either in the time, frequency, or code domain.

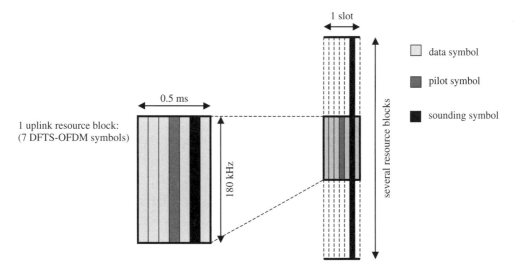

Figure 5-15   Channel sounding in the uplink

The latter assigns different Zadoff–Chu sequences to different users. The insertion of sounding symbols is illustrated in Figure 5-15.

5.2.6.3   Multi-Antenna Transmission

The mobile terminals transmit only with a single antenna. Multi-user MIMO reception is possible at the base station. Optionally, a closed-loop transmit antenna selection diversity can be implemented for the FDD mode.

5.2.7   Physical Layer Procedures

5.2.7.1   Cell Search

The cell search procedure allows the mobile terminal (UE) to find and identify a potential cell that it can use for communications after switching on the mobile terminal. For this purpose a primary and a secondary synchronization signal are transmitted in the downlink on a regular basis. The synchronization signals are inserted in the last two OFDM symbols of the first slot of the first and sixth sub-frame of a 10 ms radio frame (see Figure 5-7). In a first step the mobile terminal synchronizes to the primary synchronization signal. This allows synchronizing on a 5 ms basis. In the second step, the ambiguity with two primary synchronization signals in one radio frame is resolved by observing the second synchronization signal. The unique combination of the primary and secondary synchro-nization signals allows the mobile terminal to finally detect the broadcasted cell specific information.

In order to simplify the cell search procedure, the bandwidth of the synchronization signals is constant and independent of the bandwidth actually used in the specific cell.

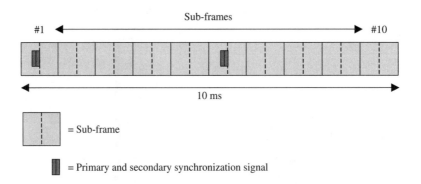

Figure 5-16   Synchronization signals for cell search

The synchronization signal occupies 72 sub-carriers with a bandwidth of about 1 MHz located in the center of the available bandwidth. Figure 5-16 illustrates the allocation of the synchronization signals within one radio frame.

The cell search procedure is also used to search for neighbor cells that might be candidates for a handover. The same procedure as described above is used for searching for neighbor cells in specific time instances. Resources have to be made available in the mobile terminal for a neighbor cell search.

5.2.7.2 Random Access Procedure

The random access procedure defines how a mobile terminal (UE) can request resources for data transmission. In LTE, a dedicated block of time and frequency resources is exclusively reserved for this procedure. Terminals that want to initiate a connection are transmitting random access preambles in these reserved resources. The preambles are based on Zadoff–Chu sequences (see Section 2.1.4.1). Each terminal selects randomly a sequence out of a set of available orthogonal Zadoff–Chu sequences. As long as no sequence is chosen twice per random access attempt, collisions can be avoided. The random access preamble has the duration of one sub-frame (1 ms) and occupies a bandwidth of about 1 MHz. Information about the allocation of the random access resources is broadcasted to all terminals. Each random access preamble of duration 0.9 ms has a 0.1 ms guard time, which is needed to compensate for the different propagation times of the different users, which at this stage are not time-aligned. After successful detection of a random access preamble, further information such as timing advance, scheduling information, and terminal identification is sent.

5.2.8 Supported Bandwidths

The bandwidths supported by LTE cover narrowband 1.4 MHz up to broadband 20 MHz. The mapping of the number of resource blocks to the channel bandwidths is given in Table 5-3. The occupied bandwidth is defined as the bandwidth containing 99 % of the total mean power of the transmitted signal.

Table 5-3 Occupied channel bandwidths and number of resource blocks

Channel bandwidth (MHz)	1.4	1.6	3	3.2	5	10	15	20
Number of resource blocks with FDD	6	N/A	15	N/A	25	50	75	100
Number of resource blocks with TDD	6	7	15	16	25	50	75	100

Table 5-4 LTE (E-UTRA) frequency bands, guard bands, and duplex modes [9]

E-UTRA band	Uplink (UL) (MHz)	Downlink (DL) (MHz)	UL-DL guard band (MHz)	Duplex mode
1	1920–1980	2110–2170	130	FDD
2	1850–1910	1930–1990	20	FDD
3	1710–1785	1805–1880	20	FDD
4	1710–1755	2110–2155	355	FDD
5	824–849	869–894	20	FDD
6	830–840	875–885	35	FDD
7	2500–2570	2620–2690	50	FDD
8	880–915	925–960	10	FDD
9	1749.9–1784.9	1844.9–1879.9	60	FDD
10	1710–1770	2110–2170	340	FDD
11	1427.9–1452.9	1475.9–1500.9	23	FDD
12	[TBD]	[TBD]	[TBD]	FDD
. . .				
33	1900–1920	1900–1920	N/A	TDD
34	2010–2025	2010–2025	N/A	TDD
35	1850–1910	1850–1910	N/A	TDD
36	1930–1990	1930–1990	N/A	TDD
37	1910–1930	1910–1930	N/A	TDD
38	2570–2620	2570–2620	N/A	TDD
39	1880–1920	1880–1920	N/A	TDD
40	2300–2400	2300–2400	N/A	TDD

5.2.9 Frequency Bands

The LTE (E-UTRA) frequency band allocations for FDD and TDD are illustrated in Table 5-4. Additionally, the uplink–downlink guard bands are shown.

Table 5-5   LTE (E-UTRA) spectrum mask for different channel bandwidths [9]

Out-of-band frequency (MHz)	Maximum power levels (dBm)								Measurement bandwidth
	1.4 MHz	1.6 MHz	3.0 MHz	3.2 MHz	5 MHz	10 MHz	15 MHz	20 MHz	
± 0–1	[TBD]	[TBD]	[TBD]	[TBD]	−15	−20	−25	−30	30 kHz
± 1–2.5	[−10]	[−10]	[−10]	[TBD]	−10	−10	−10	−10	1 MHz
± 2.5–5	[−25]	[−25]	[−10]	[−10]	−10	−10	−10	−10	1 MHz
± 5–6			[−25]	[−25]	−13	−13	−13	−13	1 MHz
± 6–10					−25	−13	−13	−13	1 MHz
± 10–15						−25	−13	−13	1 MHz
± 15–20							−25	−13	1 MHz
± 20–25								−25	1 MHz

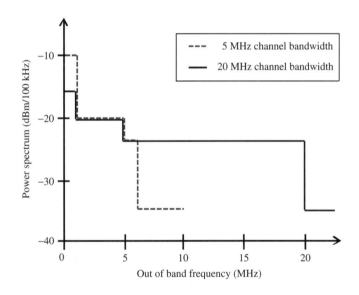

Figure 5-17   LTE spectrum mask for 5 MHz and 20 MHz channel bandwidths [9]

5.2.10   Spectrum Masks

The spectrum masks, which have to be satisfied by any mobile terminal station for different channel bandwidths, are defined in Table 5-5. The mask for the out-of-band emission for 5 MHz and 20 MHz channel bandwidths is exemplarily shown in Figure 5-17.

Table 5-6 Downlink LTE peak data rates in FDD and TDD [4]

	Peak data rate (Mbit/s in 20 MHz channel)	Spectrum efficiency (bit/s/Hz)
Requirement	100	5.0
2 × 2 MIMO, 64-QAM	172.8	8.6
4 × 4 MIMO, 64-QAM	326.4	16.3

Table 5-7 Uplink LTE peak data rates in FDD and TDD [4]

	Peak data rate (Mbit/s in 20 MHz channel)	Spectrum efficiency (bit/s/Hz)
Requirement	50	2.5
1 Tx antenna, 16-QAM	57.8	2.9
1 Tx antenna, 64-QAM	86.4	4.3

5.2.11 Performance

5.2.11.1 Peak Data Rate

The peak data rates achievable with LTE in the FDD and TDD modes are summarized in Table 5-6 for the downlink and in Table 5-7 for the uplink. Reductions in the data rate due to reference signals and the control channel are taken into account in the presented figures. It can be observed that the achievable data rates and spectrum efficiency exceeds the minimum requirements for the LTE standardization.

5.2.11.2 User Throughput

Table 5-8 and Table 5-9 show the mean and cell-edge user throughput for the LTE downlink with inter-site distances of 500 m and 1732 m respectively. The mobility of the mobile user is 3 km/h. The results are presented relative to baseline UTRA (Release 6). It can be observed that the required increase of mean user throughput of 3–4 times compared to that of UTRA and of cell-edge user throughput of 2–3 times compared to that of UTRA is met. The results are presented for single-user MIMO (SU-MIMO) schemes. Further throughput improvements can be expected with multi-user MIMO (MU-MIMO).

The mean and cell-edge user throughput for the LTE uplink are presented for inter-site distances of 500 m and 1732 m in Table 5-10 and Table 5-11 respectively. The results are presented relative to baseline UTRA. It can be observed that the required increase of mean and cell user throughput of 2–3 times compared to that of UTRA is met. Results for single-antenna transmission with two and four receive antennas and 2 × 2 SU-MIMO are shown.

5.2.11.3 Spectrum Efficiency per Cell

The spectrum efficiency achievable with LTE per cell is shown in Table 5-12 and Table 5-13 for the downlink with inter-site distances of 500 m and 1732 m respectively.

Table 5-8 Downlink LTE user throughput in a scenario with 500 m inter-site distance [4]

	Mean user throughput (relative to UTRA)	Cell-edge user throughput (relative to UTRA)
Requirement	3–4 × UTRA	2–3 × UTRA
2 × 2 SU-MIMO	3.2 × UTRA	2.7 × UTRA
4 × 2 SU-MIMO	3.5 × UTRA	3.0 × UTRA
4 × 4 SU-MIMO	5.0 × UTRA	4.4 × UTRA

Table 5-9 Downlink LTE user throughput in a scenario with 1732 m inter-site distance [4]

	Mean user throughput (relative to UTRA)	Cell-edge user throughput (relative to UTRA)
Requirement	3–4 × UTRA	2–3 × UTRA
2 × 2 SU-MIMO	3.0 × UTRA	2.3 × UTRA
4 × 2 SU-MIMO	3.6 × UTRA	2.8 × UTRA
4 × 4 SU-MIMO	4.6 × UTRA	4.8 × UTRA

Table 5-10 Uplink LTE user throughput in a scenario with 500 m inter-site distance [4]

	Mean user throughput (relative to UTRA)	Cell-edge user throughput (relative to UTRA)
Requirement	2–3 × UTRA	2–3 × UTRA
1 × 2 SIMO	2.2 × UTRA	2.5 × UTRA
1 × 4 SIMO	3.3 × UTRA	5.5 × UTRA
2 × 2 SU-MIMO	2.3 × UTRA	1.1 × UTRA

Table 5-11 Uplink LTE user throughput in a scenario with 1732 m inter-site distance [4]

	Mean user throughput (relative to UTRA)	Cell-edge user throughput (relative to UTRA)
Requirement	2–3 × UTRA	2–3 × UTRA
1 × 2 SIMO	2.2 × UTRA	2.0 × UTRA
1 × 4 SIMO	3.3 × UTRA	4.2 × UTRA

Table 5-12 Downlink LTE spectrum efficiency per cell in a scenario with 500 m inter-site distance [4]

	Spectrum efficiency (bit/s/Hz/cell)	Spectrum efficiency (relative to UTRA)
Requirement	0.53	3–4 × UTRA
2 × 2 SU-MIMO	1.69	3.2 × UTRA
4 × 2 SU-MIMO	1.87	3.5 × UTRA
4 × 4 SU-MIMO	2.67	5.0 × UTRA

Table 5-13 Downlink LTE spectrum efficiency per cell in a scenario with 1732 m inter-site distance [4]

	Spectrum efficiency (bit/s/Hz/cell)	Spectrum efficiency (relative to UTRA)
Requirement	0.52	3–4 × UTRA
2 × 2 SU-MIMO	1.56	3.0 × UTRA
4 × 2 SU-MIMO	1.85	3.6 × UTRA
4 × 4 SU-MIMO	2.41	4.6 × UTRA

Table 5-14 Uplink LTE spectrum efficiency per cell in a scenario with 500 m inter-site distance [4]

	Spectrum efficiency (bit/s/Hz/cell)	Spectrum efficiency (relative to UTRA)
Requirement	0.332	2–3 × UTRA
1 × 2 SIMO	0.735	2.2 × UTRA
1 × 4 SIMO	1.103	3.3 × UTRA
2 × 2 SU-MIMO	0.776	2.3 × UTRA

It can be seen that the minimum requirements for LTE are fulfilled. The mobility of the mobile user is 3 km/h.

Table 5-14 and Table 5-15 present the spectrum efficiency per cell for the uplink for different inter-site distances. The tables show that the minimum requirements for LTE are met.

5.3 WiMAX

5.3.1 Scope

The *World-Wide Interoperability for Microwave Access* (WiMAX) forum was initiated by a small number of companies in 2001 for promoting originally the IEEE 802.16

Table 5-15 Uplink LTE spectrum efficiency per cell in a
scenario with 1732 m inter-site distance [4]

	Spectrum efficiency (bit/s/Hz/cell)	Spectrum efficiency (relative to UTRA)
Requirement	0.316	2–3 × UTRA
1 × 2 SIMO	0.681	2.2 × UTRA
1 × 4 SIMO	1.038	3.3 × UTRA

standard specifications covering the carrier frequencies between 10 and 66 GHz for LOS applications. Later in 2003 they re-scoped the goal to cover carrier frequencies below 11 GHz as well, enabling support of NLOS reception with OFDM technology. From that time the WiMAX forum made rapid progress. The forum is a large nonprofit organization and currently made of more than 350 broadband wireless access (BWA) system manufacturers, components suppliers (silicon, RF, antenna), and operators.

Furthermore, there has been a close collaboration between the European ETSI *Broadband Radio Access Network* (BRAN) standardization body and the WiMAX forum. This collaboration has resulted in full harmonization between ETSI and IEEE and is reflected in the interoperability standards below 11 GHz: PHY- (ETSI TS 102 177) and MAC- (ETSI TS 102 178) protocol specifications. In addition, the activities of the conformance testing, used in the WiMAX forum certification process, are being done in close co-operation between the ETSI BRAN and WiMAX forum.

As illustrated in Figure 5-18, originally the WiMAX forum had the goal to cover the following applications [33]:

- broadband wireless access (BWA) for the last miles of residential properties where xDSL has no coverage;
- BWA for small and medium sized enterprises (SME)/business; and
- Wi-Fi backhauling, interconnecting the Wi-Fi base stations,

and in later steps to support nomadic and future mobility applications.

WiMAX follows an evolutionary strategy. WiMAX has the intention to cover first the last mile applications (residential and business) [28] where at the present time no xDSL services are available at these geographical places. For this application the IEEE 802.16d 256-OFDM mode will be used. Later on the support of nomadic applications without roaming/handover for dense spots networks is foreseen. Here the utilization of the so-called scalable-OFDMA (S-OFDMA) mode with antenna diversity is envisaged. Mobility with handover possibility is planned as migration strategy, where the IEEE 802.16e S-OFDMA mode, together with antenna diversity, will play an important role. The main services to be covered are: Voice over IP (VoIP), video conferencing, web./internet access, file transfer, interactive gaming, etc.

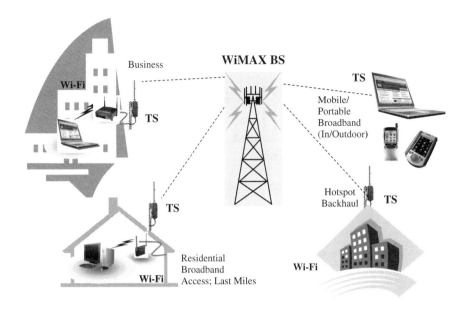

Figure 5-18   WiMAX application fields and vision

Hence the WiMAX forum missions can be summarized as follows [33, 34]:

- Ensure *interoperability* between BS and TS and build upon IEEE 802.16x standard specifications, where a concrete set of *Profiles* for each application are defined and approved by the forum.
- Enable certification, i.e. provide a WiMAX certificated stamp for compliant equipment.
- Promote the introduction of IP-based BWA services.

5.3.2   From IEEE 802.16x and ETSI BRAN HIPERMAN Towards WiMAX

Within IEEE 802.16 and ETSI BRAN standardization sub-groups, several specifications have been released since 2000 (see Figure 5-19) [19, 23, 24]. The first standards published were dated late 2000/2001 (ETSI BRAN HIPERACCESS in 2000 and IEEE 802.16 in 2001) and covered the 10–66 GHz carrier frequencies with only line of sight (LOS) applications. Early 2003 the extension was made to cover below 11 GHz (IEEE 802.16a and ETSI BRAN HIPERMAN) with the main target to cover non-LOS conditions (e.g. fixed/residential). The extension to support unlicensed bands (e.g. ISM) was made later (IEEE 802.16b). A complete revision of the standard specifications within IEEE was published in 2004 (IEEE 802.16d, also known as IEEE 802.16–2004).

Since 2004 several further specifications have been published that enable standards to offer future mobility migration (IEEE 802.16e). To cover mobility, it was quite crucial to

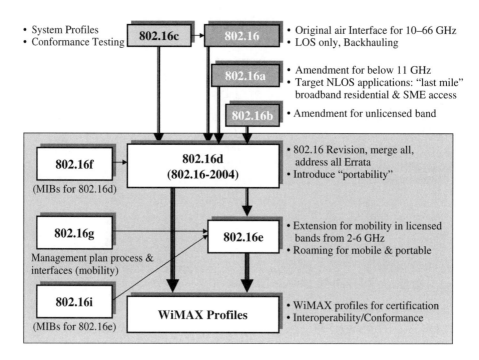

Figure 5-19 WiMAX and IEEE 802.16

specify and release in addition the roaming (handover) and network management specifications and the management information base (MIB) (IEEE 802.16f/g/i).

In parallel with the IEEE 802.16x and ETSI BRAN standardization activities, WiMAX has specified the conformance test specifications based on a predefined set of profiles (see Section 5.3.6) agreed among all forum members. The definition of these profiles enabled the system vendors and the silicon suppliers to start to develop the WiMAX products. The first chips based on IEEE 802.16d (supporting only fixed applications) were ready in 2005. Based on these components, since 2005 several field trials and WiMAX installations have been started.

In Table 5-16 a summary of the related IEEE 802.16x specification and its applications [23, 24] is given. Note that the main focus of the WiMAX forum is the so-called below 11 GHz carrier frequencies (2–11 GHz). As we see here, to increase capacity and coverage, besides the OFDM/OFDMA technology, additional implementation options such as adaptive antenna system (AAS), space time coding (STC), point to point mesh, and automatic repeat request (ARQ) are also part of these specifications.

Hence, the main technical features of WiMAX air interface specifications (MAC & PHY) versus existing networks can be summarized as follows [33]:

- Multi-carrier transmission, OFDM and OFDMA
  - Robust against multi-path fading
  - Power optimization in terminal station due to the use of OFDMA

Table 5-16 Summary of the IEEE 802.16 related applications

Designation	Applicability	Technical solutions/options	Duplex
WirelessMAN-SC (802.16/a/d)	10–66 GHz, licensed, fixed	Single carrier (SC) PMP	TDD, FDD
WirelessMAN-SCa (802.16a/d)	2–11 GHz, licensed, fixed	Single carrier PMP Adaptive antenna system (AAS) Automatic repeat request (ARQ) Space time coding (STC)	TDD, FDD
WirelessMAN-OFDM (802.16a/d)	2–11 GHz, licensed, fixed and nomadic	Multi-carrier, OFDM PMP and PTP mesh AAS ARQ STC	TDD, FDD
WirelessMAN-OFDMA (802.16e)	2–11 GHz, licensed, mobile	Multi-carrier, OFDM/OFDMA PMP AAS Hybrid ARQ STC	TDD, FDD
WirelessHUMAN (802.16b/d)	2–11 GHz, unlicensed	SC and multi-carrier, OFDM PMP and mesh AAS ARQ Mesh STC Dynamic frequency selection (DFS)	TDD

- Adaptive modulation
  - Adaptation on-the-fly
  - Maximizing capacity per sector
- Managed quality of service (QoS)
  - QoS can be guaranteed
  - Optimum share of the capacity among users
- Open to the use of intelligent antennas
- Offers nomadic and future mobile broadband access.

The OFDMA mode was adopted as a mandatory feature in IEEE 802.16e, supporting especially mobile application. Besides its large amount of sub-carriers (2048), to combat long echoes, the use of OFDMA is quite important, especially for mobile handsets. Due to its FDMA nature, the power consumption can be highly reduced for the mobile terminal (see also Chapter 3).

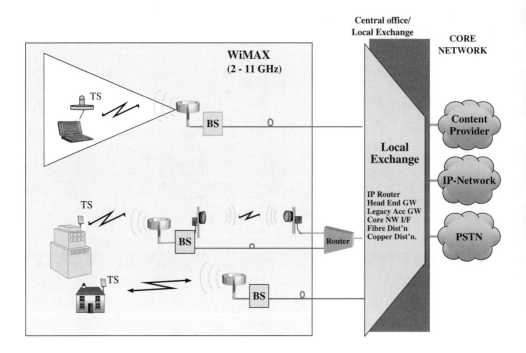

Figure 5-20 WiMAX core architecture

5.3.3 System Architecture

As stated earlier, WiMAX will cover different kinds of wireless applications. Packet IP transport is considered to be the main service to be transported. Both cellular point to multi-point and mesh point to point topologies are supported (see Figure 5-20). Furthermore, the transport of the classical switched TDM traffic (e.g. POTS, E1/T1) is also supported. The physical connection of the WiMAX base station (BS) to the core network equipments can be done by using optical fiber or a high speed point to point wireless connection at high carrier frequencies (above 11 GHz). Self-backhauling is also offered, but due to the limited bandwidth availability at frequencies between 2 and 11 GHz, this option may not be practicable.

As licensed bandwidths, the primary targets are the 2.3–2.7 GHz and 3.3–3.6 GHz. Here the main application would be both data and voice at the first stage fixed/indoor and outdoor (802.16d) and later mobile broadband (802.16e), offered by large data and voice operators. Supported main carrier frequencies for unlicensed applications is 4.9–5.85 GHz, where it encourages the Wi-Fi type of isolated spots with primarily fixed outdoor installation offered by small and medium data operators.

The WiMAX core components and network structure are illustrated in Figure 5-21. The air interface interoperability, i.e. compatibility between different terminals (TS) and base stations (BS) from different vendors, and the roaming interoperability, i.e. enabling portable and mobile applications, are considered to be the main challenges for the WiMAX networks.

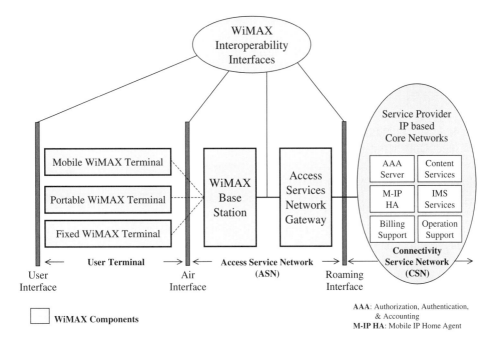

Figure 5-21 WiMAX components and interfaces

Note that the IEEE 802.16x specifications [23, 24] are covered basically by the above interoperability interfaces through several sub-specifications, namely: (a) the medium access control (MAC) layer, covering also a specific convergence sub-layer and security issues, (b) the PHY transport layer, and (c) the management plane.

5.3.4 Broadband Wireless Access Standards: HIPERMAN and IEEE 802.16x

The initial aim of the broadband wireless access (BWA) standardization activities within ETSI BRAN and IEEE 802.16 [19, 22] was to provide fixed wireless high speed access with a data rate up to 100 Mbps at carrier frequencies between 10 and 42 (66) GHz, to e.g. support broadband wireless Ethernet for fixed positioned residential customer premises and to small offices/home offices (SOHO) with high coverage. To maintain reasonably low RF costs for the residential market as well as good penetration of the radio signals, an extension of the standards was made to cover the employment of below 11 GHz carrier frequencies, e.g. the MMDS band (2.5–2.7 GHz) in the USA or around 5 GHz band in Europe and other countries. Later on interest toward nomadic and future mobility has increased.

Advantages of BWA include rapid deployment, high scalability, and lower maintenance and upgrade costs compared to the wireline network, e.g. cable or fiber. Nevertheless, the main goal of a future-proof BWA system for the residential market has to be the increase in spectral efficiency, in coverage, in flexibility for the system/network deployment, and in simplification of the installation. Above all, a reliable communication even in non line of site (NLOS) conditions has to be guaranteed. Note that in a typical urban or suburban deployment scenario at least 30 % of the subscribers are experiencing NLOS connection to the base station. In addition, for most of the remaining users LOS is obtained through rooftop positioning of the antenna, which requires very accurate pointing, thereby making the installation both time- and skill-consuming. Therefore, a system operating in NLOS conditions enabling self-installation will play an important role, especially in the success of BWA for the residential market.

In response to these trends, under the ETSI BRAN project HIPERMAN (HIgh PERformance Metropolitan Area Networks) (HM) [19] and the IEEE 802.16-project WirelessMan (Wireless Metropolitan Area Networks, WMAN) [23], several specifications have been published. Both standards offer at the first stage a wide range of data services (especially packet-based IP with a peak data rate > 50 Mbps at 10 MHz) for residential customers (i.e. single- or multi-dwelling households) and for small to medium sized enterprises by adopting multi-carrier transmission for carrier frequencies (RF) below 11 GHz with link coverages up to 20 km in the LOS case and 3–5 km in NLOS conditions. In a later stage these standards will also offer nomadic and mobility applications [24]. The promotion and the certification of these two standards are taking place in the WiMAX forum [34].

The aim of this section is to highlight the most important features of these published standards and to give a deeper understanding of the WiMAX systems.

5.3.4.1 Network Topology and Reference Model

As shown in Figure 5-22, the BWA system will be deployed to connect user network interfaces (UNIs) physically positioned in customer premises to a service node interface (SNI) of a broadband core network (e.g. IP), i.e. offering last mile connectivity. The base station typically manages communications of more than one carrier or sector. For each base station sector one antenna or more is positioned to cover the deployment region. The terminal station antenna can be directional or omni-directional. At the terminal station side the network termination (NT) interface connects the terminal station with the local user network (i.e. LAN/WLAN).

The BWA network deployments will potentially cover large areas (i.e. cities, rural areas) [19, 22]. Due to large capacity requirements of the network, large amounts of spectrum with high transmission ranges (up to 20 km) are needed. For instance, a typical network may therefore consist of some cells each covering part of the designated deployment area. Each cell will operate in a point to multi-point (PMP) or mesh manner.

Two duplex schemes can be used: (a) frequency division duplex (FDD) and (b) time division duplex (TDD). The channel size is between 1.5 and 28 MHz wide in both the FDD and the TDD case. The downlink data stream transmitted to different terminal stations is multiplexed in the time domain by MC-TDM (multi-carrier-time division multiplexing) by using OFDM or OFDMA transmission. In the uplink case, MC-TDMA (multi-carrier-time division multiple access) with OFDM or OFDMA will be used.

In Figure 5-23 the IEEE 802.16x reference model is depicted [23]. The IEEE 802.16 standard covers mainly the PHY and the MAC layers. The MAC layer in addition to the classical functions of the MAC comprises also the specific convergence sub-layer and the security sub-layer specifications. The goal of the specific convergence sub-layer (CS) specification is to enable the adaptation of the IEEE 802.16 standard specification for the transport of the higher layer services (e.g. IP, Ethernet, ATM). The PHY layer deals with the wireless transport medium, Layer 1.

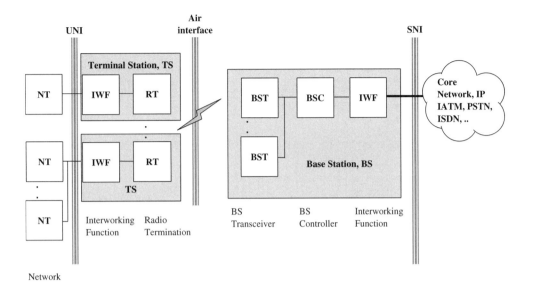

Figure 5-22   Simplified BWA reference model for standardization

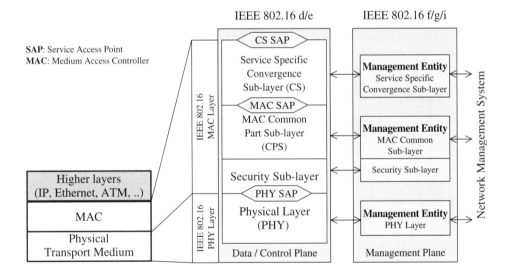

Figure 5-23   IEEE 802.16 layering model and specifications

5.3.4.2   Specific Convergence Sub-Layer

The specific convergence sub-layer (CS) covers the following functionalities [23]:

– reception of the packet data units (PDUs) from the higher layer, i.e. from the CS
   service access point (CS-SAP);

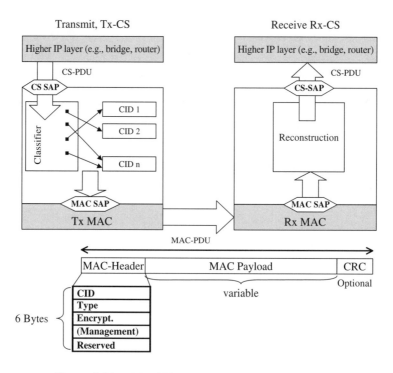

Figure 5-24   IEEE 802.16 packet convergence sub-layer

- classification of the PDUs according to their traffic content and assignment to the appropriate virtual channel connections (CID, connection identification);
- packet header compression/suppression (optional);
- forwarding and reception of the PDUs to/from the MAC service access point (MAC-SAP);
- reception of the PDUs from the peer MAC SAP; and
- re-building of any suppressed payload header information (optional).

Its exact functionality depends strongly on the convergence sub-layer type: IP, Ethernet, or ATM. The IP CS supports both IPv4 and IPv6.

In Figure 5-24 the IEEE 802.16 packet-based CS is illustrated. The packet CS is used for transport of all packet-based protocols such as the Internet Protocol (IP), Point-to-Point Protocol (PPP), and IEEE 802.3 (Ethernet). Note that the MAC PDU has a 6 bytes header and variable length payload field and an optional CRC-byte field (see Figure 5-24). The MAC header transports the following information: CID, PDU-type, MAC-PDU length, encryption information, etc.

5.3.4.3   Common MAC Sub-Layer

In addition to the convergence sub-layer as described above, the main classical functions of the IEEE 802.16 MAC control plane are depicted in Figure 5-25, where each function is briefly described [23] in the following:

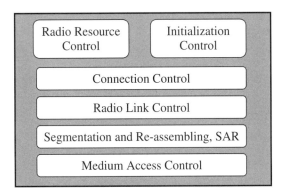

Figure 5-25   IEEE 802.16 common MAC layer functions

- *Radio resource control.* Its main goal is to supervise the QoS assigned for a given service. This function has a very close relation with the MAC (see below).
- *Initialization control.* It covers all protocol processes for any initialization, i.e. new registration, new link set-up, link recovery after link down, etc.
- *Connection control.* It deals with all functions related to set-up and release of a given connection.
- *Radio link control.* The microwave link below 11 GHz suffers from multi-path propagation, attenuation, and interferences. Control and monitoring of the link quality by automatic transmit power control (ATPC), adaptive modulation and coding (AMC), and ARQ are covered here.
- *Segmentation and re-assembling.* To be in line with the MAC packet format the long messages emanating from the control plane protocol functions will be segmented at the transmitter side and re-assembled at the reception side.
- *Medium access control.* This is the most important protocol function, which requires some real-time implementation. Its main goal is to share the available capacity of the radio resources in an efficient manner, where the required QoS for every service will be guaranteed.

Quality of Service (QoS) in IEEE 802.16 and HIPERMAN

The 802.16x/HIPERMAN MAC protocol is connection oriented; i.e. first a radio connection is established before starting any data transmission. The QoS of each connection can be managed through the so-called 'service-flow' control. The QoS parameters associated with the service flow define the radio transmission ordering and scheduling on the air interface. Note that a connection-oriented QoS control strategy ensures an accurate control over the air interface, being the main capacity bottleneck due to its limited radio bandwidth. The service flow parameters can be dynamically managed through MAC messages to accommodate the dynamic service demand. This service flow based QoS mechanism is applied to both the downlink and uplink. The supported data service and its application with varied QoS requirements are listed in Table 5-17.

Table 5-17   Applications and QoS requirements

Service category	Transported service	QoS requirements
rtPS Real-time polling service	Streaming audio or video	– Maximum reserved rate – Maximum sustained rate – Maximum latency tolerance – Traffic priority
UGS Unsolicited grant service	VoIP	– Maximum sustained rate – Maximum latency tolerance – Jitter tolerance
ErtPS Extended real-time polling service	Voice with activity detection (VoIP)	– Maximum reserved rate – Maximum sustained rate – Maximum latency tolerance – Jitter tolerance – Traffic priority
nrtPS Non-real-time polling service	File transfer protocol (FTP)	– Maximum reserved rate – Maximum sustained rate – Traffic priority
BE Best effort service	Data transfer, web, browsing, etc.	– Maximum sustained rate – Traffic priority

MAC Frame Structure

The basic IEEE 802.16 OFDM frame format [23], supporting adaptive modulation, is sketched in Figure 5-26. The same frame format is used for the FDD and TDD mode, where for the latter case the TDM portion can be extended to support the uplink TDMA mode. Each frame starts with a preamble of known sequences and a frame control header (FCH) which is protected by the most robust PHY mode (BPSK1/2). The FCH transmits the OFDM burst profile. The first burst after the FCH is made of two parts, broadcast and data. The broadcast part is made of the downlink map (DL map), uplink map (UL map), and downlink and uplink channel descriptors (DCD and UCD). The DL map transmits information regarding the contents of the DL frame with respect to the employed modulation scheme. Each frame starts with the most robust modulation and ends up with the most spectral efficient one. Note that the UL map transmits the uplink grant information assigned for each terminal station. Within the so-called downlink and uplink channel descriptors several additional pieces of information are broadcasted, which deal with the base station and the terminal station dedicated information, such as BS-identifier, TS-identifier, supported frame length (2.5, 8, and 20 ms), PHY type (SC or OFDM), TDD or FDD, modulation and FEC supported, burst concatenation, etc.

The frame format in the case of TDD for an OFDMA system [24] is illustrated in Figure 5-27. As in the case of the generic OFDM frame, each OFDMA frame is divided into downlink and uplink sub-frames, separated by transmit/receive (respectively receive/transmit) time gaps TTG (respectively RTG). This is necessary for a TDD scheme for avoiding downlink and uplink transmission collisions. Note that each burst transmits a given number of MAC-PDUs, as sketched in Figure 5-26.

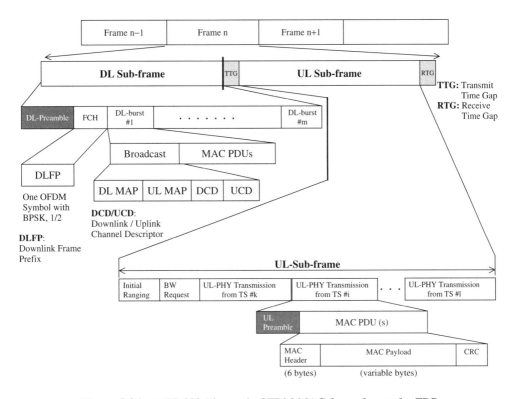

Figure 5-26   IEEE 802.16 generic OFDM MAC frame format for TDD

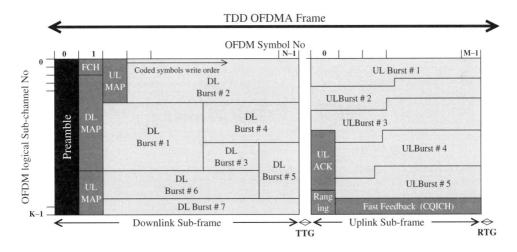

Figure 5-27   S-OFDMA (mobile) TDD frame format (time and sub-channel presentation)

Note that for the uplink three important fields for (a) ranging, (b) channel state information reporting, and (c) ARQ acknowledgment are reserved:
- *UL ranging.* It is allocated for a terminal station to perform the control of timing advance, frequency, and power measurements in a closed loop manner (measured reports are transmitted via the feedback channel to the base station).
- *UL CQICH.* This uplink channel-quality indicator field represents a feedback reporting channel in order to transmit the measured channel state information (e.g. SNR) from the terminal station to the base station.
- *UL ACK.* This is the feedback channel for an ARQ acknowledgment.

Note that the preamble used for synchronization is the first OFDM symbol of the frame. The frame control header (FCH) follows the preamble. It provides the frame configuration information, such as the MAP message length, coding scheme and usable sub-channels. The DL-MAP and UL-MAP provide sub-channel allocation and other control information for the downlink and uplink sub-frames respectively.

Terminal Station Installation and Registration Protocols
The process of a new terminal station installation is a good example to understand the different protocol processes [23]. It is depicted in Figure 5-28. The primary gaol of the WiMAX and IEEE 802.16 standard is that the equipment shall be self-installed without large outside configuration. Except for the mounting operation (indoor or outdoor) and eventually the pointing of the antenna toward the base station (in the case of using a directional antenna) all other processes will be done automatically, i.e. search of the right carrier frequency, synchronization to the right base station and extracting the transmitted downlink parameters, the initial ranging process, the negotiation of the capability of the installed terminal station, authentication, and finally the set-up of an IP data link connection. Some of these functions are detailed below. Note that major parts of these functions are quite similar to a first registration of a new mobile phone:

- *Search of downlink frequency and acquisition.* The process of searching and finding the right carrier frequency of the corresponding base station is illustrated in

Figure 5-28   WiMAX/IEEE 802.16 terminal station general installation process functions

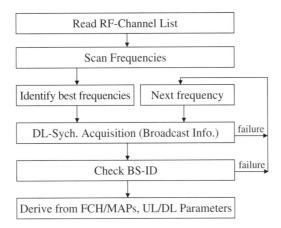

Figure 5-29 Terminal station downlink frequency scanning

Figure 5-29. The idea is to scan all available carrier frequencies and then from the transmitted downlink broadcast FCH and DL-MAP zone retrieve the right base station ID. If it is successful, further information related to the new terminal station is extracted from both downlink and uplink MAPs, i.e. ranging invitation grant.

– *Ranging process.* After receiving the corresponding invitation for transmitting the ranging messages, the terminal station transmits its long burst, called the 'ranging burst', to the base station with a given power setting. The main goal in the base station is to measure the so-called *timing advance* (propagation time between the terminal station and the base station) which will be considered in the MAC before transmitting any data burst from a terminal station. This is to avoid any burst collision at the base station. In the case where the base station does not receive a ranging burst from this new terminal station (due to a possible collision or weak received power), a new grant will be assigned and the procedure will be repeated with a higher terminal station transmit power until an exact timing advance measurement is done (see Figure 5-30). After completing the ranging process, the terminal station waits for an assignment of a normal data grant.

– *Capability negotiations.* Based on the standards and the supported profiles (see Section 5.3.6) each terminal station may implement some extra features/option that differ from the main profiles. These extra terminal station capabilities will be negotiated with the base station. They cover the following parameters:

- PHY parameters
  - OFDM parameters (guard time, modulation)
  - Coding (e.g. block Turbo codes)
  - Transmit power
- Terminal type
  - Full duplex (FDD or TDD)
  - Half duplex (H-FDD)

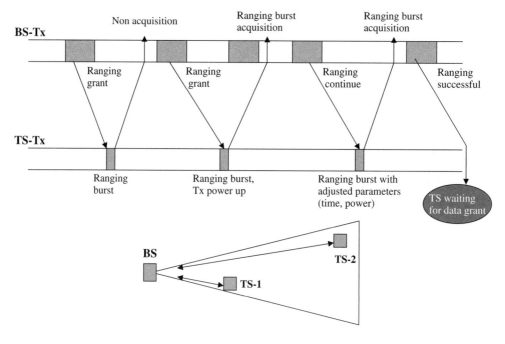

Figure 5-30   Ranging process

- Security
  - Triple DES
- QoS
  - Connection types
  - ARQ

Finally, after a successful authentication the connection can be set up and the data communication can be started.

Mobility Management [24]

Most critical issues for mobile applications are battery life time and handover/roaming. Mobile WiMAX supports both sleep and idle modes to enable low power consumption. Seamless handoff is also supported, which enables the terminal station to switch from one base station to another.

- *Power management.* Sleep and idle modes are supported. The sleep mode is a state in which the terminal station takes absence from the serving base station for a pre-negotiated period. This results in power consumption and radio resource minimization. The sleep mode also provides the possibility for the terminal station to scan other base stations in order to collect information to assist a possible handoff during the sleep mode. The idle mode provides a mechanism for the terminal station to become periodically available for the downlink broadcast traffic messaging without registration at a specific base station. This mode benefits the terminal station

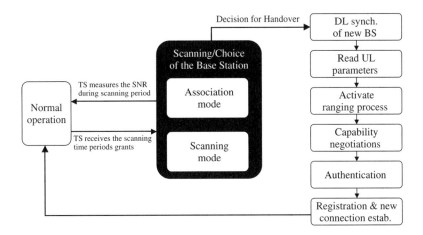

Figure 5-31   Hard handover process

by removing the requirement for a handoff and other normal operations, hence economizing battery and radio resources while still offering a simple and timely method (paging) for alerting the terminal station about the current downlink traffic.

– *Handover.* Three handover possibilities are supported within IEEE 802.16e: (a) hard handoff (HHO), (b) fast base station switching, and (c) macro diversity/soft handover. The hard handoff is mandatory and the others are optional. The requirement for an HHO to switch from one base station to another is less than 50 ms. The process of a hard handover is illustrated in Figure 5-31.

Note that a decision for a handover will result in a similar process as described above for a new registration. In the normal operation mode, the terminal station asks the base station through dedicated signalling information for a scanning period grant. After receiving the scanning grant, the terminal station initiates a frequency scanning mode, leaves the normal service mode and transmits a signal to the neighbor base station. The parameters of the neighbor base station are transmitted by the base station through the dedicated grant ACK messages in the frequency scanning period. If a new base station is found, the terminal station tries to synchronize to the downlink of this new base station and to estimate the actual reception quality in the form of an SNR. Note that during a scanning interval the so-called 'association' process (optional) can be started, where an initial ranging with the new base station can be agreed. The terminal station can make a decision on handover by comparing the SNR value of the new base station with the actual SNR. Then a similar process as for a registration can be started (see Figure 5-31). Note that the decision of a handover can also be triggered by the base station.

Security

WiMAX supports the following security managements [23]: (a) user and device authentication, (b) key management protocol, (c) traffic encryption, (d) control and management plane message protection, and (e) security protocol optimization for fast handovers.

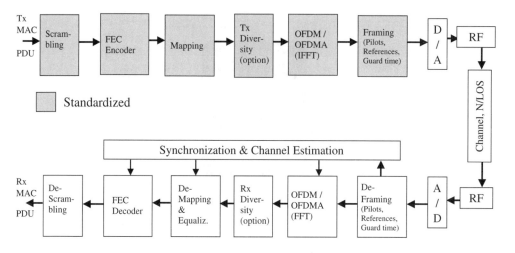

Figure 5-32 PHY layer block diagram overview

5.3.4.4 PHY Layer

An overview of the PHY layer transmission and reception functions is depicted in Figure 5-32. The transmit MAC PDUs after scrambling and FEC coding are mapped to the transmitted modulation constellation. Depending on the transmission mode (OFDM and OFDMA), the modulated data are assigned to the sub-carriers [23].

Allocated Carrier Frequencies

In Table 5-18, some target carrier frequency bands below 11 GHz are listed [23]. The channel bandwidths depend on the used carrier frequency.

The use of these radio bands provides a physical environment where, due to its wavelength characteristics, line of sight (LOS) is not necessary but multi-path may be significant (the delay spread is similar to DVB-T with up to 0.2 ms). The Doppler effects are negligible for a fixed positioned terminal station; however, for a mobile terminal station Doppler shifts will affect the link quality.

Multi-Carrier Transmission Schemes

The physical layer of both HIPERMAN and WirelessMan standards support multi-carrier transmission modes. The basic transmission mode is OFDM. Depending on the selected time/frequency parameters, the system can support TDMA as well as OFDMA. This flexibility ensures that the system can be optimized for (a) short burst type of applications as well as more streaming type oriented applications, and for (b) mobile and fixed reception conditions. The main advantage of using OFDMA with high numbers of sub-carriers with the same data rate as the OFDM mode is to provide higher coverage, i.e. larger guard time and a reduction of transmitted power for a mobile terminal station (longer battery life).

In the pure OFDM mode, in total 256 sub-carriers will be transmitted at once. The downlink applies time division multiplexing (TDM) and the uplink uses time division multiple access (TDMA). In the pure OFDMA mode, the channel bandwidth is divided into up to 2048 sub-carriers, where each user is assigned to a given group of sub-carriers. However,

Table 5-18 Example of some target carrier frequencies below 11 GHz for BWA

Frequency bands (GHz)	Allocated channel spacing	Recommendations
2.150–2.162 2.500–2.690	125 kHz to ($n \times 6$) MHz	USA CFR 47 Part 21.901, part 74.902 (MDS/MMDS)
2.305–2.320 2.345–2.360	1 or $2 \times (5 + 5)$ MHz or 1×5 MHz	USA CFR 47 Part 27 (WCS)
2.150–2.160 2.500–2.596 2.686–2.688	1 MHz to ($n \times 6$) MHz	Canada SRSP-302.5 (MCS)
2.400–2.483.5 (ISM, license-exempt)	Frequency hopping or direct sequence spread spectrum	CEPT/ERC/REC 70-03 USA CFR 47 Part 15, sub-part E
3.410–4.200	1.75 to 30 MHz paired with 1.75 to 30 MHz (FDD)	Rec. ITU-R F.1488 Annex II ETSI EN 301 021, CEPT/ERC Rec. 14-03 E, CEPT/ERC
3.400–3.700	$n \times 25$ MHz (single or paired) (FDD or TDD)	Rec. ITU-R F.1488 Annex I CITEL PCC.III/REC.47 (XII-99) Canada SRSP-303.4 (BWA)
5.150–5.850 (license-exempt)	$n \times 20$ MHz	CEPT/ERC Rec.70-03
10.000–10.680	3.5–28 MHz paired with 3.5–28 MHz (FDD)	CEPT/ERC/REC. 12-05 ETSI EN 301 021

in the scalable OFDMA mode (mobile applications) the number of sub-carriers is scalable with the used bandwidth; therefore, the number of sub-carriers varies from 256 to 2048.

OFDM Mode
This mode is used for a fixed positioned terminal station. In Figure 5-33 its sub-carrier allocation is illustrated. There are several sub-carrier types:

- Data sub-carriers
- Pilot sub-carriers (boosted and used for channel estimation purposes)
- Null sub-carriers (used for guard band and DC sub-carrier)

Detailed parameters of the sub-channel allocation for the OFDM mode are summarized in Table 5-19 and Table 5-20.

The downlink is a TDM transmission. Every downlink frame starts with a preamble. The preamble is used for synchronization purposes. It is followed by a control channel zone and downlink data bursts. Each burst uses different physical modes and each downlink burst consists of an integer number of OFDM symbols.

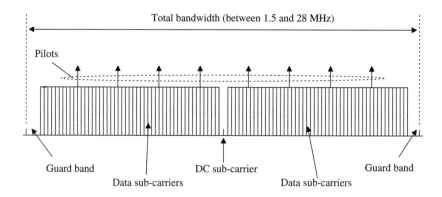

Figure 5-33   OFDM sub-carrier allocation in IEEE 802.16d

Table 5-19   OFDM mode parameters

Parameter	Value
Number of DC sub-carriers	1
Number of guard sub-carriers, left/right	28/27
Number of used sub-carriers	192
Total number of sub-carriers	256
Number of fixed located pilots	8

Table 5-20   OFDM general parameters for ETSI channelization

Bandwidth (MHz)	T_s (µs)	T_g (µs)			
1.75	128	4	8	16	32
3.5	64	2	4	8	16
7	32	1	2	4	8
14	16	1/2	1	2	4
28	8	1/4	1/2	1	2

The uplink is a TDMA transmission. Every uplink burst emanating from each terminal is preceded by a preamble. Each uplink burst, independent of the channel coding and modulation, transmits an integer number of OFDM symbols as well.

The uplink preamble consists of 2×128 samples with guard time ($=$ one OFDM symbol). The downlink preamble is made of two OFDM symbols: the first one carries 4×64 samples and the second one transmits 2×128 samples. These reference samples

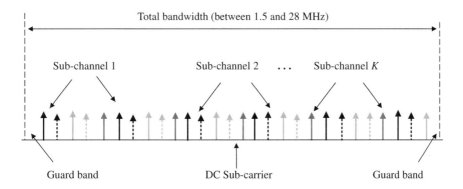

Figure 5-34   Example of OFDMA frequency allocation for K users

have good correlation properties, which eases the synchronization tasks. The power of the uplink and the downlink preambles is boosted by 3 dB compared to the data part.

OFDMA Mode

OFDMA in IEEE 802.16d was originally specified as an optional mode and later became mandatory for mobile application in IEEE 802.16e. In the following the basic parameters as described in IEEE 802.16d are presented.

As described in Chapter 3, in OFDMA only a part of all sub-carriers may be used for data transmission. A set of sub-carriers, called a sub-channel, will be assigned to each user (see Figure 5-34). For both uplink and downlink the used sub-carriers are allocated to pilot and data sub-carriers. However, there is small difference between the uplink and the downlink sub-carrier allocation. In the downlink, there is one set of common pilot carriers spread over all the bandwidth, whereas in the uplink each sub-channel contains its own pilot sub-carriers. This is because the downlink is broadcasted to all terminal stations while in the uplink each sub-channel is transmitted from a different terminal station. The goal of these pilot sub-carriers is to estimate the channel characteristics.

For OFDMA with FDD, the frame duration is an integer number of three OFDM symbols, where the actual frame duration is nearest to the nominal frame duration. In addition to the sub-channel dimension (set of sub-carriers), OFDMA uses the time dimension for data transmission. An uplink or downlink burst in OFDMA has a two- dimensional allocation. A transmit burst is mapped on to a group of contiguous sub-channels and contiguous OFDM symbols. Each data packet is first segmented into blocks sized to fit into one FEC block. Then, each FEC block spans one OFDMA sub-channel in the sub-channel axis and three OFDM symbols in the time axis. The FEC blocks are mapped such that the lowest numbered FEC block occupies the lowest numbered sub-channel in the lowest numbered OFDM symbol. The mapping is continued such that the OFDMA sub-channel index is increased for each FEC block mapped. When the edge of the data region is reached, the mapping will be continued again from the lowest numbered OFDMA sub-channel in the next OFDM symbol (see Figure 5-35).

For the uplink transmission, a number of sub-channels over a number of OFDM symbols is assigned per terminal station. The number of OFDM symbols is equal to $1 + 3N$, where

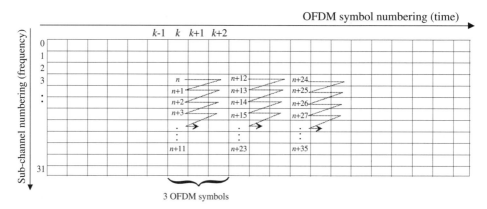

Note: The numbers n, $n+1$, ... in the boxes indicate indices in the FEC block.

Figure 5-35   Mapping of FEC blocks to OFDMA sub-channels and symbols

N is a positive integer. The smallest number of allocated sub-channels per terminal station is one sub-channel for a duration of four OFDM symbols, where the first OFDM symbol is a *preamble*.

OFDMA Downlink Sub-Carrier Allocation

As shown in Figure 5-36, for the downlink the pilots will have both fixed and variable positions [23]. The variable pilot location structure is repeated every four symbols. The allocated data sub-carriers are partitioned into groups of contiguous sub-carriers. The number of groups is therefore equal to the number of sub-carriers per sub-channel. In Table 5-21, the basic OFDMA downlink parameters are given.

OFDMA Uplink Sub-Carrier Allocation

The total number of used sub-carriers is first partitioned into sub-channels (see Figure 5-37 [23]). Within each sub-channel, there are 48 data sub-carriers, 1 fixed located pilot sub-carrier, and 4 variable located pilot sub-carriers. The fixed located pilot is always at sub-carrier 26 within each sub-channel. The variable located pilot sub-carriers are repeated every 13 symbols. The fixed and the variable positioned pilots will never coincide.

In Table 5-22 and 5-23 the OFDMA uplink parameters and guard times respectively are given. Note that for OFDMA with 2048 sub-carriers the symbol duration and guard times will be four times longer than in the case of the OFDM mode with 256 sub-carriers.

Scalable OFDMA

The IEEE 802.16e OFDMA mode for mobile applications is based on the concept of scalable OFDMA. This means that it supports a wide range of bandwidths in order to cope with different channel allocations. The scalability is supported by adjusting the FFT size while keeping the sub-carrier frequency spacing constant at 10.94 kHz. Therefore, the so-called 'resource unit' (sub-carrier bandwidth and symbol duration) is constant. Hence, the impact of a higher protocol layer is minimal when scaling the bandwidth. The

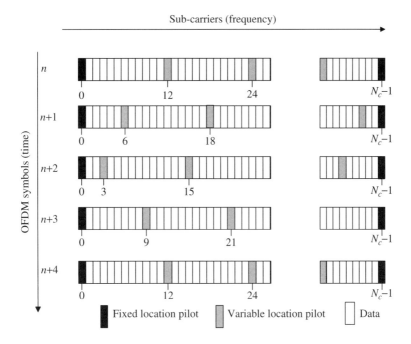

Figure 5-36   Sub-carrier allocation in the downlink

Table 5-21   OFDMA downlink sub-carrier allocation

Parameter	Value
Number of DC sub-carriers	1
Number of guard sub-carriers, left/right	173/172
Number of used sub-carriers	1702
Total number of sub-carriers	2048
Number of variable located pilots	142
Number of fixed located pilots	32
Total number of pilots	166 (where 8 fixed and variable pilots coincide)
Number of data sub-carriers	1536
Number of sub-channels	32
Number of sub-carriers per sub-channel	48

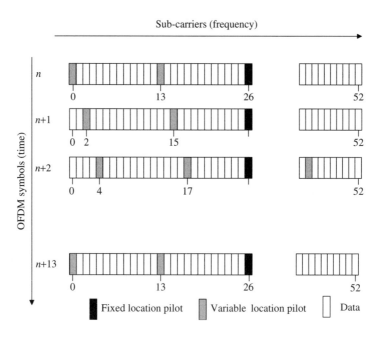

Figure 5-37   Sub-carrier allocation in the uplink

Table 5-22   OFDMA uplink sub-carrier allocation

Parameter	Value
Number of DC sub-carriers	1
Number of guard sub-carriers, left/right	176/175
Number of used sub-carriers	1696
Total number of sub-carriers	2048
Number of sub-channels	32
Number of data sub-carriers per sub-channel	48
Number of pilot sub-carriers per sub-channel	5

parameters of S-OFDMA with a guard interval of 1/8 are listed in Table 5-24. Channel bandwidths from 1.25 MHz up to 20 MHz are supported. The sampling frequency is proportional to the selected bandwidth. The number of OFDMA sub-channels increases as the bandwidth increases, leading to a better exploitation of frequency diversity.

In Table 5-25 the sub-channel allocation parameters in the case of 5 ms frame duration for 5 and 10 MHz channel allocations are given. Note that S-OFDMA supports sub-channelization in both DL and UL. The minimum frequency–time resource unit of sub-channelization is one slot, which is equal to 48 data tones (sub-carriers).

Table 5-23 General OFDMA parameters for ETSI
channelization with 2048 sub-carriers

Bandwidth (MHz)	T_s (µs)	T_g (µs)			
1.75	1024	32	64	128	256
3.5	512	16	32	64	128
7	256	8	16	32	64
14	128	4	8	16	32
28	64	2	4	8	16

Table 5-24 S-OFDMA parameters

Parameters (ANSI)	Values			
Channel bandwidth in MHz	1.25	5	10	20
Sampling frequency in MHz	1.4	5.6	11.2	22.4
FFT size	128	512	1024	2048
Number of OFDMA sub-channels	2	8	16	32
Sub-carrier spacing	10.94 kHz			
Symbol duration (without guard time)	91.4 µs			
Guard time (1/8)	11.4 µs			
OFDMA symbol duration	102.8 µs			
Number of OFDMA symbols per 5 ms frame	48			

Table 5-25 S-OFDMA sub-channel allocations

Bandwidth	5 MHz		10 MHz	
Transmission direction	DL	UL	DL	UL
FFT size	512		1024	
Number of null sub-carriers	92	104	184	184
Number of pilot sub-carriers	60	136	120	280
Number of data sub-carriers	360	272	720	560
OFDM symbols / 5 ms	48			
Data OFDM symbols / 5 ms	44			

Table 5-26 FEC coding and modulation parameters for uplink and downlink (OFDM)

PHY mode	Modulation	Inner coding	Outer coding	Overall coding rate	Efficiency (bit/s/Hz)
0	BPSK	CC 1/2	No	1/2	0.5
1	QPSK	CC 2/3	RS(32,24,4)	1/2	1.0
2	QPSK	CC 5/6	RS(40,36,2)	3/4	1.5
3	16-QAM	CC 2/3	RS(60,48,8)	1/2	2.0
4	16-QAM	CC 5/6	RS(80,72,4)	3/4	3.0
5 (optional)	64-QAM	CC 3/4	RS(108,96,6)	2/3	4.0
6 (optional)	64-QAM	CC 5/6	RS(120,108,6)	3/4	4.5

Forward Error Correcting (FEC) Coding and Modulation

The FEC employed consists of the concatenation of a Reed Solomon (RS) outer code and a punctured convolutional inner code. Block and convolutional Turbo codes can also be used. The low density parity check code (LDPC) is supported optionally in IEEE 802.16e. Different modulation schemes with Gray mapping (QPSK, 16-QAM, and 64-QAM) are employed.

The outer RS code can be shortened and punctured to enable variable block sizes and variable error-correction capability. The RS mother code is an RS (N, k, t) code, where N is the code length, k represents the number of information bytes, and t is the number of correctable error bytes. The inner convolutional code can be punctured as well to provide several inner code rates. The mother convolutional code is based on memory 6, rate 1/2, with zero tail biting. Eight tail bits are introduced at the end of each allocation. In the RS encoder, the redundant bits are sent before the input bits, keeping the tail bits at the end of the allocation.

Table 5-26 and Table 5-27 show the detailed FEC parameters for the OFDM and OFDMA modes of IEEE 802.16d. The resulting number of bytes per FEC block matches an integer number of OFDM symbols. As 64-QAM is optional, the codes for this modulation will only be implemented if the modulation is implemented. As these tables show, different coding and modulation schemes are supported. The lowest concatenated coding scheme with code rate 1/2 will be used for control information.

For S-OFDMA, support for QPSK, 16-QAM, and 64-QAM are mandatory in the downlink. In the uplink, 64-QAM is optional. Both the convolutional code (CC) with RS and the convolutional Turbo code (CTC) with a variable code rate and repetition coding are supported. The block Turbo code and low density parity check code (LDPC) are supported as optional features. Table 5-28 summarizes the coding and modulation schemes supported in the mobile WiMAX profile.

Hybrid-ARQ is supported by IEEE 802.16e. HARQ is enabled using the N channel 'stop and wait' protocol, which provides a fast response to packet errors and improves cell-edge coverage. Both combining and incremental redundancy are supported to improve the

Table 5-27 FEC coding and modulation parameters for uplink and downlink (OFDMA)

PHY mode	Modulation	Inner coding	Outer coding	Overall coding rate	Efficiency (bit/s/Hz)
0	BPSK	CC 1/2	No	1/2	0.5
1	QPSK	CC 2/3	RS(24,18,3)	1/2	1.0
2	QPSK	CC 5/6	RS(30,26,2)	~3/4	~1.5
3	16-QAM	CC 2/3	RS(48,36,6)	1/2	2.0
4	16-QAM	CC 5/6	RS(60,54,3)	3/4	3.0
5 (optional)	64-QAM	CC 3/4	RS(80,72,4)	2/3	4.0
6 (optional)	64-QAM	CC 5/6	RS(90,82,4)	~3/4	~4.5

Table 5-28 FEC coding and modulation parameters for uplink and downlink (S-OFDMA)

Link		Downlink	Uplink
Modulation		QPSK, 16-QAM, 64-QAM	QPSK, 16-QAM, 64-QAM (optional)
Code rate	CC+RS	1/2, 2/3, 3/4, 5/6	1/2, 2/3, 5/6
	CTC	1/2, 2/3, 3/4, 5/6	1/2, 2/3, 5/6
	Repetition	1/2, 1/4, 1/6	1/2, 1/4, 1/6

reliability of the re-transmission. A dedicated ACK channel is provided in the uplink for HARQ ACK/NACK signaling. At the cost of higher overheads, the HARQ combined with adaptive coding and modulation provides robust link adaptation in mobile environments at vehicular speeds up to 120 km/h.

5.3.5 Transmit Diversity / MIMO in WiMAX

The implementation of transmit diversity with MIMO is not mandatory [23, 24]. Three schemes are supported [29]: (a) space–time coding (STC), (b) spatial multiplexing (SM), and (c) adaptive antenna systems (AAS).

- *Space–time coding (STC)*. The procedure specified here is the Alamouti STC scheme. STC is foreseen as an option for both WirelessMAN-OFDM and WirelessMAN-(S)-OFDMA schemes (see Figure 5-38). Two transmit antennas are foreseen. Depending on the receiver implementation one or two antennas could be used. The diversity combining gain will be larger if two Rx antennas are implemented.

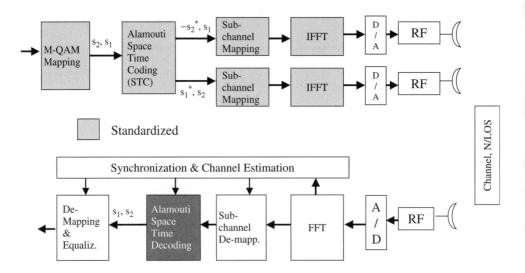

Figure 5-38   Alamouti STC in WiMAX

The encoder of the Alamouti STC can be represented by the following matrix:

$$A = \begin{pmatrix} s_1 & -s_2^* \\ s_2 & s_1^* \end{pmatrix}. \tag{5.1}$$

This means that over antenna 1 the pair of M-QAM symbols s_1 and $-s_2^*$ and over antenna 2 the pair s_2 and s_1^* will be transmitted. At the receiver side the state of each MIMO channel is estimated and combined. For more details about the Alamouti space–time coding and decoding please refer to Chapter 6. Note that the STC provides only higher system gain, the transmitted data rate remains unchanged (versus the uncoded STC system).

– *Space diversity in conjunction with STC* In WirelessMAN-(S)-OFDMA a combination of space diversity and STC is defined as an option. Compared to STC, this combination allows an increase of the data rate. In the case of pure space diversity with two antennae, the data rate is split into two independent sub-streams (without adding any redundancy), where each stream is transmitted over a given antenna. In this case the encoder (no redundancy) will be represented by the matrix B:

$$B = \begin{pmatrix} s_1 \\ s_2 \end{pmatrix}. \tag{5.2}$$

In WiMAX, up to four transmit antennae for the downlink are foreseen (see Figure 5-39), where for the uplink only up to two transmit antennas can be used. In the case of employing four Tx antennas for the downlink, three combinations are possible: (a) pure STC, (b) a combination of STC with space diversity, and (c) pure space diversity. Each of these combinations can be chosen according to the channel condition and the demanded QoS.

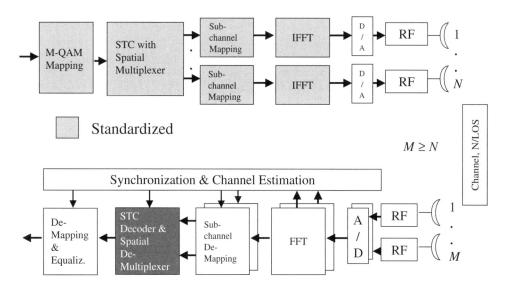

Figure 5-39   Space diversity in WiMAX

Similar to two antennas, in the case of four antennas we can define the following matrices:

$$
A = \begin{pmatrix} s_1 & -s_2^* & 0 & 0 \\ s_2 & s_1^* & 0 & 0 \\ 0 & 0 & s_3 & -s_4^* \\ 0 & 0 & s_4 & s_3^* \end{pmatrix}, B = \begin{pmatrix} s_1 & -s_2^* & s_5 & -s_7 \\ s_2 & s_1^* & s_6 & -s_8^* \\ s_3 & -s_4^* & s_7 & s_5^* \\ s_4 & s_3^* & s_8 & s_6^* \end{pmatrix}, C = \begin{pmatrix} s_1 \\ s_2 \\ s_3 \\ s_4 \end{pmatrix}. \quad (5.3)
$$

With matrix A, the encoder becomes a pure STC with full diversity (gain of 4) and without an increase of the data rate. With matrix B we have a combination of STC with a diversity gain of 2 and a data rate doubling. Matrix C allows a pure spatial multiplexing offering up to 4 times more data rate without any diversity gain.

Note that in the last case the number of used receive antennas is at least the same as at the transmitter side. The main advantage of this scheme is the increase of the transmit data rate. Here, as in the STC case, the receiver estimates for each MIMO channel the corresponding channel state information that is needed for the diversity combiner.

– *Adaptive antenna system (AAS).* AAS is foreseen as an option for both WirelessMAN-OFDM and WirelessMAN-(S)-OFDMA modes. The idea behind the AAS is to concentrate for a given period of time all radiated power toward a given user by using, for instance, the so-called phased array antennas, where the antenna gain can be adjusted towards a given direction. Several antennas will be used at the transmitter (or receiver) side. Note that the combination or the steering of the antennas can be done at the transmission or reception side. The example given in Figure 5-40 takes into account the fact that beamforming is done at the

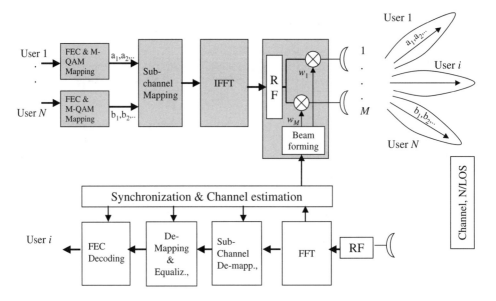

Figure 5-40   Adaptive antenna system in WiMAX

transmission side, where the transmitted signal before transmission is weighted by the corresponding weighting factor w_i. In order to steer the antennas in the optimum direction, knowledge of the channel state information is necessary at the transmitter side for instantaneous derivation of the weighting factors. Note that these weighting factors not only modify the antenna phases but also have an influence on the transmitted signal amplitude. By these means not only the antenna main direction but also the antenna opening angle and the null on the neighboring directions can be monitored, leading to a higher antenna gain toward a desired direction. The total adjustment freedom depends strongly on the number of antennas present.

At the cost of slightly higher complexity, the throughput can also be increased if the steering of the antennas is done jointly for all users. Since each user has its own position within a sector, the geographical separation of individual receivers is straightforward. In other words, for a given user, a given antenna can be directed such that its neighboring directions can be minimized, while in its main direction the gain is maximized.

By concentrating all radiated power toward a given user, two advantages can be derived: (a) increase of the system gain and (b) reduction of the interference, especially those co-channel interferences emanating from the frequency re-use. Therefore, by using AAS the coverage is highly increased.

Furthermore, in order to support AAS, the basic frame structure shall be extended (see Figure 5-41). As shown in the basic frame, the AAS part is appended. This kind of extension offers compatibility between the terminal station with or without AAS and the base station with or without AAS. The same frame structure is used for both TDD and FDD duplex schemes. The activation of the AAS mode to the corresponding TS

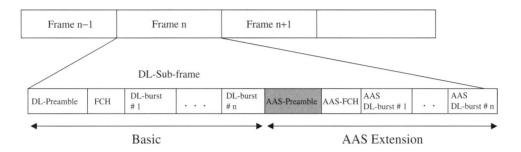

Figure 5-41 Modified frame structure to support AAS

will be done by sending corresponding signaling from the base station via a dedicated management channel.

When comparing AAC with STC or the spatial diversity scheme, the main drawback of the AAS implementation is its complexity, which resides mainly in the RF part, and is usually not an easy task. Furthermore, in the case of FDD the problem especially in the mobile environment might be the adaptation of the weighting factor for the antenna adjustment, where the real instantaneous channel state information (CSI) is usually delayed at the transmission side (due to the closed loop, i.e. transmission of the measured CSI from each terminal station to the base station).

– *Adaptive combination of all transmit diversity schemes*. In order to use the same hardware components (e.g. IFFT, D/A, RF, and antennas), to simplify the implementation of AAS and to adapt dynamically the transmit diversity on-the-fly depending on the channel condition, a functional combination of all these transmit diversity schemes is depicted in Figure 5-42. Note that with up to four transmit antennas and with the corresponding pre-coding scheme (STC, spatial multiplexing and beamforming weighting matrix) all these combinations can be realized and dynamically adapted to the channel condition. In this scheme the AAS is realized in the digital domain; weighting of the antennas is done before the IFFT operation, which is used for each antenna branch.

5.3.6 WiMAX Profiles

The IEEE 802.16x standard specifications are quite flexible and support a lot of options [23, 24]. To reduce the number of options and make them implementable and testable, a set of predefined profiles for both the MAC and PHY layer [34] has been specified.

The implementation of these profiles is quite important, because based on this set of profiles the WiMAX certification tests will be done and the WiMAX conformance label will be agreed. The basic profiles for IEEE 802.16d are listed in Table 5-29. The favorite duplex scheme for WiMAX is TDD. The initial release of WiMAX certification profiles will only include TDD. With ongoing releases, FDD profiles will be considered by the WiMAX forum.

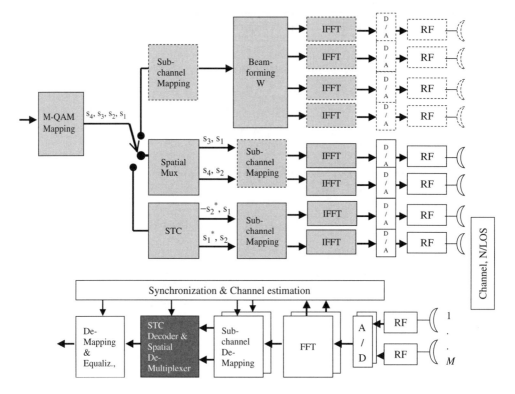

Figure 5-42   Functional combination of all transmit diversities

5.3.6.1   MAC Profiles

In Table 5-30 the basic WirelessMAN MAC profiles for OFDM and OFDMA modes are detailed. The IP packet convergence sub-layer is mandatory, while the ATM convergence sub-layer is optional. On the QoS level side, best effort and non-real-time polling are considered to be the main applications.

5.3.6.2   PHY Profiles

The baseband PHY basic profiles are given in Table 5-31, Table 5-32 and Table 5-33. For all used modulation schemes the minimum receiver sensitivity levels are given. Note that 64-QAM is mandatory only for unlicensed carrier frequencies.

5.3.6.3   RF Profiles

One of the important RF parameters is the spectrum mask, which is illustrated in Figure 5-43.

Table 5-29 WirelessMAN-OFDM and OFDMA basic profiles

	IEEE 802.16d OFDM		IEEE 802.16d OFDMA	
	Profile name	Application	Profile name	Application
MAC	profM3_PMP	PMP, packet	profM1_PMP	PMP, packet
	profM3_Mesh	Mesh, packet	N/A	N/A
PHY	profP3_1.75	1.75 MHz bandwidth	profP1_1.25	1.25 MHz bandwidth
	profP3_3.5	3.5 MHz bandwidth	profP2_3.5	3.5 MHz bandwidth
	profP3_7	7 MHz bandwidth	profP3_7	7 MHz bandwidth
	profP3_3	3 MHz bandwidth	profP4_14	14 MHz bandwidth
	profP3_5,5	5.5 MHz bandwidth	profP5_28	28 MHz bandwidth
	profP3_10	10 MHz bandwidth	N/A	N/A
RF, max. Tx power class	profC3_0	$P_{max} < 14$ dBm	N/A	N/A
	profC3_14	$14 \leq P_{max} < 17$ dBm	profC1	$17 \leq P_{max} < 20$ dBm
	profC3_17	$17 \leq P_{max} < 20$ dBm	profC2	$20 \leq P_{max} < 23$ dBm
	profC3_20	$20 \leq P_{max} < 23$ dBm	profC3	$23 \leq P_{max} < 30$ dBm
	profC3_23	$P_{max} \geq 23$ dBm	profC4	$P_{max} \geq 30$ dBm
Duplex	TDD, FDD			

Table 5-30 MAC WirelessMAN-OFDM and OFDMA basic profiles

Properties	Mandatory or not
Packet convergence sub-layer	Yes
• Payload head suppression	No
• IPv4 over 802.3/Ethernet	Yes
• 802.3/Ethernet	Yes
ATM convergence sub-layer	No
Multicast polling	Yes
CRC	Yes
Unsolicited grant services	No
• Real-time polling	No
• Non-real-time polling	Yes
• Best effort	Yes
ARQ	No
Adaptive antennas	No

Table 5-31 Example of minimum PHY requirements for the OFDM mode

Parameter	profP3_1.75	profP3_3.5	profP3_7	profP3_5.5	profP3_10
B (MHz)	1.75	3.5	7	5.5	10
Frame length	BS selects one from 5 ms, 10 ms, and 20 ms; TS supports all				
Receiver sensitivity at BER $= 10^{-6}$					
QPSK1/2	−91 dBm	−88 dBm	−85 dBm	−86 dBm	−83 dBm
QPSK3/4	−89 dBm	−86 dBm	−83 dBm	−84 dBm	−81 dBm
16-QAM1/2	−84 dBm	−81 dBm	−78 dBm	−79 dBm	−76 dBm
16-QAM3/4	−82 dBm	−79 dBm	−76 dBm	−77 dBm	−74 dBm
64-QAM2/3	−77 dBm	−74 dBm	−71 dBm	−72 dBm	−69 dBm
64-QAM3/4	−76 dBm	−73 dBm	−70 dBm	−71 dBm	−68 dBm

Table 5-32 Example of minimum PHY requirements for the OFDMA mode

Parameter	profP1_1.25	profP2_3.5	profP3_7	profP4_14	profP5_28
B (MHz)	1.25	3.5	7	14	28
Frame length	BS selects a frame duration and TS supports all				
Receiver sensitivity at BER $= 10^{-6}$					
QPSK1/2	−90 dBm	−87 dBm	−84 dBm	−81 dBm	−78 dBm
QPSK3/4	−87 dBm	−84 dBm	−81 dBm	−78 dBm	−75 dBm
16-QAM1/2	−84 dBm	−80 dBm	−77 dBm	−74 dBm	−71 dBm
16-QAM3/4	−80 dBm	−77 dBm	−74 dBm	−71 dBm	−68 dBm
64-QAM2/3	N/A	−73 dBm	−71 dBm	−67 dBm	−64 dBm
64-QAM3/4	N/A	−71 dBm	−68 dBm	−65 dBm	−62 dBm

Table 5-33 PHY features for OFDM and OFDMA

Properties	Mandatory or not
64-QAM	Yes for unlicensed bands
Block Turbo code	No
Convolutional Turbo code	No
Sub-channelization	No
Space time coding	No
Guard time	Base station supports at least one and terminal station all

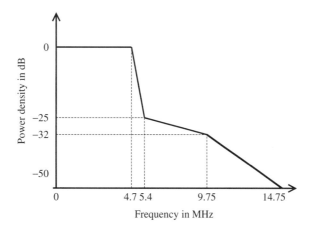

Figure 5-43   WiMAX RF profile: spectrum mask, 10 MHz

Table 5-34   WiMAX OFDM certifications

Frequency band	Duplex	PHY profile	Bandwidth	MAC profile
3400–3600 MHz	TDD	Prof3_3.5	3.5 MHz	profM3_PMP
		Prof3_7	7 MHz	
	FDD	Prof3_3.5	3.5 MHz	
		Prof3_7	7 MHz	
5725–5850 MHz	TDD	Prof3_10	10 MHz	

5.3.6.4   WiMAX Certification Profiles

As mentioned earlier, the WiMAX systems will be introduced first for fixed positioned equipment and later for mobile applications. The certification instance will use the pre-defined set of profiles for testing any WiMAX-compliant product and will then deliver the certification label. The profiles for fixed services (WirelessMAN OFDM) are given in Table 5-34 for both TDD and FDD duplexes. TDD is the preferred WiMAX duplexing scheme.

Release-1 mobile WiMAX OFDMA profiles will cover 5, 7, 8.75, and 10 MHz channel bandwidths for licensed spectrum allocations in the 2.3, 2.5, 3.3, and 3.5 GHz frequency bands [33]. The IEEE 802.16e S-OFDMA is designed to be scaled to different channel-izations from 1.25 to 20 MHz in order to comply with varied bandwidth requirements.

5.3.7   Performance

5.3.7.1   Peak Transmission Capacity

The peak data rate achievable for both the OFDM and OFDMA mode depends on the used channel bandwidth, the PHY mode, and the guard time. In Table 5-35 and Table 5-36

Table 5-35 Peak data rate for the OFDM mode versus modulation (7 MHz, 256 sub-carriers, no diversity)

T_g/T_s	QPSK 1/2	QPSK 3/4	16-QAM 1/2	16-QAM 3/4	64-QAM 2/3	64-QAM 3/4
1/32	5.94	8.91	11.88	17.82	23.76	26.73
1/4	4.9	7.35	9.80	14.70	19.60	22.05

Table 5-36 Peak data rate for the OFDMA mode versus modulation (7 MHz, 2048 sub-carrier, no diversity)

T_g/T_s	QPSK 1/2	QPSK 3/4	16-QAM 1/2	16-QAM 3/4	64-QAM 2/3	64-QAM 3/4
1/32	5.82	8.73	11.64	17.45	23.27	26.18
1/4	4.8	7.2	9.6	14.40	19.20	21.6

the maximum achievable data rate for the OFDM and OFDMA mode for different PHY modes in a 7 MHz ETSI channelization raster is given. Both systems provide similar performance. All MAC overheads are included in the tables.

The achievable peak rates versus M-QAM modulation for S-OFDMA are given in Table 5-37 for both 5 and 10 MHz FDD channel bandwidths by considering a frame duration of 5 ms. In this table the parameters for the uplink and downlink are given. In the case of the 10 MHz channel bandwidth, data rates up to 32 Mbps and 23 Mbps for downlink and uplink can be achieved respectively. The penalty in data rate for the uplink is due to the large amount of overhead spent for pilots and guard times. The repetition code is used only in the case of QPSK.

The peak data rates found by employing the most efficient mandatory modulation schemes (64-QAM for DL and 16-QAM for UL) with MIMO for S-OFDMA are given in Table 5-38 for 10 MHz TDD channel bandwidths by considering a frame duration of 5 ms. The DL/UL ratio defines the ratio between the DL and the UL allocated time frame for each direction.

5.3.7.2 Link Budget, Data Rate, and Coverage

For the link budget evaluation the thresholds for a bit error rate of BER $= 10^{-6}$ are considered. The receiver power thresholds are taken from Table 5-31 and 5-32. The receiver E_s/N_0 values are listed in Table 5-39.

Fixed Positioned Terminal Sation

The real WiMAX coverage and capacity performance for a fixed positioned terminal station depend strongly on the terminal equipment configuration: indoor or outdoor, use of directional antenna or not, use of transmit diversity, etc. The use of a short frame length and long guard time has a negative impact on the total data rate efficiency. In Table 5-40 the link budget for a 7 MHz FDD channel allocation is given. Note that the system gain for the uplink is lower than the downlink. This is due to the higher BS

Table 5-37 Peak data rate versus modulation and bandwidth for S-OFDMA; no diversity

Bandwidth		5 MHz		10 MHz	
Modulat.	FEC, rate	DL	UL	DL	UL
QPSK	CTC1/2, 6 ×	0.53 Mbps	0.38 Mbps	1.06 Mbps	0.78 Mbps
	CTC1/2, 4 ×	0.79 Mbps	0.57 Mbps	1.58 Mbps	1.18 Mbps
	CTC1/2, 2 ×	1.58 Mbps	1.14 Mbps	3.127 Mbps	2.35 Mbps
	CTC1/2, 1 ×	3.17 Mbps	2.28 Mbps	6.34 Mbps	4.7 Mbps
	CTC3/4	4.75 Mbps	3.43 Mbps	9.50 Mbps	7.06 Mbps
16-QAM	CTC1/2	6.34 Mbps	4.75 Mbps	12.67 Mbps	9.41 Mbps
	CTC3/4	9.50 Mbps	6.85 Mbps	19.01 Mbps	14.11 Mbps
64-QAM	CTC1/2	9.50 Mbps	6.85 Mbps	19.01 Mbps	14.11 Mbps
	CTC2/3	12.67 Mbps	9.14 Mbps	25.34 Mbps	18.82 Mbps
	CTC3/4	14.16 Mbps	10.28 Mbps	28.51 Mbps	21.17 Mbps
	CTC5/6	15.84 Mbps	11.42 Mbps	31.68 Mbps	23.52 Mbps

Table 5-38 Peak data rate versus bandwidth for S-OFDMA with MIMO, for TDD

DL/UL ratio		1:0	2:1	1:1	0:1
SIMO (1 × 2)	DL	31.68 Mbps	20.16 Mbps	15.84 Mbps	0
	UL	0	5.04 Mbps	7.06 Mbps	14.11 Mbps
MIMO (2 × 2)	DL	63.36 Mbps	40.32 Mbps	31.68 Mbps	0
	UL	0	10.08 Mbps	14.12 Mbps	28.22 Mbps

Table 5-39 Estimated SNR in AWGN for concatenated RS+CC FEC

Modulation	Coding rate	Receiver E_s/N_o (dB)
QPSK	1/2	9.4
	3/4	11.2
16-QAM	1/2	16.4
	3/4	18.2
64-QAM	2/3	22.7
	3/4	24.4

Table 5-40   Link budget estimation for LOS and NLOS fixed positioned terminal without MIMO

Transmission modes	Outdoor LOS (directional TS antenna)		Outdoor NLOS (omni-directional TS antenna)	
Topology/Profile	PMP, four sectors (OFDM profile Prof3_7), FDD			
Carrier frequency	3.5 GHz			
Bandwidth	7 MHz (UL and DL): Full usage of bandwidth for the UL			
Antenna location	fixed		portable	
UL modulation E_s/N_0	QPSK3/4 11.2 dB	16-QAM3/4 18.2 dB	QPSK1/2 9.4 dB	QPSK3/4 11.2 dB
UL receiver sensitivities	QPSK3/4 −83 dBm	16-QAM −76 dBm	QPSK1/2 −85 dBm	QPSK3/4 −83 dBm
DL modulation E_s/N_0	QPSK3/4 11.2 dB	64-QAM3/4 24.4 dB	QPSK1/2 9.4 dB	16-QAM3/4 18.2 dB
DL receiver sensitivities	QPSK3/4	64-QAM	QPSK1/2	16-QAM3/4
	−83 dBm	−70 dBm	−85 dBm	−76 dBm
TS transmit power	23 dBm			
TS antenna gain	17 dBi (directional)		3 dBi (∼ omni-directional)	
BS transmit power	30 dBm			
BS antenna gain	12 dBi (90° opening)			
BS antenna height	30 m			
Fading, interference margin	5 dB		10 dB	
UL system gain	QPSK 130 dB	16-QAM 123 dB	QPSK1/2 113 dB	QPSK3/4 109 dB
DL system gain	QPSK 137 dB	64-QAM 124 dB	QPSK1/2 120 dB	16-QAM 111 dB
Propagation models	Free space		See Section 1.1.5, WiMAX	
Coverage	QPSK ∼ 21 km	16-QAM ∼ 9 km	QPSK1/2 ∼ 1 km	QPSK3/4 ∼ 0.6 km
Mean UL sector capacity	12 Mbps		7 Mbps	
Mean DL sector capacity	18 Mbps		12 Mbps	

Table 5-41 DL link budget estimation for mobile terminal station

Transmission modes	Outdoor NLOS (omni-directional TS antenna)	
Topology/profile	PMP, four sectors (OFDMA profile Prof3_10)	
Carrier frequency	2.5 GHz	
Bandwidth	10 MHz	
DL modulation and FEC scheme	QPSK-CTC1/2	16-QAM-CTC1/2
Receiver sensitivity per sub-carrier	QPSK1/2: −123 dBm	16-QAM1/2: −118 dBm
BS transmit power/antenna	30 dBm	
BS antenna configuration	Tx/Rx: 2/2	
BS antenna gain	12 dBi (90° opening)	
BS antenna height	30 m	
TS transmit power	23 dBm	
TS antenna gain	3 dBi (∼ omni-directional)	
TS antenna configuration	Tx/Rx: 2/1	
Fading, interference margin	10 dB	
Propagation models	See Section 1.1.5, WiMAX [26]	
Distance between two BS	2.8 km	
TS distributions	60 % pedestrian (3 km/h), 30 % 30 km/h speed, and 10 % mobile with 120 km/h	

transmit power possibility. In this example, for the uplink a full usage of all OFDM sub-carriers is considered. We notice that the coverage depends strongly on the used antenna and the terminal location. By offering an average data rate of 18 Mbps/sector up to 21 km coverage with line of sight (LOS) and directional antennas is achievable.

However, in the non line of sight (NLOS) condition using an omni-directional antenna the coverage will be highly reduced to 1 km. In order to overcome this issue, OFDMA (OFDMA_profile_Prof3_7) can be used for the uplink, where at the cost of a lower peak data rate per terminal station, a higher coverage can be obtained. For instance, if we consider eight sub-channels for the OFDMA system, the uplink system gain increases by 9 dB. This leads to an increase of the uplink coverage by about 50–60 %.

Mobile Terminal Station

The parameter assumptions for mobile reception using the S-OFDMA mode are given in Table 5-41. Based on these assumptions, the estimated data rate per sector with MIMO is given in Table 5-42 [33]. Note that by using the MIMO system (two Tx and two Rx antennas), the spectral efficiency can be increased by about 60 %.

Table 5-42 Estimated data rate in mobile environment, TDD, 10 MHz [33]

Cases		DL: 28 data symbols UL: 9 data symbols		DL: 22 data symbols UL: 15 data symbols	
Antenna	Link	Sector throughput	Spectral efficiency	Sector throughput	Spectral efficiency
SIMO	DL	8.8 Mbps	1.19 bit/s/Hz	6.6 Mbps	1.07 bit/s/Hz
	UL	1.38 Mbps	0.53 bit/s/Hz	2.20 Mbps	0.57 bit/s/Hz
MIMO	DL	13.60 Mbps	1.84 bit/s/Hz	10.63 Mbps	1.73 bit/s/Hz
	UL	1.83 Mbps	0.70 bit/s/Hz	3.05 Mbps	0.79 bit/s/Hz

5.4 Future Mobile Communications Concepts and Field Trials

5.4.1 Objectives

Besides the introduction of new technologies to cover the need for higher data rates and new services, integration of the existing technologies in a common platform is an important objective of the next-generation wireless systems, referred to as IMT-Advanced.

The design of a generic multiple access scheme for various wireless systems is challenging. The new multiple access scheme should (a) enable the integration of existing technologies, (b) provide higher data rates in a given spectrum, i.e. maximize the spectral efficiency, (c) support different cell configurations and automatic adaptation to the channel conditions, (d) provide simple protocol and air interface layers, and, finally, (e) enable a seamless adaptation of new standards and technologies in the future.

Especially for the downlink of a mobile communications system, data rates of 1 Gbit/s are targeted with IMT-Advanced. Therefore, new physical layer and multiple access technologies are needed to provide high speed data rates with flexible bandwidth allocation. A next-generation low cost generic radio interface has to be operational in various mixed-cell environments with a scalable bandwidth and data rate.

5.4.2 Network Topology and Basic Concept

An adaptive and flexible concept for a cellular system based on multi-carrier transmission has been proposed by NTT DOCOMO. The concept has been proven by several field trials. The generic architecture allows a capacity optimization with seamless transition from a single-cell to a multi-cell environment.

5.4.3 Experiments and Field Trials

A series of successful experiments and field trials based on innovative OFDM schemes with challenging data rate and throughput values have been performed by NTT DOCOMO. The series of experiments and field trials ranged from 100 Mbit/s in 2002 up to about 5 Gbit/s in 2006. The achieved maximum data rate in the field trials together with the characteristic system parameters are summarized in Table 5-43.

Table 5-43 OFDM-based experiments and field trials performed by NTT DOCOMO

Maximum data rate	100 Mbit/s	1 Gbit/s	2.5 Gbit/s	4.92 Gbit/s
Access scheme	VSF-OFCDM	OFDM	OFDM	OFDM
Bandwidth	101.5 MHz	101.5 MHz	101.4 MHz	101.4 MHz
Symbol mapping	QPSK	16-QAM	64-QAM	64-QAM
Channel coding	Rate 3/4 Turbo code	Rate 8/9 Turbo code	Rate 8/9 Turbo code	Rate 8/9 Turbo code
Antenna scheme	1 spatial stream	4x4 MIMO	6x6 MIMO	12x12 MIMO
Date of field trial	07-2003	05-2005	12-2005	12-2006
Reference	[27]	[32]	[31]	[30]

The experiments with variable spreading factor orthogonal frequency and code division multiplexing (VSF-OFCDM) [10–12] were performed in cellular environments and applied spreading codes of variable length. For isolated cell scenarios without inter-cell interference, a spreading factor of one can be chosen, resulting in an OFDM transmission without spreading. In scenarios with high inter-cell interference the spreading factor can be adapted accordingly. While the first experiments targeted scenarios with cellular interference, the later ones focused on maximum data rates in isolated cells exploiting OFDM in combination with multiple antenna transmission (MIMO). The VSF-OFCDM access scheme is explained in detail in the following section.

5.4.4 VSF-OFCDM Access Scheme

Figure 5-44 illustrates the generic architecture of a broadband packet based air interface with VSF-OFCDM with two-dimensional spreading in the downlink and MC-DS-CDMA for the uplink proposed by NTT DOCOMO [10–12]. The use of a two-dimensional variable spreading code together with adaptive channel coding and M-QAM modulation in an MC-CDMA system allows an automatic adaptation of the radio link parameters to different traffic, channel, and cellular environment conditions. Furthermore, by appropriate selection of the transmission parameters (FEC, constellation, frame length, FFT size, duplex, i.e. TDD/FDD, etc.), this concept can support different multi-carrier or spread spectrum based transmission schemes. For instance, by choosing a spreading factor of one in both time and frequency directions, one obtains a pure OFDM transmission system. If the spreading factor in frequency direction and the number of sub-carriers are set to one, we can configure the system to a classical DS-CDMA scheme. Hence, such a flexible architecture can be seen as a *basic platform* for the integration of existing technologies as well.

The proposal mainly focuses on FDD in order to avoid the necessity of inter-cell synchronization in multi-cell environments and to accommodate an independent traffic assignment in the up- and downlinks according to the respective traffic. An application

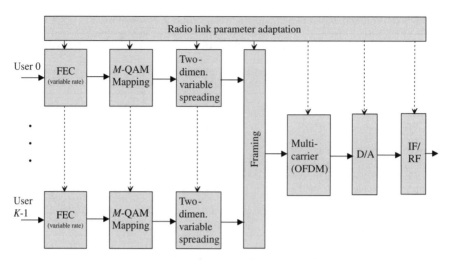

Figure 5-44   Generic VSF-OFCDM concept

of TDD for special environments is also foreseen. In both duplex cases (FDD and TDD) the same air interface is used.

5.4.5   System Parameters

5.4.5.1   Downlink

As depicted in Figure 5-45, the principle of VSF-OFCDM for the downlink is to apply variable spreading code lengths L and different spreading types. In multi-cell environments spreading codes of length $L > 1$ are chosen in order to achieve a high link capacity by using a frequency re-use factor of one. Two-dimensional spreading has a total spreading code length of

$$L = L_{time} L_{freq}. \tag{5.4}$$

The two-dimensional spreading with priority for time domain spreading rather than frequency domain spreading is used. The motivation is that in frequency selective fading channels it is easier to maintain orthogonality among the spread user signals by spreading in the time direction than in the frequency direction. The concept of two-dimensional spreading is described in detail in Section 2.1.4.3. Additional frequency domain spreading in combination with interleaving along with time domain spreading is used for channels that have low SNR such that additional frequency diversity can enhance the transmission quality. The spreading code lengths L_{time} and L_{freq} are adapted to the radio link conditions like delay spread, Doppler spread, and inter-cell interference as well as to the link parameters such as symbol mapping. In isolated areas (hot-spot areas or indoor offices) only one-dimensional spreading in the time direction is used in order to maintain the orthogonality between the spread user signals. Finally, spreading can be completely switched off with $L = 1$ if a single user operates in an isolated cell with a high data rate.

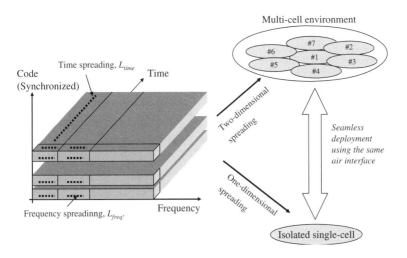

Figure 5-45   Downlink transmission based on VSF-OFCDM

For the channel estimation two different frame formats have been defined. The first format is based on a time-multiplexed pilot structure where two subsequent OFDM symbols with reference data are transmitted periodically in predefined distances. The second format applies a code-multiplexed pilot structure where the reference data are spread by a reserved spreading code and multiplexed with the spread data symbols so that no explicit pilot symbols or carriers are required. The assumption for this channel estimation method is that the whole spreading codes are faded flat and the different spreading codes remain orthogonal.

Table 5-44 summarizes the downlink system parameters. For signal detection at the terminal station side, single-user detection with MMSE equalization is proposed before de-spreading, which is a good compromise between receiver complexity and performance achievement. Furthermore, the high order modulation like 16-QAM or 64-QAM is used without frequency or time domain spreading [10]. In a dense cellular system with high interference and frequency selectivity the lowest order modulation QPSK with the highest spreading factor in both directions is employed.

The throughput of a VSF-OFCDM system in the downlink is shown in Figure 5-46 [10]. The throughput in Mbit/s versus the SNR per symbol in a Rayleigh fading channel is plotted. The system applies a spreading code length of $L = 16$, where 12 codes are used. The symbol timing is synchronized using a guard interval correlation and the channel estimation is realized with a time-multiplexed pilot channel within a frame. It can be observed from Figure 5-46 that an average throughput over 100 Mbit/s can be achieved at an SNR of about 13 dB when using QPSK with rate 1/2 Turbo coding.

5.4.5.2   Uplink

In contrast to the downlink, a very low number of sub-carriers in an asynchronous MC-DS-CDMA has been chosen. This guarantees a low power mobile terminal since it has a lower PAPR, reducing the back-off of the amplifier compared to MC-CDMA or OFDM.

Table 5-44 System parameters for the VSF-OFCDM downlink

Parameters	Characteristics/values
Multiple access	VSF-OFCDM
Bandwidth B	101.5 MHz
Data rate objective	> 100 Mbit/s
Spreading code	Walsh–Hadamard codes
Spreading code length L	1–256
Number of sub-carriers N_c	768
Sub-carrier spacing F_s	131.8 kHz
OFDM symbol duration T_s	7.585 μs
Guard interval duration T_g	1.674 μs
Total OFDM symbol duration T'_s	9.259 μs
Number of OFDM symbols per frame N_s	54
OFDM frame length T_{fr}	500 μs
Symbol mapping	QPSK, 16-QAM, 64-QAM
Channel code	Convolutional Turbo code, memory 4
Channel code rate R	1/3–8/9

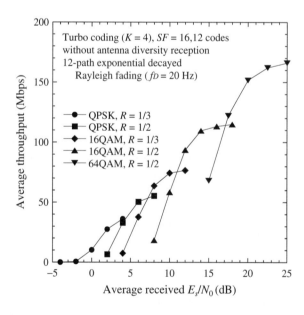

Figure 5-46 Throughput with VSF-OFCDM in the downlink [10]

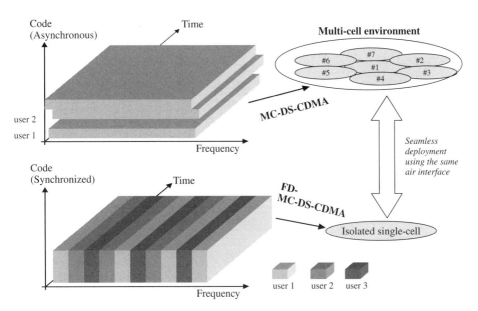

Figure 5-47 Uplink transmission based on MC-DS-CDMA and with an FD-MC-DS-CDMA option

A code-multiplexed pilot structure is applied for channel estimation based on the principle described in the previous section. To combat the multiple access interference, a rake receiver with an interference cancellation in conjunction with an adaptive array antenna at the base station is proposed. As shown in Figure 5-47, the capacity can be optimized for each cell configuration.

In a multi-cell environment MC-DS-CDMA with complex interference cancellation at the base station is used, where in a single-cell environment an orthogonal function in the frequency (FD-MC-DS-CDMA) or time direction (TD-MC-DS-CDMA) is introduced into MC-DS-CDMA. In addition, this approach allows a seamless deployment from a multi-cell to a single cell with the same air interface. The basic system parameters for the uplink are summarized in Table 5-45.

High order modulation like 16-QAM or 64-QAM is used even with no spreading in a single cell with good reception conditions [10]. However, in a dense cellular system with high frequency selectivity and high interference the lowest order modulation QPSK with the highest spreading factor will be deployed.

In Figure 5-48, the throughput of an MC-DS-CDMA system in the uplink is shown [10]. The throughput in Mbit/s versus the SNR per symbol in a Rayleigh fading channel is plotted. The system applies a spreading code length of $L = 4$, where three codes are used. Receive antenna diversity with two antennas is exploited. The channel estimation is realized with a code-multiplexed pilot channel within a frame. It can be observed from Figure 5-48 that an average throughput over 20 Mbit/s can be achieved at an SNR of about 9 dB when using QPSK with rate 1/2 Turbo coding.

Table 5-45 System parameters for the MC-DS-CDMA uplink

Parameters	Characteristics/values
Multiple access	MC-DS-CDMA
Bandwidth B	40 MHz
Data rate objective	> 20 Mbit/s
Spreading code length L	1–256
Number of sub-carriers N_c	2
Sub-carrier spacing F_s	20 MHz
Chip rate per sub-carrier	16.384 Mcps
Roll-off factor	0.22
Total OFDM symbol duration T_s'	9.259 µs
Number of chips per frame	8192
Frame length T_{fr}	500 µs
Symbol mapping	QPSK, 16-QAM, 64-QAM
Channel code	Convolutional Turbo code, memory 4
Channel code rate R	1/16–3/4

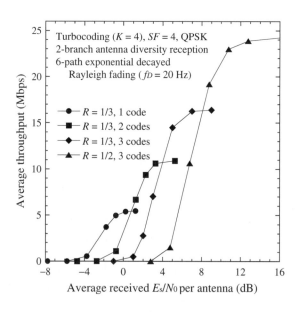

Figure 5-48 Throughput with MC-DS-CDMA in the uplink [10]

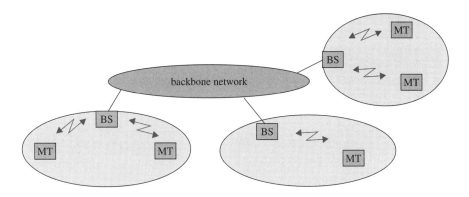

Figure 5-49   WLAN as an infrastructure network

5.5   Wireless Local Area Networks

Local area networks typically cover a story or building and their wireless realization should avoid complex installation of a wired infrastructure. WLANs are used in public and private environments and support high data rates. They are less expensive than wired networks for the same data rate, are simple and fast to install, offer flexibility and mobility, and are cost efficient due to the possibility of license exempt operation.

5.5.1   Network Topology

WLANs can be designed for infrastructure networks, ad hoc networks, or combinations of both. The mobile terminals (MTs) in infrastructure networks communicate via the base stations (BSs), which control the multiple access. The base stations are linked to each other by a wireless (e.g. FWA) or wired backbone network. Infrastructure networks have access to other networks like the Internet. The principle of an infrastructure network is illustrated in Figure 5-49.

In ad hoc networks, the mobile terminals communicate directly with each other. These networks are more flexible than infrastructure networks but require a higher complexity in the mobile terminals since they have to control the complete multiple access as in a base station. The communication within ad hoc networks is illustrated in Figure 5-50.

5.5.2   Channel Characteristics

WLAN systems often use the license exempt 2.4 GHz and 5 GHz frequency bands, which have strict limitations on the maximum transmit power since these frequency bands are also used by many other communications systems. This versatile use of the frequency band results in different types of narrowband and wideband interference, such as from a microwave oven, which the WLAN system has to cope with.

The WLAN cell size is up to several 100 m and multi-path propagation typically results in maximum delays of less than 1 μs. The mobility in WLAN cells is low and corresponds to a walking speed of about 1 m/s. The low Doppler spread in the order of 10–20 Hz makes OFDM very interesting for high rate WLAN systems.

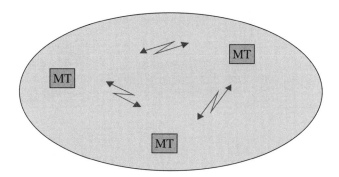

Figure 5-50   WLAN as an ad hoc network

5.5.3   IEEE 802.11a

The physical layer of the OFDM-based WLAN standards IEEE 802.11a, HIPERLAN/2, and MMAC has been harmonized, which enables the use of the same chip set for products of different standards. However, today only products according to IEEE 802.11a and extensions of it are widely used. IEEE 802.11a and 802.11g operate in the 2.4 GHz frequency band while IEEE 802.11h operates in the 5 GHz frequency band.

IEEE 802.11a applies carrier sense multiple access with collision avoidance (CSMA/CA) and share the channel in the time direction for user separation within one channel and apply FDMA for cell separation. The basic OFDM parameters of IEEE 802.11a are summarized in Table 5-46 [21].

Table 5-46   OFDM parameters of IEEE 802.11a

Parameter	Value
IFFT/FFT length	64
Sampling rate	20 MHz
Sub-carrier spacing	312.5 kHz (= 20 MHz/64)
Useful OFDM symbol duration	3.2 µs
Guard duration	0.8 µs
Total OFDM symbol duration	4.0 µs
Number of data sub-carriers	48
Number of pilot sub-carriers	4
Total number of sub-carriers	52

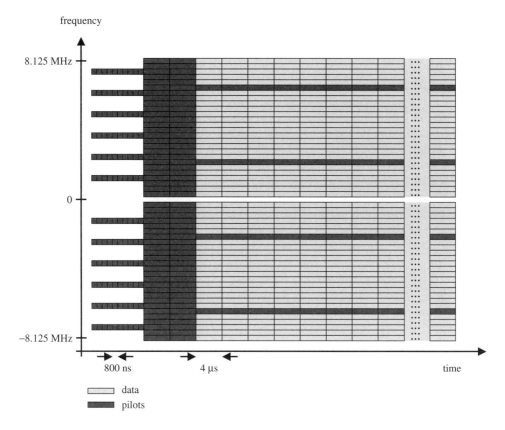

Figure 5-51   OFDM frame of IEEE 802.11a

5.5.3.1   Frame Structure

The OFDM frame structure specified by IEEE 802.11a is shown in Figure 5-51. The frame starts with up to 10 short pilot symbols, depending on the frame type. These pilot symbols are used for coarse frequency synchronization, frame detection, and automatic gain control (AGC). The following two OFDM symbols contain pilots used for fine frequency synchronization and channel estimation. The OFDM frame has four pilot sub-carriers, which are the sub-carriers -21, -7, 7, and 21. These pilot sub-carriers are used for compensation of frequency offsets. The sub-carrier 0 is not used to avoid problems with DC offsets.

5.5.3.2   FEC Coding and Modulation

The IEEE 802.11a standard supports the modulation schemes BPSK, QPSK, 16-QAM, and 64-QAM in combination with punctured convolutional codes (CC) with rates in the range of 1/2 up to 3/4. The different FEC and modulation combinations supported by IEEE 802.11a are shown in Table 5-47. This flexibility offers a good tradeoff between the coverage and data rate.

5.5.4 Transmission Performance

5.5.4.1 Transmission Capacity

As shown in Table 5-48, the use of flexible channel coding and modulation in the IEEE 802.11a standard provides eight physical modes (PHY modes), i.e. combinations of FEC and modulation. The data rates that can be supported are in the range of 6 Mbit/s up to 54 Mbit/s and depend on the coverage and channel conditions.

It should be emphasized that the overall data rate in a cellular system is limited by the coverage distance and the amount of interference due to a dense frequency re-use. A global capacity optimization per cell (or per sector) can be achieved if the PHY mode

Table 5-47 FEC and modulation parameters of IEEE 802.11a

Modulation	Code rate R	Coded bits per sub-channel	Coded bits per OFDM symbol	Data bits per OFDM symbol
BPSK	1/2	1	48	24
BPSK	3/4	1	48	36
QPSK	1/2	2	96	48
QPSK	3/4	2	96	72
16-QAM	1/2	4	192	96
16-QAM	3/4	4	192	144
64-QAM	2/3	6	288	192
64-QAM	3/4	6	288	216

Table 5-48 Data rates of IEEE 802.11a

PHY mode		Data rate (Mbit/s)
1	BPSK, CC1/2	6
2	BPSK, CC3/4	9
3	QPSK, CC1/2	12
4	QPSK, CC3/4	18
5	16-QAM, CC1/2	24
6	16-QAM, CC3/4	36
7	64-QAM, CC2/3	48
8	64-QAM, CC3/4	54

is adapted to each terminal station link condition individually. Results in Reference [20] show that compared to a single PHY mode, the area spectral efficiency can be at least doubled if adaptive PHY modes are employed.

5.5.4.2 Link Budget

The transmit power, depending on the coverage distance, is given by

$$P_{T_x} = Pathloss + P_{Noise} - G_{Antenna} + FadeMargin + Rx_{loss} + \frac{C}{N}, \quad (5.5)$$

where

$$Pathloss = 10\log_{10}\left(\frac{4\pi f_c d^{\eta/2}}{c}\right)^2 \quad (5.6)$$

is the propagation path loss, d represents the distance between the transmitter and the receiver, f_c is the carrier frequency, and c is the speed of light. In the case of WLANs, η can be estimated to be in the order of 3 to 5.

$$P_{Noise} = F\,N_{Thermal} = F\,K\,T\,B \quad (5.7)$$

is the noise power at the receiver input, where F is the receiver noise factor (about 6 dB), K is the Boltzman constant ($K = 1.38 \times 10^{-23}$ J/K), T is the temperature in K, and B is total occupied Nyquist bandwidth. The noise power is expressed in dBm.

$G_{Antenna}$ is the sum of the transmit and receive antenna gains, expressed in dBi. In WLANs, the terminal station antenna can be omni-directional with 0 dBi gain, but the base station antenna may have a gain of about 14 dBi. The *fade margin* is the margin needed to counteract the fading and is about 5 to 10 dB. Rx_{loss} is the margin for all implementation losses and all additional uncertainties such as interference. This margin can be about 5 dB. C/N is the carrier-to-noise power ratio (equivalent to E_s/N_0) for BER $= 10^{-6}$. By considering a transmission power of about 23 dBm and following the above parameters, the maximum coverage for the robust PHY mode at 2.4 GHz carrier frequency can be estimated to be about 300 m for an omni-directional antenna.

The minimum receiver sensitivity thresholds for IEEE 802.11a, depending on the PHY mode, i.e. the data rate for a BER of 10^{-6}, are given in Table 5-49. The receiver sensitivity threshold Rx_{th} is defined by

$$Rx_{th} = P_{Noise} + \frac{C}{N} + Rx_{loss}. \quad (5.8)$$

5.6 Interaction Channel for DVB-T: DVB-RCT

The VHF and the UHF frequency bands (typically from 120 to 860 MHz) are reserved especially for broadcast TV services and similarly to convey uplink information from the subscriber premises, i.e. terminal stations, to the base station for interactive services. In addition to the uplink channel, interactive services also require a downlink channel

Table 5-49 Minimum receiver sensitivity
thresholds for IEEE 802.11a

Nominal bit rate (Mbit/s)	Minimum sensitivity
6	−85 dBm
9	−83 dBm
12	−81 dBm
18	−79 dBm
24	−76 dBm
36	−73 dBm
48	−70 dBm
54	−68 dBm

for transmitting messages, uplink channel access control commands, and other data. The downlink interactive information can be either embedded in the broadcast channels or defined as a standalone channel specifically devoted to interaction with subscribers.

5.6.1 Network Topology

A DVB-T interactive network architecture is illustrated in Figure 5-52 [17]. This could be seen as a *point to multi-point* (PMP) network topology consisting of a base station and several subscribers (terminal stations). Like other digital access networks, DVB-RCT networks are also intended to offer a variety of services requiring different data rates. Therefore, the multiple access scheme needs to be flexible in terms of data rate assignment to each subscriber. Furthermore, another major constraint for the choice of the parameters of the DVB-RCT specification is to employ the existing infrastructure already used for broadcast DVB-T services.

As shown in Figure 5-53, the interactive downlink path is embedded in the broadcast channel, exploiting the existing DVB-T infrastructure [18]. The access for the uplink interactive channels carrying the return interaction path data is based on a combination of OFDMA and TDMA type of multiple access scheme [17].

The downlink interactive information data are made up of MPEG-2 transport stream packets with a specific header that carries the medium access control (MAC) management data. The MAC messages control the access of the subscribers, i.e. terminal stations, to the shared medium. These embedded MPEG-2 transport stream packets are carried in the DVB-T broadcast channel (see Figure 5-53).

The uplink interactive information is mainly made up of ATM cells mapped on to physical bursts. ATM cells include application data messages and MAC management data.

To allow access by multiple users, the VHF/UHF radio frequency return channel is partitioned both in the frequency and time domain, using frequency division and time division. Each subscriber can transmit his or her data for a given period of time on a

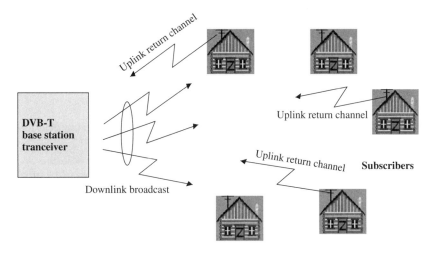

Figure 5-52   DVB-RCT network architecture

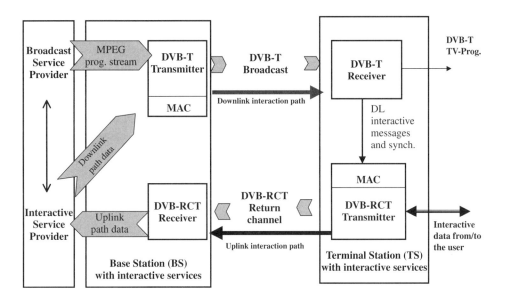

Figure 5-53   Overview of the DVB-RCT standard

given sub-carrier, resulting in a combination of OFDMA and TDMA type of multiple access.

A global synchronization signal, required for the correct operation of the uplink demodulator at the base station, is transmitted to all users via global DVB-T timing signals. Time synchronization signals are conveyed to all users through the broadcast channel, either within the MPEG-2 transport stream or via global DVB-T timing signals. In other words, the DVB-RCT frequency synchronization is derived from the broadcast DVB-T signal while the time synchronization results from the use of MAC management packets conveyed through the broadcast channel. Furthermore, the so-called periodic *ranging signals* are transmitted from the base station to individual terminal stations for timing misalignment adjustment and power control purposes.

The DVB-RCT OFDMA based system employs either 1024 (1 k) or 2048 (2 k) sub-carriers and operates as follows:

- Each terminal station transmits one or several low bit rate modulated sub-carriers toward the base station.
- The sub-carriers are frequency locked and power ranged and the timing of the modulation is synchronized by the base station. In other words, the terminal stations derive their system clock from the DVB-T downstream. Accordingly, the transmission mode parameters are fixed in a strict relationship with the DVB-T downstream.
- On the reception side, the uplink signal is demodulated, using an FFT process, like the one performed in a DVB-T receiver.

5.6.2 Channel Characteristics

As in the downlink terrestrial channel, the return channels suffer especially from high multi-path propagation delays.

In the DVB-RCT system, the downlink interaction data and the uplink interactive data are transmitted in the same radio frequency bands, i.e. VHF/UHF bands III, IV, and V. Hence, the DVB-T and DVB-RCT systems may form a bi-directional FDD communication system that shares the same frequency bands with sufficient duplex spacing. Thus, it is possible to benefit from common features in regard to the RF devices and parameters (e.g. antenna, combiner, propagation conditions). The return channel (RCT) can also be located in any free segment of an RF channel, taking into account existing national and regional analogue television assignments, interference risks, and future allocations for DVB-T.

5.6.3 Multi-Carrier Uplink Transmission

The method used to organize the DVB-RCT channel is inspired by the DVB-T standard. The DVB-RCT RF channel provides a grid of time–frequency slots, each slot usable by any terminal station. Hence, the concept of the DVB uplink channel allocation is based on a combination of OFDMA with TDMA. Thus, the uplink is divided into a number of time slots. Each time slot is divided in the frequency domain into groups of sub-carriers, referred to as sub-channels. The MAC layer controls the assignment of sub-channels and time slots by resource requests and grant messages.

The DVB-RCT standard provides two types of sub-carrier shaping, of which only one will be used at a certain time. The shaping functions are:

- *Nyquist shaping* in the time domain on each sub-carrier to provide immunity against both ICI and ISI. A square root raised cosine pulse with a roll-off factor $\alpha = 0.25$ is employed. The total symbol duration is 1.25 times the inverse of the sub-carrier spacing.
- *Rectangular shaping* with guard interval T_g that has a possible value of $T_s/4$, $T_s/8$, $T_s/16$, $T_s/32$, where T_s is the useful symbol duration (without guard time).

Table 5-50 DVB-RCT targeted sub-carrier spacing for the 8 MHz channel

Sub-carrier spacing	Targeted sub-carrier spacing
Sub-carrier spacing 1	$\approx 1\,kHz$ (symbol duration $\approx 1000\,\mu s$)
Sub-carrier spacing 2	$\approx 2\,kHz$ (symbol duration $\approx 500\,\mu s$)
Sub-carrier spacing 3	$\approx 4\,kHz$ (symbol duration $\approx 250\,\mu s$)

Table 5-51 DVB-RCT transmission mode parameters for the 8 and 6 MHz DVB-T systems

Parameters	8 MHz DVB-T system		6 MHz DVB-T system	
Total number of sub-carriers	2048 (2 k)	1024 (1 k)	2048 (2 k)	1024 (1 k)
Used sub-carriers	1712	842	1712	842
Useful symbol duration	896 µs	896 µs	1195 µs	1195 µs
Sub-carrier spacing	1.116 kHz	1.116 kHz	0.837 kHz	0.837 kHz
RCT channel bandwidth	1.911 MHz	0.940 MHz	1.433 MHz	0.705 MHz
Useful symbol duration	448 µs	448 µs	597 µs	597 µs
Sub-carrier spacing	2.232 kHz	2.232 kHz	1.674 kHz	1.674 kHz
RCT channel bandwidth	3.821 MHz	1.879 MHz	2.866 MHz	1.410 MHz
Useful symbol duration	224 µs	224 µs	299 µs	299 µs
Sub-carrier spacing	4.464 kHz	4.464 kHz	3.348 kHz	3.348 kHz
RCT channel bandwidth	7.643 MHz	3.759 MHz	5.732 MHz	2.819 MHz

5.6.3.1 Transmission Modes

The DVB-RCT standard provides six transmission modes characterized by a dedicated combination of the maximum number of sub-carriers used and their sub-carrier spacings [17]. Only one transmission mode will be implemented in a given RCT radio frequency channel, i.e. transmission modes should not be mixed.

The sub-carrier spacing governs the robustness of the system in regard to the possible synchronization misalignment of any terminal station. Each value implies a given maximum transmission cell size and a given resistance to the Doppler shift experienced when the terminal station is in motion, i.e. in the case of portable receivers. The three targeted DVB-RCT sub-carrier spacing values are defined in Table 5-50.

Table 5-51 gives the basic DVB-RCT transmission mode parameters applicable for the 8 MHz and 6 MHz radio frequency channels with 1024 or 2048 sub-carriers. Due to the combination of the above parameters, the DVB-RCT final bandwidth is a function of the sub-carrier spacing and of the FFT size. Each combination will have a specific tradeoff between frequency diversity and time diversity, and between coverage range and portability/mobility capability.

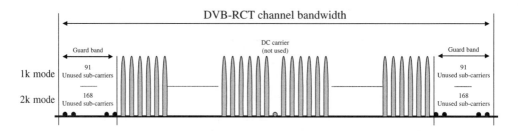

Figure 5-54 DVB-RCT channel organization for the 1 k and 2 k mode

Table 5-52 Sub-carrier organization for the 1 k and 2 k mode

Parameters	1 k mode structure	2 k mode structure
Number of FFT points	1024	2048
Overall usable sub-carriers	842	1712
Overall used sub-carriers – with burst structures 1 and 2 – with burst structure 3	 840 841	 1708 1711
Lower and upper channel guard band	91 sub-carriers	168 sub-carriers

5.6.3.2 Time and Frequency Frames

Depending on the transmission mode in operation, the total number of allocated sub-carriers for uplink data transmission is 1024 carriers (1 k mode) or 2048 carriers (2 k mode) (see Figure 5-54). Table 5-52 shows the main parameters.

Two types of transmission frames (TFs) are defined:

- *TF1*. The first frame type consists of a set of OFDM symbols, which contains several data sub-channels, a null symbol, and a series of synchronization/ranging symbols.
- *TF2*. The second frame type is made up of a set of general purpose OFDM symbols, which contains either data or synchronization/ranging sub-channels.

Furthermore, three different burst structures are specified as follows:

- *Burst structure 1* uses one unique sub-carrier to carry the total data burst over time, with an optional frequency hopping law applied within the duration of the burst.
- *Burst structure 2* uses four sub-carriers simultaneously, each carrying a quarter of the total data burst over time.
- *Burst structure 3* uses 29 sub-carriers simultaneously, each carrying 1/29 of the total data burst over time.

These three burst structures provide a pilot aided modulation scheme to allow coherent detection in the base station. The defined pilot insertion ratio is approximately 1/6,

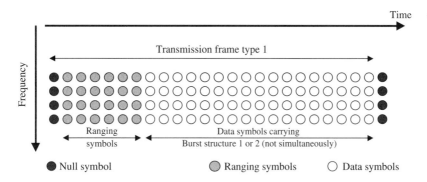

Figure 5-55 Organization of the TF1 frame

which means one pilot carrier is inserted for approximately every five data sub-carriers. Furthermore, they result in various combinations of time and frequency diversity, thereby providing various degrees of robustness, burst duration, and a wide range of bit rates to the system.

Each burst structure makes use of a set of sub-carriers called a sub-channel. One or several sub-channels can be used simultaneously by a given terminal station depending on the allocation performed by the MAC process.

Figure 5-55 depicts the organization of a TF1 frame in the time domain. It should be noticed that the burst structures are symbolized regarding their duration and not regarding their occupancy in the frequency domain. The corresponding sub-carrier(s) of burst structures 1 and 2 are spread over the whole RCT channel.

Null symbol and ranging symbols always use the rectangular shaping. The user symbols of TF1 can use either rectangular shaping or Nyquist shaping. If the user part employs rectangular shaping, the guard interval value will be identical for any OFDM symbol embedded in the whole TF1 frame. If the user part performs Nyquist shaping, the guard interval value to apply on to the null symbol and ranging symbols will be $T_s/4$. The user part of the TF1 frame is suitable to carry one burst structure 1 or four burst structure 2. Both burst structures should not be mixed in a given DVB-RCT channel.

The time duration of a transmission frame depends on the number of consecutive OFDM symbols and on the time duration of the OFDM symbol. The time duration of an OFDM symbol depends on

- the reference downlink DVB-T system clock;
- the sub-carrier spacing; and
- the rectangular filtering of the guard interval (1/4, 1/8, 1/16, 1/32 times T_s).

In Table 5-53, the values of the frame durations in seconds for TF1 using burst structure 1 are given.

Figure 5-56 depicts the organization of the TF2 frame in the time domain. The corresponding sub-carrier(s) of burst structures 2 and 3 are spread on the whole RCT channel. TF2 is used only in the rectangular pulse shaping case. The guard interval applied on any OFDM symbols embedded in the whole TF2 is the same (i.e. either 1/4, 1/8, 1/16 or

Table 5-53 Transmission frame duration in seconds with burst structure 1 and with rectangular filtering with $T_g = T_s/4$ or Nyquist filtering and for reference clock 64/7 MHz

Shaping scheme	Number of consecutive OFDM symbols	Sub-carrier spacing 1	Sub-carrier spacing 2	Sub-carrier spacing 3
Rectangular	187	0.20944	0.10472	0.05236
Nyquist without FH	195	0.2184	0.1092	0.0546
Nyquist with FH	219	0.24528	0.12264	0.06132

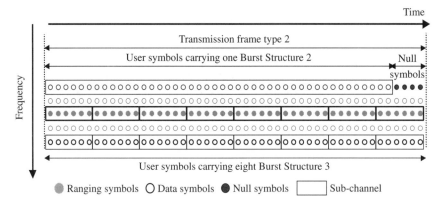

Figure 5-56 Organization of the TF2 frame structure (time domain)

1/32 of the useful symbol duration). The user part of the TF2 allows the usage of burst structure 3 or, optionally, burst structure 2. When one burst structure 2 is transmitted, it should be completed by a set of four null modulated symbols to have a duration equal to the duration of eight burst structure 3.

5.6.3.3 FEC Coding and Modulation

The channel coding is based on a concatenation of a Reed Solomon outer code and a rate-compatible convolutional inner code. Convolutional Turbo codes can also be used. Different modulation schemes (QPSK, 16-QAM, and 64-QAM) with Gray mapping are employed.

Whatever FEC is used, the data bursts produced after the encoding and mapping processes have a fixed length of 144 modulated symbols. Table 5-54 defines the original sizes of the useful data payloads to be encoded in relation to the selected physical modulation and encoding rate.

Under the control of the base station, a given terminal station can use different produced successive bursts having different combinations of encoding rates. Here, the use of adaptive coding and modulation is aimed to provide flexible bit rates to each terminal station, in relation to the individual reception conditions encountered in the base station.

Table 5-54 Number of useful data bytes per burst

Parameters	QPSK		16-QAM		64-QAM	
FEC encoding rate	$R = 1/2$	$R = 3/4$	$R = 1/2$	$R = 3/4$	$R = 1/2$	$R = 3/4$
Number of data bytes in 144 symbols	18	27	36	54	54	81

Table 5-55 Overall encoding rates

Outer RS encoding rate R_{outer}	Inner CC encoding rate R_{inner}	Overal code rate $R_{total} = R_{outer}R_{inner}$
3/4	2/3	1/2
9/10	5/6	3/4

Table 5-56 Coding parameters for combinations of coding rate and modulation

Modulation code rate	RS input	CC input	Number of CC output bits
QPSK 1/2	144 bits = 24 RS symbols	32 RS symbols = 192 bits	288
QPSK 3/4	216 bits = 36 RS symbols	40 RS symbols = 240 bits	288
16-QAM 1/2	288 bits = 48 RS symbols	2×32 RS symols = 384 bits	576
16-QAM 3/4	432 bits = 72 RS symbols	2×40 RS symbols = 480 bits	576
64-QAM 1/2	432 bits = 72 RS symbols	3×32 RS symbols = 576 bits	864
64-QAM 3/4	648 bits = 108 RS symbols	3×40 RS symbols = 720 bits	864

The outer Reed Solomon encoding process uses a shortened systematic RS(63, 55, $t = 4$) encoder over a Galois field GF(64); i.e. each RS symbol consists of 6 bits. Data bits issued from the Reed Solomon encoder are fed to the convolutional encoder of constraint length 9. To produce the two total coding rates expected (1/2 and 3/4) the RS and convolutional encoder have implemented the coding rates defined in Table 5-55.

The terminal station uses the modulation scheme determined by the base station through MAC messages. The encoding parameters defined in Table 5-56 are used to produce the desired coding rate in relation with the modulation schemes. It should be noted that the number of symbols per burst in all combinations remains constant, i.e. 144 modulated symbols per burst.

Pilot sub-carriers are inserted into each data burst in order to constitute the burst structure and are modulated according to their sub-carrier location. Two power levels are used for these pilots, corresponding to +2.5 dB or 0 dB relative to the mean useful symbol power. The selected power depends on the position of the pilot inside the burst structure.

Table 5-57 Net bit rate in kbit/s per sub-carrier for burst structure 1 using rectangular shaping

Channel spacing, modulation, coding parameters			Rectangular shaping with/without FH		Nyquist shaping without FH
			$T_G = 1/4T_s$	$T_G = 1/32T_s$	$\alpha = 0.25$
Channel spacing 1	4-QAM	1/2	0.66	0.69	0.83
		3/4	0.99	1.03	1.25
	16-QAM	1/2	1.32	1.37	1.67
		3/4	1.98	2.06	2.50
	64-QAM	1/2	1.98	2.06	2.50
		3/4	2.97	3.09	3.75
Channel spacing 3	4-QAM	1/2	2.63	2.75	3.33
		3/4	3.95	4.12	5.00
	16-QAM	1/2	5.27	5.50	6.67
		3/4	7.91	8.25	10.00
	64-QAM	1/2	7.91	8.25	10.00
		3/4	11.87	12.38	15.00

5.6.4 Transmission Performance

5.6.4.1 Transmission Capacity

The transmission capacity depends on the used M-QAM modulation density, the error control coding, and the used mode with Nyquist or rectangular pulse shaping. The net bit rate per sub-carrier for burst structure 1 is given in Table 5-57 with and without frequency hopping (FH).

5.6.4.2 Link Budget

The service range given for the different transmission modes and configurations can be calculated using the RF figures derived from the DVB-T implementation and propagation models for rural and urban areas. In order to limit the terminal station RF power to reasonable limits, it is recommended that the complexity be put on the base station side by using high gain sectorized antenna schemes and optimized reception configurations.

To define mean service ranges, Table 5-58 details the RF configurations for carrier spacing 1 and QPSK 1/2 modulation levels for 800 MHz in transmission modes with burst structures 1 and 2. The operational C/N is derived from Reference [18] while considering a +2 dB implementation margin, +1 dB gain due to block Turbo code/concatenated RS and convolutional codes, and +1 dB gain when using time interleaving in Rayleigh channels.

Table 5-58 Parameters for service range simulations

Transmission modes	Outdoor	Indoor
Antenna location	Rural/fixed	Indoor urban/portable
Frequency	800 MHz	800 MHz
Sub-carrier spacing	1 kHz	1 kHz
Modulation scheme C/N [18] Operational C/N	4-QAM 1/2 3.6 dB 5 dB	4-QAM 1/2 3.6 dB 5 dB
BS receiver antenna gain	16 dBi (60 degree)	16 dBi (60 degree)
Antenna height (user side)	Outdoor 10 m	Indoor 10 m (second floor)
TS antenna gain	13 dBi (directive)	3 dBi (~ omni-directional)
Cable loss	4 dB	1 dB
Duplexer loss	4 dB	1 dB (separate antennas/switch)
Indoor penetration loss	/	10 dB (mean second floor)
Propagation models	ITU-R 370	OKUMURA-HATA suburban
Standard deviation for location variation	− 10 dB for BS1 − 5 dB for BS-2 and BS-3 (spread multi-carrier)	− 10 dB for BS1 − 5 dB for BS-2 and BS-3 (spread multi-carrier)

Reasonable dimensioning of the output amplifier in terms of bandwidth and inter-modulation products (linearity) indicates that a transmit power in the order of 25 dBm could be achievable at low cost. It is shown in Reference [17] that with 24 dBm transmit power indoor reception would be possible up to a distance of 15 km, while outdoor reception would be offered up to 40 km or higher.

References

[1] 3GPP (TS 25.308), "High speed downlink packet access (HSDPA): overall description; Stage 2," *Technical Specification*, Sophia Antipolis, France, 2004.

[2] 3GPP (TS 25.309), "FDD enhanced uplink: overall description; Stage 2," *Technical Specification*, Sophia Antipolis, France, 2006.

[3] 3GPP (TS 25.401), "UTRAN overall description," *Technical Specification*, Sophia Antipolis, France, 2002.

[4] 3GPP (TR 25.912), "Feasibility study for evolved universal terrestrial radio access (UTRA) and universal terrestrial radio access network (UTRAN) (Release 7)," *Technical Report*, Sophia Antipolis, France, 2007.

[5] 3GPP (TR 25.913), "Requirements for evolved UTRA (E-UTRA) and evolved UTRAN (E-UTRAN)," *Technical Report*, Sophia Antipolis, France, 2006.

[6] 3GPP (TS 36.211), "Evolved universal terrestrial radio access (UTRA); physical channels and modulation (Release 8)," *Technical Specification*, Sophia Antipolis, France, 2007.

[7] 3GPP (TS 36.212), "Evolved universal terrestrial radio access (UTRA); multiplexing and channel coding (Release 8)," *Technical Specification*, Sophia Antipolis, France, 2007.

[8] 3GPP (TS 36.300), "Evolved universal terrestrial radio access (UTRA) and evolved universal terrestrial radio access network (UTRAN); overall description; Stage 2 (Release 8)," *Technical Specification*, Sophia Antipolis, France, 2007.

[9] 3GPP (TS 36.803), "Evolved universal terrestrial radio access (UTRA); user equipment (UE) radio transmission and reception; (Release 8)," *Technical Specification*, Sophia Antipolis, France, 2007.

[10] Atarashi H., Maeda N., Abeta S., and Sawahashi M., "Broadband packet wireless access based on VSF-OFCDM and MC-DS-CDMA," in *Proc. IEEE International Symposium on Personal, Indoor and Mobile Radio Communications (PIMRC 2002)*, Lisbon, Portugal, pp. 992–997, Sept. 2002.

[11] Atarashi H., Maeda N., Kishiyama Y., and Sawahashi M., "Broadband wireless access based on VSF-OFCDM and MC-DS-CDMA and its experiments," *European Transactions on Telecommunications (ETT)*, vol. 15, pp. 159–172, May/June 2004.

[12] Atarashi H. and Sawahashi M., "Variable spreading factor orthogonal frequency and code division multiplexing (VSF-OFCDM)," in *Proc. International Workshop on Multi-Carrier Spread Spectrum and Related Topics (MC-SS 2001)*, Oberpfaffenhofen, Germany, pp. 113–122, Sept. 2001.

[13] Burow R., Fazel K., Höher P., Kussmann H., Progrzeba P., Robertson P., and Ruf M., "On the Performance of the DVB-T system in mobile environments," in *Proc. IEEE Global Telecommunications Conference (GLOBECOM '98), Communication Theory Mini Conference*, Sydney, Australia, Nov. 1998

[14] Dahlman E., Parkvall S., Sköld J., and Beming P., *3G Evolution: HSPA and LTE for Mobile Broadband*, Oxford: Academic Press, 2007.

[15] ETSI DAB (EN 300 401), "Radio broadcasting systems; digital audio broadcasting (DAB) to mobile, portable and fixed receivers," Sophia Antipolis, France, April 2000.

[16] ETSI DVB-H (EN 302 304), "Digital video broadcasting (DVB); transmission system for handheld terminals (DVB-H)," Sophia Antipolis, France, Nov. 2004.

[17] ETSI DVB RCT (EN 301 958), "Interaction channel for digital terrestrial television (RCT) incorporating multiple access OFDM," Sophia Antipolis, France, March 2001.

[18] ETSI DVB-T (EN 300 744), "Digital video broadcasting (DVB); framing structure, channel coding and modulation for digital terrestrial television," Sophia Antipolis, France, July 1999.

[19] ETSI HIPERMAN (TS 102 177), "High performance metropolitan area network, Part-1: physical layer," Feb. 2004.

[20] Fazel K., Decanis C., Klein J., Licitra G., Lindh L., and Lebret Y. Y., "An overview of the ETSI-BRAN HA physical layer air interface specification," *in Proc. IEEE International Symposium on Personal, Indoor and Mobile Radio Communications (PIMRC 2002)*, Lisbon, Portugal, pp. 102–106, Sept. 2002.

[21] IEEE 802.11 (P802.11a/D6.0), "LAN/MAN specific requirements – Part 2: wireless MAC and PHY specifications – high speed physical layer in the 5 GHz band," IEEE 802.11, May 1999.

[22] IEEE 802.16ab-01/01, "Air interface for fixed broadband wireless access systems – Part A: systems between 2 and 11 GHz," IEEE 802.16, June 2000.

[23] IEEE 802.16d, "Air interface for fixed broadband wireless access systems," IEEE 802.16, May 2004.

[24] IEEE 802.16e, "Air interface for fixed and mobile broadband wireless access systems," IEEE 802.16, Feb. 2005.

[25] IST FP6 Project WINNER II, "Final WINNER II system requirements," *Deliverable D6.11.4*, June 2007.

[26] ITU-R Recommendations M.1225, "Guidelines for evaluation of radio transmission technologies for ITM-2000," ITU-R, 1997.

[27] Kishiyama Y., Maeda N., Higuchi K., Atarashi H., and Sawahashi M., "Transmission performance analysis of VSF-OFCDM broadband packet wireless access based on field experiments in 100-MHz forward link," in *Proc. IEEE Vehicular Technology Conference (VTC2004-Fall)*, Los Angeles, USA, pp. 1113–1117, Sept. 2004.

[28] Maucher J. and Furrer J., "WiMAX, der IEEE 802.16 Standard: Technik, Anwendung, Potential," Hannover: Heise Publisher, 2007.

[29] Salvekar A., Sandhu S., Li Q., Vuong M., and Qian X., "Multiple-antenna technology in WiMAX systems," *Intel Technology Journal*, vol. 08, Aug. 2004.

[30] Taoka H., Dai K., Higuchi K., and Sawahashi M., "Field experiments on ultimate frequency efficiency exceeding 30 bit/second/Hz using MLD signal detection in MIMO-OFDM broadband packet radio access," in *Proc. IEEE Vehicular Technologies Conference (VTC 2007-Spring)*, Dublin, Ireland, April 2007.

[31] Taoka H., Dai K., and Sawahashi M., "Field experiments on 2.5-Gbps packet transmission using MLD-based signal detection in MIMO-OFDM broadband packet radio access," in *Proc. International Symposium on Wireless Personal Multimedia Communications (WPMC 2006)*, San Diego, USA, Sept. 2006.

[32] Taoka H., Higuchi K., and Sawahashi M., "Field experiments on real-time 1-Gbps high-speed packet transmission in MIMO-OFDM broadband packet radio access," in *Proc. IEEE Vehicular Technologies Conference (VTC 2006-Spring)*, Melbourne, Australia, pp. 1812–1816, May 2006.

[33] WiMAX forum, "Mobile WiMAX –Part I: a technical overview and performance evaluation," White Paper, Aug. 2006.

[34] WiMAX forum, www.wimaxforum.org.

6

Additional Techniques for Capacity and Flexibility Enhancement

6.1 Introduction

As shown in Chapter 1, wireless channels suffer from attenuation due to the destructive addition of multi-path propagation paths and interference. Severe attenuation makes it difficult for the receiver to detect the transmitted signal unless some additional, less-attenuated replicas of the transmitted signal are provided. This principle is called diversity and is the most important factor in achieving reliable communications. Examples of diversity techniques are:

- *Time diversity.* Time interleaving in combination with channel coding or spreading provides replicas of the transmitted signal in the form of redundancy in the temporal domain to the receiver.
- *Frequency diversity.* The signal transmitted on different frequencies induces different structures in the multi-path environment. Replicas of the transmitted signal are provided to the receiver in the form of redundancy in the frequency domain. The best examples of how to exploit the frequency diversity are the technique of multi-carrier spread spectrum and coding in the frequency direction.
- *Spatial diversity.* Spatially separated antennas provide replicas of the transmitted signal to the receiver in the form of redundancy in the spatial domain. This can be provided with no penalty in spectral efficiency.

Exploiting all forms of diversity in future systems (e.g. B3G, 4G) will ensure the highest performance in terms of capacity and spectral efficiency.

Furthermore, the future generation of broadband mobile/fixed wireless systems will aim to support a wide range of services and bit rates. The transmission rate may vary from voice to very high rate multi-media services requiring data rates up to 1 Gbit/s. Communication channels may change in terms of their grade of mobility, cellular infrastructure, required symmetrical or asymmetrical transmission capacity, and whether they are indoor or outdoor.

Multi-Carrier and Spread Spectrum Systems   Second Edition   K. Fazel and S. Kaiser
© 2008 John Wiley & Sons, Ltd

Hence, air interfaces with the highest flexibility are demanded in order to maximize the area spectrum efficiency in a variety of communication environments. The adaptation and integration of existing and new systems to emerging new standards would be feasible if both the receiver and the transmitter are re-configurable using software-defined radio (SDR) approaches.

The aim of this last chapter is to look at new antenna diversity techniques (e.g. *space time coding* (STC), *space frequency coding* (SFC), and spatial multiplexing) and at the concept of *software-defined radio* (SDR), which will all play a major role in the realization of 4G.

6.2 MIMO Overview

In conventional wireless communications, spectral and power efficiency are achieved by exploiting time and frequency diversity techniques. However, the spatial dimension, so far only exploited for cell sectorization, will play a more important role in future wireless communication systems. In the past most of the work was concentrated on the design of intelligent antennas, applied for *space division multiple access* (SDMA) [5]. In the meantime, more general techniques have been introduced where arbitrary antenna configurations at the transmit and receive sides are considered.

If we consider M transmit antennas and L receive antennas, the overall transmission channel defines the so-called *multiple input/multiple output* (MIMO) channel (see Figure 6-1). If the MIMO channel is assumed to be linear and time-invariant during one symbol duration, the channel impulse response $h(t)$ can be written as

$$h(t) = \begin{bmatrix} h_{0,0}(t) & \cdots & h_{0,L-1}(t) \\ \vdots & \ddots & \vdots \\ h_{M-1,0}(t) & \cdots & h_{M-1,L-1}(t) \end{bmatrix}, \tag{6.1}$$

where $h_{m,l}(t)$ represents the impulse response of the channel between the transmit (Tx) antenna m and the receive (Rx) antenna l.

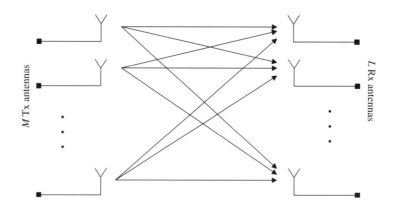

Figure 6-1 MIMO channel

From the above general model, two cases exist: (a) case $M = 1$, resulting in a single input/multiple output (SIMO) channel, and (b) case $L = 1$, resulting in a multiple input/single output (MISO) channel. In the case of SIMO, conventional receiver diversity techniques such as MRC can be realized, which can improve power efficiency, especially if the channels between the Tx and the Rx antennas are independently faded paths (e.g. Rayleigh distributed), where the multi-path diversity order is identical to the number of receiver antennas [21].

With diversity techniques, a frequency- or time-selective channel tends to become an AWGN channel. This improves the power efficiency. However, there are two ways to increase the spectral efficiency. The first one, which is the trivial way, is to increase the symbol alphabet size and the second one is to transmit different symbols in parallel in space by using the MIMO properties.

The capacity of MIMO channels for an uncoded system in flat fading channels with perfect channel knowledge at the receiver is calculated by Foschini [12] as

$$C = \log_2 \left[\det \left(I_L + \frac{E_s/N_o}{M} h(t) h^{*T}(t) \right) \right],$$ (6.2)

where 'det' means determinant, I_L is an $L \times L$ identity matrix, and $(\cdot)^{*T}$ is the conjugate complex of the transpose matrix. Note that this formula is based on the Shannon capacity calculation for a simple AWGN channel.

Two approaches exist to exploit the capacity in MIMO channels. The information theory shows that with M transmit antennas and $L = M$ receive antennas, M independent data streams can be simultaneously transmitted, hence reaching the channel capacity. As an example, the BLAST (Bell-Labs layered space time) architecture can be referred to [12, 28]. Another approach is to use an MISO scheme to obtain diversity, where in this case sophisticated techniques such as space–time coding (STC) can be realized. All transmit signals occupy the same bandwidth, but they are constructed such that the receiver can exploit spatial diversity, as in the Alamouti scheme [2]. The main advantage of STC, especially for mobile communications, is that they do not require multiple receive antennas.

6.2.1 BLAST Architecture

The basic concept of the BLAST architecture is to increase channel capacity by increasing the data rate through simultaneous transmission of independent data streams over M transmit antennas. In this architecture, the number of receive antennas should at least be equal to the number of transmit antennas $L \geq M$ (see Figure 6-1).

For m-array modulation, the receiver has to choose the most likely out of m^M possible signals in each symbol time interval. Therefore, the receiver complexity grows exponentially with the number of modulation constellation points and the number of transmit antennas. Consequently, sub-optimum detection techniques such as those proposed in BLAST can be applied. Here, in each step only the signal transmitted from a single antenna is detected, whereas the transmitted signals from the other antennas are canceled using the previously detected signals or suppressed by means of zero-forcing or MMSE equalization.

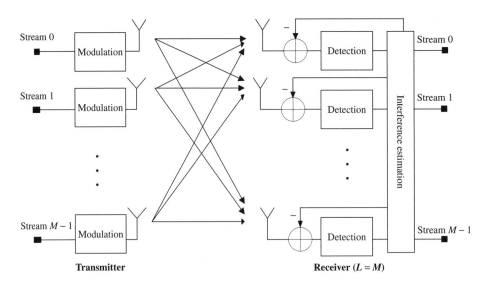

Figure 6-2   V-BLAST transceiver

Two basic variants of BLAST are proposed [12, 28]: D-BLAST (diagonal BLAST) and V-BLAST (vertical BLAST). The only difference is that in V-BLAST transmit antenna m corresponds all the time to the transmitted data stream m, while in D-BLAST the assignment of the antenna to the transmitted data stream is hopped periodically. If the channel does not vary during transmission, in V-BLAST the different data streams may suffer from asymmetrical performance. Furthermore, in general the BLAST performance is limited due to the error propagation issued by the multi-stage decoding process.

As illustrated in Figure 6-2, for detection of data stream 0, the signals transmitted from all other antennas are estimated and suppressed from the received signal of the data stream 0. In References [3] and [4] an iterative decoding process for the BLAST architecture is proposed, which outperforms the classical approach.

However, the main disadvantage of the BLAST architecture for mobile communications is the need for high numbers of receive antennas, which is not practical in a small mobile terminal. Furthermore, high system complexity may prohibit the large-scale implementation of such a scheme.

6.2.2 Space–Time Coding

An alternative approach is to obtain transmit diversity with M transmit antennas, where the number of received antennas is not necessarily equal to the number of transmit antennas. Even with one receive antenna the system operates. This approach is more suitable for mobile communications.

The basic philosophy with STC is different from the BLAST architecture. Instead of transmitting independent data streams, the same data stream is transmitted in an appropriate manner over all antennas. This could be, for instance, a downlink mobile communication, where in the base station M transmit antennas are used while in the terminal station only one or few antennas might be applied.

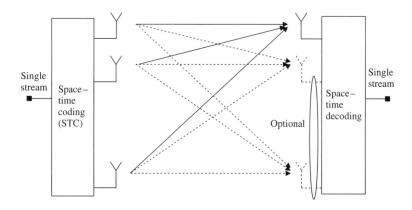

Figure 6-3 General principle of space–time coding (STC)

The principle of STC is illustrated in Figure 6-3. The basic idea is to provide through coding *constructive superposition* of the signals transmitted from different antennas. Constructive combining can be achieved, for instance, by modulation diversity, where orthogonal pulses are used in different transmit antennas. The receiver uses the respective matched filters, where the contributions of all transmit antennas can be separated and combined with MRC.

The simplest form of modulation diversity is delay diversity, a special form of space–time trellis codes. The other alternative of STC are space–time block codes. Both spatial coding schemes are described in the following.

6.2.2.1 Space–Time Trellis Code (STTC)

The simplest form of STTCs is the delay diversity technique (see Figure 6-4). The idea is to transmit the same symbol with a delay of iT_s from transmit antenna $i = 0, \ldots, M - 1$.

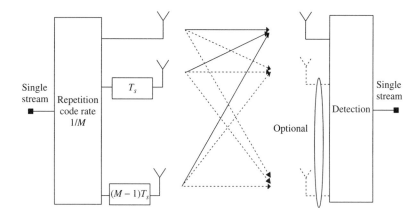

Figure 6-4 Space–time trellis code with delay diversity technique

The delay diversity can be viewed as a rate $1/M$ repetition code. The detector could be a standard equalizer. Replacing the repetition code by a more powerful code, additional coding gain on top of the diversity advantage can be obtained [23]. However, there is no general rule of how to obtain good space–time trellis codes for arbitrary numbers of transmit antennas and modulation methods. Powerful STTCs are given in Reference [25] and obtained from an exhaustive search. However, the problem of STTCs is that the detection complexity measured in the number of states grows exponentially with m^M.

In Figure 6-5, an example of an STTC for two transmit antennas $M = 2$ in the case of QPSK $m = 2$ is given. This code has four states with spectral efficiency of 2 bit/s/Hz. Assuming ideal channel estimation, the decoding of this code at the receive antenna j can be performed by minimizing the following metric:

$$D = \sum_{j=0}^{L-1} \left| r_j - \sum_{i=0}^{M-1} h_{i,j} x_i \right|^2, \tag{6.3}$$

where r_j is the received signal at receive antenna j and x_i is the branch metric in the transition of the encoder trellis. Here, the Viterbi algorithm can be used to choose the best path with the lowest accumulated metric. The results in Reference [25] show the coding advantages obtained by increasing the number of states as the number of received antennas is increased.

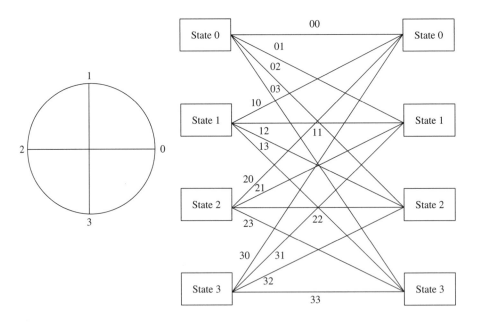

Figure 6-5   Space–time trellis code with four states

6.2.2.2 Space–Time Block Code (STBC)

A simple transmit diversity scheme for two transmit antennas using STBCs was introduced by Alamouti in Reference [2] and generalized to an arbitrary number of antennas by Tarokh *et al.* [24]. Basically, STBCs are designed as pure diversity schemes and provide no additional coding gain as with STTCs. In the simplest Alamouti scheme with $M = 2$ antennas, the transmitted symbols x_i are mapped to the transmit antenna with the mapping

$$B = \begin{bmatrix} x_0 & x_1 \\ -x_1^* & x_0^* \end{bmatrix}, \qquad (6.4)$$

where the row corresponds to the time index and the column to the transmit antenna index. In the first symbol time interval x_0 is transmitted from antenna 0 and x_1 is transmitted from antenna 1 simultaneously, while in the second symbol time interval antenna 0 transmits $-x_1^*$ and simultaneously antenna 1 transmits x_0^*.

The coding rate of this STBC is one, meaning that no bandwidth expansion takes place (see Figure 6-6). Due to the orthogonality of the space–time block codes, the symbols can be separated at the receiver by a simple linear combining (see Figure 6-7). The spatial diversity combining with block codes applied for multi-carrier transmission is described in more detail in Section 6.3.4.

6.2.3 Achievable Capacity

For STBCs of rate R the channel capacity is given by [3]

$$C = R \log_2 \left[\left(1 + \frac{E_s/N_o}{M} \sum_{i=0}^{M-1} \sum_{j=0}^{L-1} |h_{i,j}|^2 \right) \right]. \qquad (6.5)$$

For $R = 1$ and $L = 1$, this is equivalent to the channel capacity of a MISO scheme. However, for $L > 1$, the capacity curve is only shifted, but the asymptotic slope is not

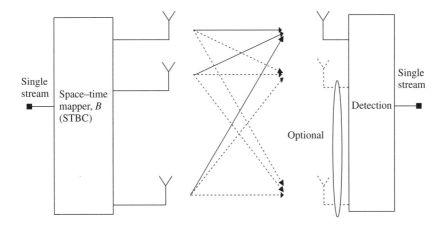

Figure 6-6   Space–time block code transceiver

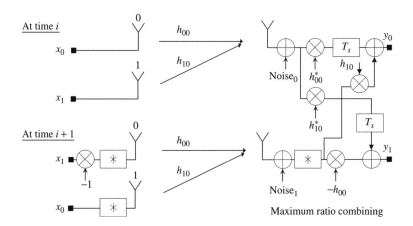

Figure 6-7   Principle of space–time block coding

increased, therefore, the MIMO capacity will not be achieved [4]. This also corresponds to results for STTCs.

From an information theoretical point of view it can be concluded that STCs should be used in systems with $L = 1$ receive antennas. If multiple receive antennas are available, the data rate can be increased by transmitting independent data from different antennas as in the BLAST architecture.

6.3   Diversity Techniques for Multi-Carrier Transmission

6.3.1   Transmit Diversity

Several techniques to achieve spatial transmit diversity in OFDM systems are discussed in this section. The number of used transmit antennas is M. OFDM is realized by an IFFT and the OFDM blocks shown in the following figures also include a frequency interleaver and a guard interval insertion/removal. It is important to note that the total transmit power Ω is the sum of the transmit power Ω_m of each antenna, i.e.

$$\Omega = \sum_{m=0}^{M-1} \Omega_m. \tag{6.6}$$

In the case of equal transmit power per antenna, the power per antenna is

$$\Omega_m = \frac{\Omega}{M}. \tag{6.7}$$

6.3.1.1   Delay Diversity

As discussed before, the principle of delay diversity (DD) is to increase the frequency selectivity of the mobile radio channel artificially by introducing additional constructive

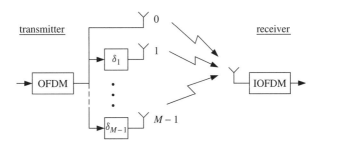

Figure 6-8   Delay diversity

delayed signals. Delay diversity can be considered a simple form of STTCs. Increased frequency selectivity can enable a better exploitation of diversity, which results in an improved system performance. With delay diversity, the multi-carrier modulated signal itself is identical on all M transmit antennas and differs only in an antenna-specific delay $\delta_m, m = 1, \ldots, M - 1$ [20]. The block diagram of an OFDM system with spatial transmit diversity applying delay diversity is shown in Figure 6-8.

In order to achieve frequency selective fading within the transmission bandwidth B, the delay has to fulfill the condition

$$\delta_m \geq \frac{1}{B}. \tag{6.8}$$

To increase the frequency diversity by multiple transmit antennas, the delay of the different antennas should be chosen as

$$\delta_m \geq \frac{km}{B}, \quad k \geq 1, \tag{6.9}$$

where k is a constant factor introduced for the system design. The factor has to be chosen large enough ($k \geq 1$) in order to guarantee a diversity gain. A factor of $k = 2$ seems to be sufficient to achieve promising performance improvements in most scenarios. This result is verified by the simulation results presented in Section 6.3.1.2.

The disadvantage of delay diversity is that the additional delays $\delta_m, m = 1, \ldots, M - 1$, increase the total delay spread at the receiver antenna and require an extension of the guard interval duration by the maximum $\delta_m, m = 1, \ldots, M - 1$, which reduces the spectral efficiency of the system. This disadvantage can be overcome by phase diversity presented in the next section.

6.3.1.2   Phase Diversity

Phase diversity (PD) transmits signals on M antennas with different phase shifts, where $\Phi_{m,n}, m = 1, \ldots, M - 1, n = 0, \ldots, N_c - 1$, is an antenna- and sub-carrier-specific phase offset [14, 15]. The phase shift is efficiently realized by a phase rotation before OFDM, i.e. before the IFFT. The block diagram of an OFDM system with spatial transmit diversity applying phase diversity is shown in Figure 6-9.

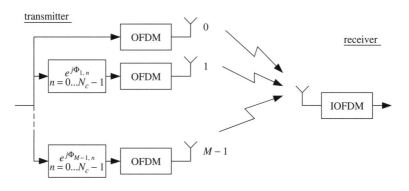

Figure 6-9 Phase diversity

In order to achieve frequency-selective fading within the transmission bandwidth of the N_c sub-channels, the phase $\Phi_{m,n}$ has to fulfill the condition

$$\Phi_{m,n} \geq \frac{2\pi f_n}{B}$$

$$\geq \frac{2\pi n}{N_c}, \tag{6.10}$$

where $f_n = n/T_s$ is the nth sub-carrier frequency, T_s is the OFDM symbol duration without a guard interval, and $B = N_c/T_s$. To increase the frequency diversity by multiple transmit antennas, the phase offset of the nth sub-carrier at the mth antenna should be chosen as

$$\Phi_{m,n} = \frac{2\pi\, kmn}{N_c}, \quad k \geq 1, \tag{6.11}$$

where k is a constant factor introduced for the system design that has to be chosen large enough ($k \geq 1$) to guarantee a diversity gain. The constant k corresponds to k introduced in Section 6.3.1.1. Since no delay of the signals at the transmit antennas occurs with phase diversity, no extension of the guard interval is necessary compared to delay diversity.

In Figure 6-10, the SNR gain to reach a BER of 3×10^{-4} with two transmit antennas applying delay diversity and phase diversity compared to a one transmit antenna scheme over the parameter k introduced in Equations (6.9) and (6.11) is shown for OFDM and OFDM-CDM. The results are presented for an indoor and outdoor scenario. The performance of delay diversity and phase diversity is the same for the chosen system parameters, since the guard interval duration exceeds the maximum delay of the channel and the additional delay due to delay diversity. The curves show that gains of more than 5 dB in the indoor scenario and of about 2 dB in the outdoor scenario can be achieved for $k \geq 2$ and justify the selection of $k = 2$ as a reasonable value. It is interesting to observe that even in an outdoor environment, which already has frequency-selective fading, significant performance improvements can be achieved.

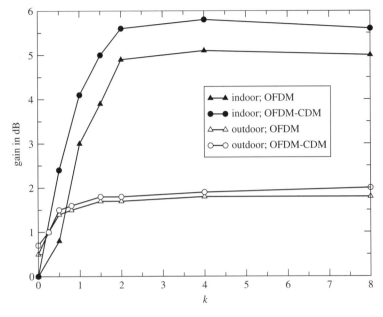

Figure 6-10   Performance gains with delay diversity and phase diversity over k: $M = 2$; BER $= 3 \times 10^{-4}$

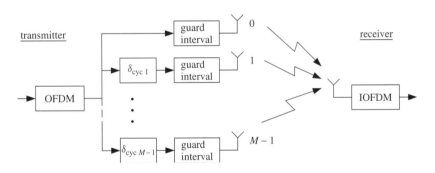

Figure 6-11   Cyclic delay diversity (CDD)

6.3.1.3   Cyclic Delay Diversity (CDD)

An efficient implementation of phase diversity is cyclic delay diversity (CDD) [7], which instead of M OFDM operations requires only one OFDM operation in the transmitter. The signals constructed by phase diversity and by cyclic delay diversity are equal. Signal generation with cyclic delay diversity is illustrated in Figure 6-11. With cyclic delay diversity, $\delta_{cycl\ m}$ denotes cyclic shifts [8]. Both phase diversity and cyclic delay diversity are performed before guard interval insertion. CDD is, for example, applied in LTE (see Chapter 5).

6.3.1.4   Time-Variant Phase Diversity/Doppler Diversity

The spatial transmit diversity concepts presented in the previous sections introduce only frequency diversity. Time-variant phase diversity (TPD), also referred to as Doppler diversity, can additionally exploit time diversity. It can be used to introduce time diversity or to introduce both time and frequency diversity. The block diagram shown in Figure 6-9 is still valid, only the phase offsets $\Phi_{m,n}$ have to be replaced by the time-variant phase offsets $\Theta_{m,n}(t)$, $m = 1, \ldots, M - 1$, $n = 0, \ldots, N_c - 1$, which are given by [15]

$$\Theta_{m,n}(t) = \Phi_{m,n} + 2\pi t F_m. \tag{6.12}$$

The frequency shift F_m at transmit antenna m has to be chosen such that the channel can be considered as time-invariant during one OFDM symbol duration, but appears time-variant over several OFDM symbols. It has to be taken into account in the system design that the frequency shift F_m introduces ICI, which increases with increasing F_m.

The gain in SNR to reach the BER of 3×10^{-4} with two transmit antennas applying time-variant phase diversity compared to time-invariant phase diversity with two transmit antennas over the frequency shift F_1 is shown in Figure 6-12. The frequency shifts F_m should be less than a few percent of the sub-carrier spacing to avoid non-negligible degradations due to ICI.

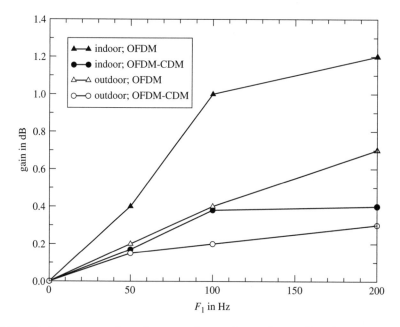

Figure 6-12   Performance gains due to time-variant phase diversity: $M = 2$; $k = 2$; BER $= 3 \times 10^{-4}$

6.3.1.5 Discontinuous Doppler Diversity (DDD)

The introduction of ICI with time-variant phase diversity/Doppler diversity can be avoided by applying discontinuous Doppler diversity (DDD). The principle of DDD is to quantize the time-variant phase offset $\Theta_{m,n}(t)$ such that it is constant during one OFDM symbol but changes between adjacent OFDM symbols [9]. This guarantees the same time diversity as introduced with TPD but avoids completely the ICI. The increased time variance of the superimposed channel at the receiver antenna has to be taken into account in the system design in the same way as with TPD. This holds especially for the pilot grid design used for channel estimation.

6.3.1.6 Sub-Carrier Diversity

With sub-carrier diversity (SCD), the sub-carriers used for OFDM are clustered in M smaller blocks and each block is transmitted over a separate antenna [6]. The principle of sub-carrier diversity is shown in Figure 6-13.

After serial-to-parallel (S/P) conversion, each OFDM block processes N_c/M complex-valued data symbols out of a sequence of N_c. Each of the M OFDM blocks maps its N_c/M data symbols on to its exclusively assigned set of sub-carriers. The sub-carriers of one block should be spread over the entire transmission bandwidth in order to increase the frequency diversity per block, i.e. the sub-carriers of the individual blocks should be interleaved.

The advantage of sub-carrier diversity is that the peak-to-average power ratio per transmit antenna is reduced compared to a single antenna implementation since there are fewer sub-channels per transmit antenna.

6.3.2 Receive Diversity

6.3.2.1 Maximum Ratio Combining (MRC)

The signals at the output of the L receive antennas are combined linearly so that the SNR is maximized. The optimum weighting coefficient is the conjugate complex of the assigned channel coefficient as illustrated in Figure 6-14.

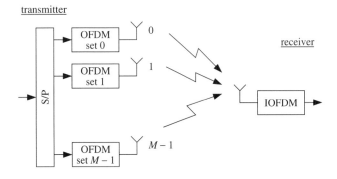

Figure 6-13   Sub-carrier diversity

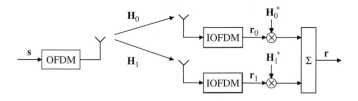

Figure 6-14   OFDM with MRC receiver; $L = 2$

With the received signals

$$\mathbf{r}_0 = \mathbf{H}_0 \mathbf{s} + \mathbf{n}_0,$$
$$\mathbf{r}_1 = \mathbf{H}_1 \mathbf{s} + \mathbf{n}_1 \tag{6.13}$$

the diversity gain achievable with MRC can be observed as follows:

$$\mathbf{r} = \mathbf{H}_0^* \mathbf{r}_0 + \mathbf{H}_1^* \mathbf{r}_1$$
$$= (|\mathbf{H}_0|^2 + |\mathbf{H}_1|^2)\mathbf{s} + \mathbf{H}_0^* \mathbf{n}_0 + \mathbf{H}_1^* \mathbf{n}_1. \tag{6.14}$$

6.3.2.2 Delay and Phase Diversity

The transmit diversity techniques delay, phase, and time-variant phase diversity presented in Section 6.3.1 can also be applied in the receiver, achieving the same diversity gains plus an additional gain due to the collection of the signal power from multiple receive antennas. A receiver with phase diversity is shown in Figure 6-15.

6.3.3 Transmit/Receive Diversity Performance Analysis

The gain in SNR due to different transmit diversity techniques to reach the BER of 3×10^{-4} with M transmit antennas compared to one transmit antenna over the number of antennas M is shown in Figure 6-16. The results are presented for a rate 1/2 coded OFDM system in an indoor environment. Except for sub-carrier diversity without interleaving, promising performance improvements have already been obtained with two transmit antennas. The optimum choice of the number of antennas M is a tradeoff between cost and performance.

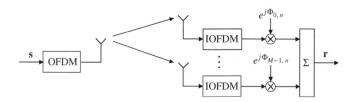

Figure 6-15   Phase diversity at the receiver

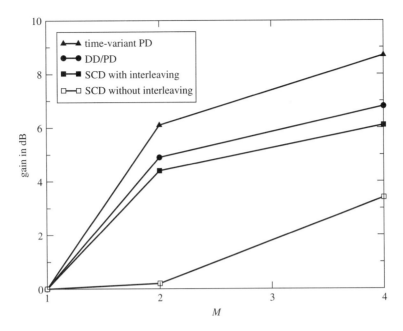

Figure 6-16   Spatial transmit diversity gain over the number of antennas M: $k = 2$; $F_1 = 100\,\text{Hz}$ for time-variant phase diversity; indoor; BER $= 3 \times 10^{-4}$

The BER performance of the presented spatial transmit diversity concepts is shown in Figure 6-17 for an indoor environment with two transmit antennas. Simulation results are shown for coded OFDM and OFDM-CDM systems. The performance of the OFDM system with one transmit antenna is given as a reference. The corresponding simulation results for an outdoor environment are shown in Figure 6-18. It can be observed that time-variant phase diversity outperforms the other investigated transmit diversity schemes and that phase diversity and delay diversity are superior to sub-carrier diversity in the indoor environment. Moreover, the performance can be improved by up to 2 dB with an additional CDM component.

Finally, some general statements should be made about spatial transmit diversity concepts. The disadvantages of spatial transmit diversity concepts are that multiple transmit chains and antennas are required, increasing the system complexity. Moreover, accurate oscillators are required in the transmitters, such that the sub-carrier patterns at the individual transmit antennas fit together and ICI can be avoided. As long as this is done in the base station, e.g. in a broadcasting system or in the downlink of a mobile radio system, the additional complexity is reasonable. Nevertheless, this complexity can also be justified in a mobile transmitter.

The clear advantage of the presented spatial transmit and receive diversity concepts is that significant performance improvements of several dB can be achieved in critical propagation scenarios. The results presented in this chapters for DD and PD are also valid for CDD. Moreover, the presented CDD technique is standard-compliant and can be applied in already standardized systems such as DAB, DVB-T, or IEEE 802.11a.

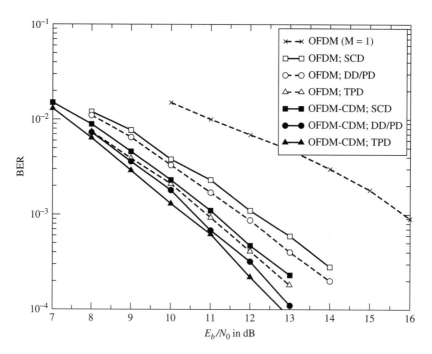

Figure 6-17   BER versus SNR: $M = 2$; $k = 2$; $F_1 = 100\,\text{Hz}$ for time-variant phase diversity; indoor

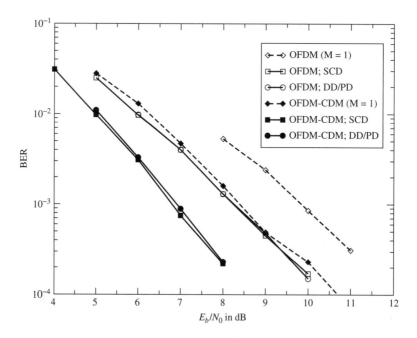

Figure 6-18   BER versus SNR: $M = 2$; $k = 2$; outdoor

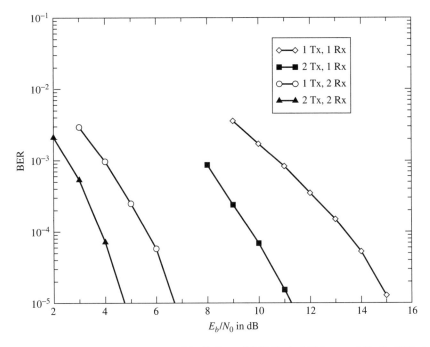

Figure 6-19   Performance of DVB-T with CDD and MRC: 4-QAM; rate 1/2; 2k FFT; indoor; $f_D = 10\,\text{Hz}$

Transmit and receive diversity techniques can easily be combined. Performance improvements with CDD in the transmitter and MRC in the receiver are shown for a DVB-T transmission in Figure 6-19 [8]. The chosen DVB-T parameters are 4-QAM, code rate 1/2, and 2k FFT in an indoor environment with maximum Doppler frequency of 10 Hz. A single antenna system is given as reference.

Further performance improvements can be obtained by combining transmit and receive diversity techniques with beamforming [10]. Beamforming reduces interference within a propagation environment and can efficiently cancel interference. Since the channel of each beam has a small delay spread, the channel appears nearly flat. The diversity techniques presented in Sections 6.3.1 and 6.3.2 can artificially introduce frequency- and time-selectivity and, thus, improve the performance.

6.3.4 Space–Frequency Block Codes (SFBC)

Transmit antenna diversity in the form of space–time block codes exploits time and space diversity and achieves a maximum diversity gain for two transmit antennas without rate loss. They have to be applied under the assumption that the channel coefficients remain constant for two subsequent symbol durations in order to guarantee the diversity gain. This is a tough precondition in OFDM systems, where the OFDM symbol duration T_s is N_c times the duration of a serial data symbols T_d. To overcome the necessity of using two successive OFDM symbols for coding, symbols belonging together can be sent on different

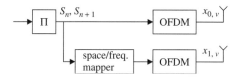

Figure 6-20   Space–frequency block coding in an OFDM transmitter

sub-carriers in multi-carrier systems. The feature of OFDM can be exploited, that two adjacent narrowband sub-channels are affected by almost the same channel coefficients. Thus, a space–frequency block code (SFBC) requires only the reception of one OFDM symbol for detection, avoids problems with coherence time restrictions, and reduces delay in the detection process.

Figure 6-20 shows an OFDM transmitter with space–frequency block coding. Sequences of N_c interleaved data symbols S_n are transmitted in one OFDM symbol. The data symbols are interleaved by the block Π before space–frequency block coding such that the data symbols combined with space–frequency mapping and, thus, affected by the same fading coefficient are not subsequent data symbols in the original data stream. The interleaver with size I performs frequency interleaving for $I \leq N_c$ and time and frequency interleaving for $I > N_c$.

The mapping scheme of the data symbols S_n for SFBC with two transmit antennas and code rate 1 is shown in Table 6-1. The mapping scheme for SFBC is chosen such that on the first antenna the original data are transmitted without any modification. Thus, the mapping of the data symbols on the sub-carriers for the first antenna corresponds to the classical inverse discrete Fourier transform with

$$x_{0,v} = \frac{1}{N_c} \sum_{n=0}^{N_c-1} S_n e^{j2\pi nv/N_c}, \tag{6.15}$$

where n is the sub-carrier index and v is the sample index of the time signal. Only the data symbol mapping on the second antenna has to be modified according to the mapping scheme for space–frequency block coding given in Table 6-1. The data symbols of the second transmit antenna are mapped on the sub-carriers as follows:

$$x_{1,v} = \frac{1}{N_c} \sum_{n=0}^{N_c/2-1} (S_{2n}^* e^{j2\pi(2n+1)v/N_c} - S_{2n+1}^* e^{j2\pi 2nv/N_c}). \tag{6.16}$$

Table 6-1   Mapping with space–frequency block codes and two transmit antennas

Sub-carrier number	Antenna 1	Antenna 2
n	S_n	$-S_{n+1}^*$
$n+1$	S_{n+1}	S_n^*

Figure 6-21   Space–frequency block decoding in an OFDM receiver

The OFDM block comprises inverse fast Fourier transform (IFFT) and cyclic extension of an OFDM symbol. A receiver with inverse OFDM operation and space–frequency block decoding is shown in Figure 6-21. The received signals on sub-channels n and $n + 1$ after guard interval removal and fast Fourier transform (FFT) can be written as

$$R_n = H_{0,n} S_n - H_{1,n} S_{n+1}^* + N_n,$$
$$R_{n+1} = H_{0,n+1} S_{n+1} + H_{1,n+1} S_n^* + N_{n+1}, \tag{6.17}$$

where $H_{m,n}$ is the flat fading coefficient of sub-channel n assigned to transmit antenna m and N_n is the additive noise on sub-carrier n. OFDM systems are designed such that the fading per sub-channel can be considered as flat, from which it can be concluded that the fading between adjacent sub-carriers can be considered as flat as well, and $H_{m,n}$ can be assumed to be equal to $H_{m,n+1}$. Thus, when focusing the analysis on an arbitrary pair of adjacent sub-channels n and $n + 1$, we can write H_m as the flat fading coefficient assigned to the pair of sub-channels n and $n + 1$ and to transmit antenna m. After using the combining scheme

$$\hat{R}_n = H_0^* R_n - H_1 R_{n+1}^*,$$
$$\hat{R}_{n+1} = -H_1 R_n^* + H_0^* R_{n+1}, \tag{6.18}$$

the received signals result in

$$\hat{R}_n = (|H_0|^2 + |H_1|^2) S_n + H_0^* N_n + H_1 N_{n+1}^*,$$
$$\hat{R}_{n+1} = (|H_0|^2 + |H_1|^2) S_{n+1} - H_1 N_n^* + H_0^* N_{n+1}. \tag{6.19}$$

After de-interleaving Π^{-1}, symbol detection, and de-mapping, the soft decided bit w is obtained, which in the case of channel coding is fed to the channel decoder. Optimum soft decision channel decoding is guaranteed when using log-likelihood ratios (LLRs) in the Viterbi decoder. The normalized LLRs for space–frequency block coded systems result in

$$\Gamma = \frac{2\sqrt{|H_0|^2 + |H_1|^2}}{\sigma^2} w. \tag{6.20}$$

6.3.5   SFBC Performance Analysis

In this section, the performance of a classical OFDM system, a space–frequency block coded OFDM system, an OFDM-CDM system, and combinations of these are compared in a Rayleigh fading channel. It is important to note that the performance of OFDM-CDM is comparable to the performance of a fully loaded MC-CDMA scheme in the downlink and that the performance of OFDM is comparable to the performance of OFDMA or MC-TDMA with perfect interleaving. The transmission bandwidth of the systems is $B = 2\,\text{MHz}$ and the carrier frequency is located at $2\,\text{GHz}$. The number of sub-carriers is 512. The guard interval duration is $20\,\mu s$. As channel codes, punctured convolutional codes with memory 6 and variable code rate R between 1/3 and 4/5 are applied. QPSK is used for symbol mapping. The depth of the interleaver Π is equal to 24 subsequent OFDM symbols, so that time and frequency interleaving is applied. SFBC is realized with $M = 2$ transmit antennas. There is no correlation between the antennas in the space–frequency block coded OFDM system. In the case of OFDM-CDM, short Hadamard codes of length $L = 8$ are applied for spreading. The mobile radio channel is modeled as an uncorrelated Rayleigh fading channel, assuming perfect interleaving in the frequency and time direction.

In Figure 6-22, the BER versus the SNR per bit is shown for space–frequency block coding in OFDM and OFDM-CDM systems. The results are presented for OFDM-CDM with different single-symbol detection techniques. No FEC coding is used. The corresponding results without SFBC are shown in Figure 2-14.

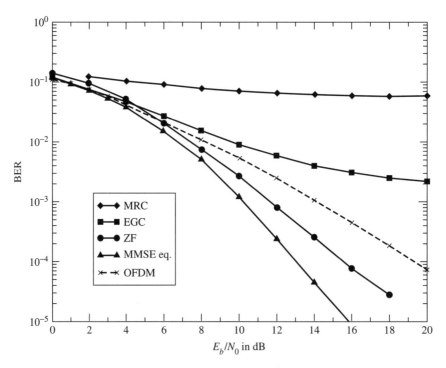

Figure 6-22 BER versus SNR for space–frequency block coding applied to OFDM and OFDM-CDM with different single-symbol detection techniques: no FEC coding

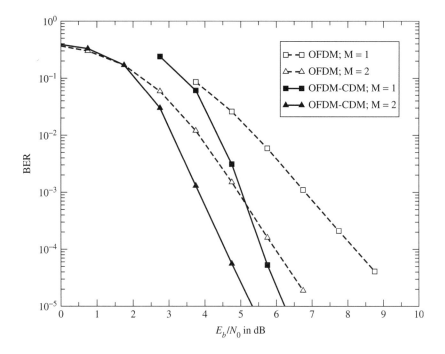

Figure 6-23   BER versus SNR for different OFDM schemes: code rate $R = 2/3$

The BER performance of various OFDM-based diversity schemes with and without SFBC is shown in Figure 6-23. The channel code rate is $R = 2/3$. OFDM-CDM applies soft interference cancellation with one iteration at the receiver.

In Figure 6-24, the gain, i.e. SNR reduction, with SFBC OFDM-CDM compared to SFBC OFDM is shown versus the channel code rate for a BER of 10^{-5}. Additionally, OFDM-CDM without SFBC is compared to SFBC OFDM. In the case of OFDM-CDM, soft interference cancellation with one iteration is applied. It can be observed that the gains due to CDM increase with an increasing code rate. This result shows that the weaker the channel code, the more diversity can be exploited by CDM.

6.4   Spatial Pre-Coding for Multi-Carrier Transmission

Spatial pre-coding schemes exploit knowledge about the channel state information at the transmitter. The signals at the multiple transmit antennas are pre-coded such that they improve the performance at the receiver. The channel knowledge available at the transmitter is typically provided by a feedback channel from the receiver to the transmitter. The spatial pre-coding schemes described in this section differ in the amount of channel knowledge required at the transmitter. An optimized system design has to find the appropriate trade-off between the amount of channel knowledge required and the performance. With increased channel knowledge the overhead for channel estimation and the feedback channel increase.

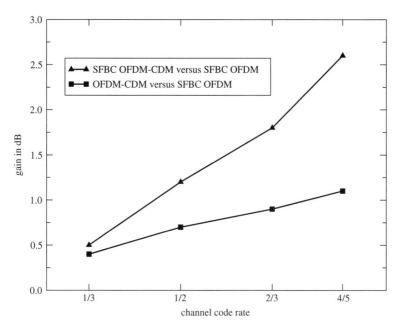

Figure 6-24   Gain with OFDM-CDM compared to OFDM with space–frequency block coding in dB versus channel code rate R: BER $= 10^{-5}$

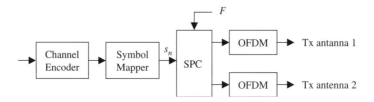

Figure 6-25   Transmitter of an OFDM system with spatial pre-coding with two Tx antennas

The block diagram of an OFDM transmitter with spatial pre-coding and two transmit antennas is shown in Figure 6-25. The spatial pre-coder block exploits the feedback information F from the receiver about the actual channel. This can be full or partial channel state information, depending on the chosen pre-coding scheme. The number of sub-carriers in one OFDM symbol is given by N_c. A data symbol on sub-carrier n is represented by s_n, $n = 0, \ldots, N_c - 1$. After spatial pre-coding and OFDM, the signal s_n is transmitted over the channel exploiting M transmit antennas. In this section, the spatial pre-coding schemes are described and analyzed for the case with $M = 2$ transmit antennas. The signals transmitted on the two antennas are $s_n^{(1)}$ and $s_n^{(2)}$, where $^{(m)}$ is the antenna index $m = 1, 2$. The pre-coding is defined as

$$\mathbf{s}_n = s_n w_n \mathbf{c}_n = (s_n^{(1)}, s_n^{(2)})^T, \tag{6.21}$$

where $\mathbf{c}_n$ is the spatial pre-coding vector and w_n the power Normalization factor.

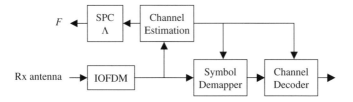

Figure 6-26   Receiver of an OFDM system with spatial pre-coding with a single Rx antenna

The channels from the two transmit antennas to the receive antenna are given by the complex-valued fading coefficients $H_n^{(1)}$ and $H_n^{(2)}$ respectively. The superimposed channel at the receiver antenna is

$$H_n = H_n^{(1)} + H_n^{(2)}. \tag{6.22}$$

The block diagram of the receiver is shown in Figure 6-26. The receiver can have multiple receive antennas. For simplification but without loss of generality, the case with one receiver antenna is analyzed in the following. At the receiver antenna, the signals from the different transmit antennas superimpose. Depending on the spatial pre-coding scheme, the receiver has to estimate all channels from the M transmit antennas to the receiver antenna or only the superimposed channel coefficient H_n.

The spatial pre-coding schemes are presented with increasing complexity, i.e. spatial phase coding (SPC), selection diversity (SD), equal gain transmission (EGT), and maximum ratio transmission (MRT) [13, 16, 18, 22]. The least complex pre-coding scheme is SPC, which requires only the estimation of the superimposed channel H_n at the receiver and a 1 bit feedback F. The most complex is MRT, which requires the estimation of the channels from each transmit antenna to the receive antenna and a feedback that provides this information to the transmitter.

6.4.1   Spatial Phase Coding (SPC)

The principle of SPC is to achieve a constructive superposition of the signals from the different transmit antennas at the receiver antenna without the necessity to estimate the two channels from the two transmit antennas to the receive antenna. Only one channel has to be estimated, which is the superimposed channel H_n.

By comparing the absolute value of the superimposed channel $|H_n|$ with a predefined threshold Λ, SPC can detect a destructive superposition at the receiver with high probability and indicates to the transmitter over a feedback channel that the phase relation of the transmitted signals should in this case be changed by π. A straightforward solution is to flip the phase of the signal at one transmit antenna by π. Thus in the subsequent transmission the channels superimpose constructively at the receive antenna. The assumption is that the phase relation between both channels is quasi constant between subsequent transmissions. This is a typical assumption in OFDM systems. If due to variations of the channel over the time $|H_n|$ falls below the threshold Λ again, the phase will be flipped again.

The spatial phase encoder has two possible states. These are termed state A and state B. Depending on the actual state, the data symbols s_n to be transmitted are pre-coded in different ways. The SPC vector $\mathbf{c}_n$ for the two states is defined as [16]

$$\mathbf{c}_n = \begin{cases} (1, 1)^T & \text{State A} \\ (1, e^{-j\pi})^T & \text{State B} \end{cases} .$$

(6.23)

The property of SPC is that the receiver evaluates the received signal and decides if the SPC should remain in the actual state or should perform a state change. The receiver does not need to know the actual state of the spatial phase pre-coder. It is sufficient to evaluate a predefined flipping criterion and based on this to indicate to the transmitter via a feedback channel that it needs to flip the phase on one antenna compared to the previous transmission, i.e. to perform a state change, or to remain in the actual state. The criterion for a state change is define as follows [16]:

$$F = \begin{cases} \text{no state change} & |H_n| \geq \Lambda \\ \text{state change} & |H_n| < \Lambda \end{cases} ,$$

(6.24)

where F is the feedback information from the receiver to the transmitter indicating if a state change is necessary or not. The principle of SPC is shown in Figure 6-27. Figure 6-27(a) shows $H_n^{(1)}$ and $H_n^{(2)}$ with a relative angle α_n such that $|\alpha_n| < \pi/2$ where they superimpose constructively. In Figure 6-27(b) the relative angle α_n between both channels is such that $\pi/2 < |\alpha_n| < \pi$ where $H_n^{(1)}$ and $H_n^{(2)}$ superimpose destructively before spatial phase pre-coding and constructively after spatial phase pre-coding.

Since it is only a binary feedback, 1 bit is sufficient for the feedback signal F. The performance of SPC depends on the choice of the threshold Λ. With SPC using two Tx antennas the power normalization factor w_n results in

$$w_n = \frac{1}{\sqrt{2}}.$$

(6.25)

This normalizes the overall transmit power proportional to the number of transmit antennas.

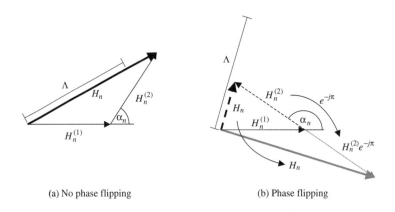

(a) No phase flipping   (b) Phase flipping

Figure 6-27   Principle of spatial phase coding

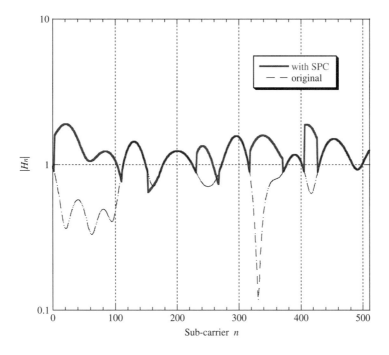

Figure 6-28   COST 207 typical urban channel with and without spatial phase coding

Exemplarily the effects of SPC on the channel transfer function for an OFDM system with 512 sub-carriers and two transmit antennas applying the COST 207 typical urban (TU) channel are shown in Figure 6-28. A snapshot of the absolute value of the superimposed channel coefficient $|H_n|$ is plotted over the 512 sub-carriers. It has to be mentioned that the channel $|H_n|$ without SPC (original) is normalized to $E\{|H_n|^2\} = 1$. The thin dashed line shows the original channel coefficients while the thick solid line shows $|H_n|$ after SPC. The threshold Λ is chosen equal to 0.9. It can be observed that in cases where the original channel is in a deep fade, the channel after SPC often is even enhancing the transmitted signal.

6.4.2   Selection Diversity (SD)

With SD the signal is transmitted over the antenna where the absolute value of the individual channel is largest, i.e.

$$\mathbf{c}_n = \begin{cases} (1, 0)^T & \text{if } |H_n^{(1)}| \geq |H_n^{(2)}| \\ (0, 1)^T & \text{otherwise} \end{cases}, \tag{6.26}$$

and the power normalization factor w_n is equal to 1. With SD the receiver has to estimate both channels in order to determine which channel has a larger absolute value and has to indicate this back to the transmitter via a feedback channel. As with SPC, 1 bit is sufficient for the feedback. However, SD needs twice the overhead for channel estimation than SPC since SD needs the estimation of $|H_n^{(1)}|$ and $|H_n^{(2)}|$ instead of $|H_n|$ only.

6.4.3 Equal Gain Transmission (EGT)

EGT is a pre-coding scheme that requires knowledge about the phase of the channels $H_n^{(1)}$ and $H_n^{(2)}$. The signals at the antennas are transmitted with a relative phase shift such that they superimpose constructively at the receiver antenna with $\alpha_n = 0$. Thus the pre-coding vector is

$$\mathbf{c}_n = (1, \, e^{j\alpha_n})^T, \tag{6.27}$$

where α_n is the relative phase difference between $H_n^{(1)}$ and $H_n^{(2)}$. The transmit power is equally split between both antennas, i.e.

$$w_n = \frac{1}{\sqrt{2}}. \tag{6.28}$$

The phase of the channels $H_n^{(1)}$ and $H_n^{(2)}$ has to be estimated at the receiver and the relative phase difference between both channels has to be fed back to the transmitter. EGT is a lower bound for SPC.

6.4.4 Maximum Ratio Transmission (MRT)

The optimum pre-coding scheme is MRT. The pre-coding coefficient of MRT is

$$\mathbf{c}_n = (H_n^{(1)*}, \, H_n^{(2)*})^T. \tag{6.29}$$

Compared to the previous schemes, MRT allows for an unequal distribution of the transmit power between both antennas. The power normalization factor results in

$$w_n = \frac{1}{\sqrt{\sum_{m=1}^{2} |H_n^{(m)}|^2}}. \tag{6.30}$$

The full channel information about $H_n^{(1)}$ and $H_n^{(2)}$ has to be estimated at the receiver. Thus, MRT requires the most complex channel estimation with the highest overhead compared to the previous schemes. Moreover, MRT requires full feedback of the amplitude and phase of each channel, which exceeds the feedback required by the previous schemes.

6.4.5 Performance Analysis

An OFDM system with two transmit antennas and one receive antenna is investigated. The transmission bandwidth is 2 MHz and the carrier frequency is 2 GHz. The multi-carrier modulation is realized by OFDM occupying $N_c = 512$ sub-carriers. An OFDM frame consists of 24 OFDM symbols. The guard interval duration is 5 μs. Results are presented for a rate 1/2 and 2/3 coded transmission using convolutional codes with memory 6 as well as uncoded transmission. QPSK is chosen for symbol mapping. In order to compare the performance differences between pre-coding schemes without side effects, disturbances due to synchronization and feedback errors are omitted by assuming that these components are perfect. The channels from different transmit antennas to the receive antenna are assumed to be uncorrelated. The SNR given in the following results always refers to E_t/N_0, where E_t is the total energy per bit at the transmitter and N_0 is the one-sided noise power spectral density.

6.4.5.1 Lower Bounds

For the performance analysis of different pre-coding schemes three different propagation scenarios are considered. The first environment corresponds to environment A, where $H_n^{(1)}$ and $H_n^{(2)}$ have constant amplitude but random independent phase to each other. It is important to note that the superimposed channel H_n at the receive antenna results in a fading channel due to constructive and destructive superposition of $H_n^{(1)}$ and $H_n^{(2)}$. The second is environment B, where $|H_n^{(1)}|$ and $|H_n^{(2)}|$ are each Rice distributed with a Rice factor of 10; i.e. there is a LOS component and multi-path propagation. The third is environment C, where $|H_n^{(1)}|$ and $|H_n^{(2)}|$ are each Rayleigh distributed, i.e. no LOS component. Each symbol s_n is faded with an independent fading coefficient, assuming perfect interleaving. The symbol mapping is QPSK. Results are presented for an uncoded transmission as well as with rate 1/2 coded transmission using convolutional codes with memory 6. Perfect channel estimation is assumed. As reference the performance of an OFDM system with one transmit antenna (1Tx) is included in the results.

Figure 6-29 presents the BER versus the SNR for different spatial pre-coding schemes in environment A. This enables the array gain to be analyzed since the diversity gain is omitted. The system is uncoded. Since the individual channels have constant amplitude the performance of SD is identical to the performance of the 1Tx scheme. The lower bound is given by EGT and MRT, which perform identically in environment A. The performance of

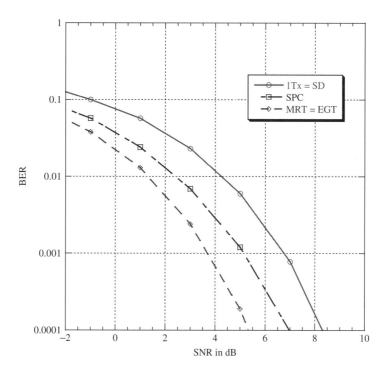

Figure 6-29 Achievable array gain by showing the BER versus SNR for environment A without channel coding

both curves shows the 3 dB array gain compared to the 1Tx scheme, which results from the coherent superposition of the signals from both Tx antennas. For the SPC scheme an array gain of about 1.5 dB compared to the 1Tx scheme can be observed. SD cannot achieve an array gain compared to the SPC schemes. The array gain obtained with SPC is the reason why SPC outperforms SD in channels with LOS, as shown in the next section.

The BER versus SNR in the Ricean fading scenario (environment B) for a rate 1/2 coded transmission is shown in Figure 6-30. SPC outperforms the 1Tx scheme as well as SD. EGT and MRT are shown as lower bounds. QEGT represents 1 bit quantized EGT. Figure 6-31 shows the corresponding results, where $|H_n^{(1)}|$ and $|H_n^{(2)}|$ are Rayleigh distributed (environment C). SPC requires a 0.6 dB higher SNR at a BER of 10^{-4} compared to SD. However, SPC requires less overheads for channel estimation compared to SD since only one channel has to be estimated at the receiver.

6.4.5.2 Effects of Channel Estimation

The system parameters are the same as in the previous section except for the channel estimation, which is no longer perfect. The filtering is performed by two-times one-dimensional Wiener filtering ($2 \times 1D$ filtering) where for the filtering in each dimension five taps are used. The pilot spacing is six sub-carriers in the frequency direction and three

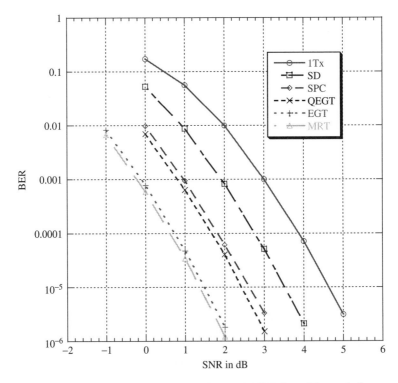

Figure 6-30 BER versus SNR for SPC, SD, EGT, and MRT in a Ricean fading environment (environment B) with channel coding, $\Lambda = 0.9$

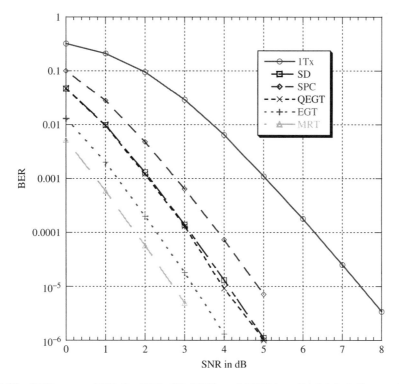

Figure 6-31   BER versus SNR for SPC, SD, EGT, and MRT in a Rayleigh fading environment (environment C) with channel coding, $\Lambda = 0.8$

OFDM symbols in the time direction. As a propagation channel the COST 207 typical urban (TU) channel model is taken. The maximum velocity of the mobile user is 30 km/h, which corresponds to a Doppler frequency of 55.6 Hz. A classical Doppler spectrum is assumed.

In Figure 6-32 the BER versus SNR for the spatial pre-coding schemes with channel estimation errors based on a two-dimensional channel estimation are shown. The same total pilot power is assigned to each scheme. The pre-coding schemes MRT, EGT, and SD need to estimate two channels at the receiver. Since SPC and 1Tx have to estimate only one channel, they have twice the pilot power available for channel estimation per channel compared to MRT, EGT, and SD. The channel estimation accuracy for MRT, EGT, and SD degrades compared to the accuracy with SPC and 1Tx. It can be observed that, when taking imperfection due to channel estimation into account, the low complex SPC scheme outperforms SD and EGT and performs similar as the optimum MRT scheme at BERs $\leq 10^{-4}$ [17].

6.4.5.3   Effects of Spreading

For spreading, Walsh–Hadamard codes of length $L = 8$ are used and eight data symbols are superimposed, i.e. code division multiplexing (CDM) is applied. For data detection

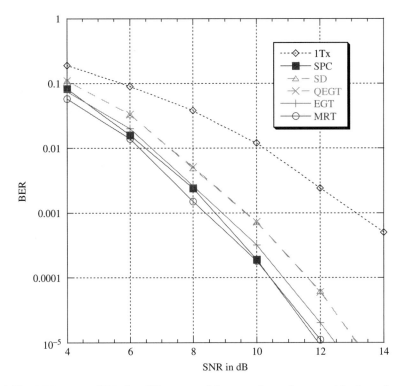

Figure 6-32 BER versus SNR for different spatial pre-coding schemes with channel estimation errors; same total pilot power per scheme

Table 6-2 Gain with CDM compared to OFDM without spreading for different spatial diversity and pre-coding schemes at a BER $= 10^{-5}$

	1Tx	CDD	SPC	EGT	MRT
Gain with CDM	1.7 dB	3.1 dB	1.5 dB	1.4 dB	1.4 dB

soft IC with one iteration is applied. The parameters are the same as in Section 6.4.5.2 except that the channel estimation is perfect.

Table 6-2 shows the gain achievable with CDM in OFDM systems for different spatial diversity and pre-coding schemes at a BER of 10^{-5}. It can be observed that OFDM-CDM outperforms OFDM without spreading in all investigated scenarios. The highest performance improvements can be obtained by combining CDM with CDD. The reason is that CDD increases the fading ratio of the channel, which can be better exploited by CDM. The other spatial pre-coding schemes result in a more flat fading channel at the Rx antenna. If the channel would be completely flat, OFDM-CDM and OFDM would perform identically.

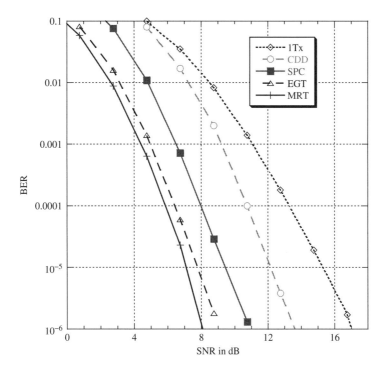

Figure 6-33   BER versus SNR for OFDM-CDM with CDD, SPC, EGT, and MRT: COST 207 TU channel, fully loaded, $R = 2/3$, QPSK, $\Lambda = 0.4$

The performance of OFDM-CDM with CDD, SPC, EGT, and MRT is shown in Figure 6-33. The system is fully loaded. The results show that CDD achieves a significant performance improvement of about 3 dB at a BER of 10^{-5} compared to the 1Tx scheme. Further investigations on OFDM-CDM with spatial diversity can be found in Reference [19].

6.5   Software-Defined Radio

The transmission rate for future wireless systems may vary from low rate messages up to very high rate data services up to 1 Gbit/s. The communication channel may change in terms of its grade of mobility, the cellular infrastructure, the required symmetrical or asymmetrical transmission capacity, and whether it is indoor or outdoor. Hence, air interfaces with high flexibility are required in order to maximize the area spectrum efficiency in a variety of communication environments. Future systems are also expected to support various types of services based on the IP protocol, which require a varying quality of services (QoS).

Recent advances in digital technology enable the faster introduction of new standards that benefit from the most advanced physical (PHY) and data link control (DLC) layers (see Table 6-3). These trends are still growing and new standards or their enhancements are being added continuously to the existing network infrastructures. As explained in

Table 6-3   Examples of current wireless communication standards

Mobile communication systems		Wireless LAN/WLL	
CDMA based	TDMA based	Multi-carrier or CDMA based	Non-MC, non-CDMA based
IS-95/-B: digital cellular standard in the USA	*GSM*: global system for mobile communications	*IEEE. 802.11a/b*: WLAN based on OFDM/CDMA	*DECT*: digital enhanced cordless telecommunications
WCDMA/UMTS: wideband CDMA	*PDC*: personal digital cellular system	*HIPERMAN*: WLL based on OFDM	*HIPERACCESS*: WLL based on single-carrier TDMA
CDMA-2000: multi-carrier CDMA based on IS-95	*IS-136*: North American TDMA system	*IEEE 802.16d*: WLL based on OFDM	*IEEE 802.16*: WLL based on single-carrier TDMA
TD-CDMA: time division synchronous CDMA	*UWC136*: universal wireless communications based on IS-136		
	GPRS: general packet radio service		
	EDGE: enhanced data rate for global evolution		

Chapter 5, the integration of all these existing and future standards in a common platform is one of the major goals of the next generation (4G) of wireless systems.

Hence, a fast adaptation/integration of existing systems to emerging new standards would be feasible if the 4G system has a generic architecture, while its receiver and transmitter parameters are both reconfigurable per software.

6.5.1   General

A common understanding of a software-defined radio (SDR) is that of a transceiver, where the *functions* are realized as programs running on suitable processors or re-programmable components [27]. On the hardware, different transmitter/receiver algorithms, which describe transmission standards, could be executed for corresponding application software. For instance, the software can be specified in such a manner that several standards can be loaded via parameter configurations. This strategy can offer a seamless change/adaptation of standards, if necessary.

The software-defined radio can be characterized by the following features:

– The radio functionality is configured per software.
– Different standards can be executed on the hardware according to the parameter lists.

A software-defined radio offers the following features:

- The radio can be used everywhere if all major wireless communication standards are supported. The corresponding standard-specific application software can be downloaded from the existing network itself.
- The software-defined radio can guarantee compatibility between several wireless networks. If UMTS is not supported in a given area, the terminal station can search for another network, e.g. GSM or IS-95.
- Depending on the hardware used, SDR is open to adopt new technologies and standards.

Therefore, SDR plays an important role in the success and penetration of 4G systems.

A set of examples of the current standards for cellular networks is given in Table 6-3. These standards, following their multi-access schemes, can be characterized as follows:

- Most of the 2G mobile communication systems are based on TDMA, while a CDMA component is adopted in 3G systems.
- In conjunction with TDMA, many broadband WLAN and WLL standards support multi-carrier transmission (OFDM).

For standards beyond 3G we may expect that a combination of CDMA with a multi-carrier (OFDM) component is a potential candidate. Hence, a generic air interface based on multi-carrier CDMA using a software-defined radio would support many existing and future standards (see Figure 6-34).

6.5.2 Basic Concept

A basic implementation concept of a software-defined radio is illustrated in Figure 6-35. The digitization of the received signal can be performed directly on the radio frequency

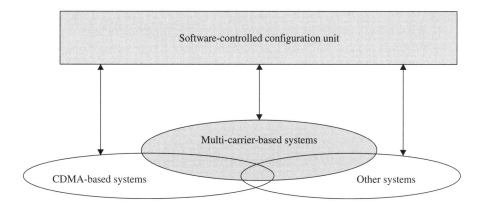

Figure 6-34   Software configured air interface

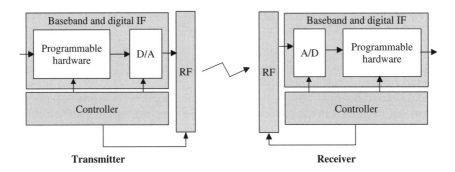

Figure 6-35   Basic concept of SDR implementation

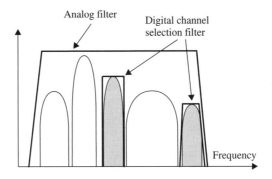

Figure 6-36   Channel selection filter in the digital domain

(RF) stage with a direct down-conversion or at some intermediate (IF) stage. In contrast to the conventional multi-hardware radio, channel selection filtering will be done in the digital domain (see Figure 6-36). However, it should be noticed that if the A/D converter is placed too close to the antenna, it has to convert a lot of useless signals together with the desired signal. Consequently, the A/D converter would have to use a resolution that is far too high for its task, therefore leading to a high sampling rate that would increase the cost. Digital programmable hardware components such as digital signal processors (DSPs) or field programmable gate arrays (FPGAs) can, besides the baseband signal processing tasks, execute some digital intermediate frequency (IF) unit functions including channel selection. Today, the use of fast programmable DSP or FPGA components allows the implementation of efficient real-time multi-standard receivers.

The SDR might be classified into the following categories [27]:

- *multi-band radio*, where the RF head can be used for a wide frequency range, e.g. from VHF (30–300 MHz) to SHF (30 GHz), to cover all services (e.g. broadcast TV to microwave FWA);
- *multi-role radio*, where the transceiver, i.e. the digital processor, supports different transmission, connection, and network protocols;

- *multi-function radio*, where the transceiver supports different multi-media services such as voice, data, and video.

The first category may require a quite complex RF unit to handle all frequency bands. However, if one concentrates on the main application, for instance, in mobile communications using the UHF frequency band (from 800 MHz/GSM/IS-95 to 2200 MHz/UMTS to even 5 GHz/IEEE 802.11h) it would be possible to cover this frequency region with a single wideband RF head [26]. Furthermore, regarding the transmission standards that use this frequency band, all parameters such as transmitted services, allocated frequency region, occupied channel bandwidth, signal power level, required SNR, coding, and modulation are known. Knowledge about these parameters can ease the implementation of the second and the third SDR categories.

6.5.3 MC-CDMA-Based Software-Defined Radio

A detailed SDR transceiver concept based on MC-CDMA is illustrated in Figure 6-37. At the transmitter side, the higher layer, i.e. the protocol layer, will support several connections at the user interface (TS), e.g. voice, data, video. At the base station it can offer several network connections, e.g. IP, PSTN, ISDN. The data link controller (DLC)/medium access controller (MAC) layer according to the chosen standard takes care of the scheduling (sharing capacity among users) to guarantee the required quality of service (QoS). Furthermore, in adaptive coding, modulation, spreading, and power leveling the task of the DLC layer is the selection of appropriate parameters such as the FEC code rate, modulation density, and spreading codes/factor. The protocol data units/packets

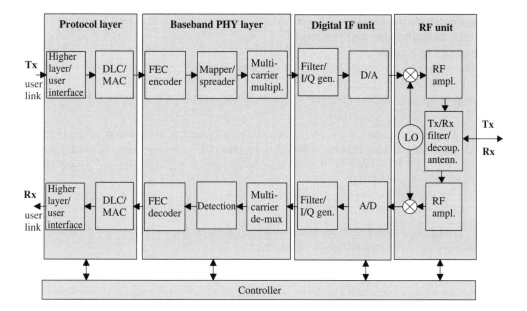

Figure 6-37   MC-CDMA-based SDR implementation

(PDUs) from the DLC layer are submitted to the baseband processing unit, consisting mainly of FEC encoder, mapper, spreader, and multi-carrier (i.e. OFDM) modulator. After digital I/Q generation (digital IF unit), the signal can be directly up-converted to the RF analogue signal, or it may have an analogue IF stage. Note that the digital I/Q generation has the advantage that only one converter is needed. In addition, this avoids problems of I and Q sampling mismatch. Finally, the transmitted analogue signal is amplified, filtered, and tuned by the local oscillator to the radio frequency and submitted to the Tx antenna. An RF decoupler is used to separate the Tx and Rx signals.

Similarly, the receiver functions, being the inverse of the transmitter functions (but more complex), are performed. In case of an analogue IF unit, it is shown in Reference [27] that the filter dimensioning and sampling rate are crucial to support several standards. The sampling rate is related to the selected wideband analogue signal, e.g. in the case of direct down-conversion [26]. However, the A/D resolution depends on many parameters: (a) the ratio between the narrowest and the largest selected channel bandwidths, (b) the used modulation, (c) the needed dynamic for different power levels, and (d) the receiver degradation tolerance.

As an example, the set of parameters that might be configured by the controller given in Figure 6-37 could be:

- higher layer connection parameters (e.g. port, services);
- DLC, MAC, multiple access parameters (QoS, framing, pilot/reference, burst formatting and radio link parameters);
- ARQ/FEC (CRC, convolutional, block, Turbo, STC, SFC);
- modulation (M-QAM, M-PSK, MSK) and constellation mapping (Gray, set partitioning, pragmatic approach);
- spreading codes (one- or two-dimensional spreading codes, spreading factors);
- multi-carrier transmission, i.e. OFDM (FFT size, guard time, guard band);
- A/D, sampling rate, and resolution;
- channel selection;
- detection scheme (single- or multi-user detection);
- diversity configuration;
- duplex scheme (FDD, TDD).

Hence, SDR offers elegant solutions to accommodate various modulation constellations, coding, and multi-access schemes. Besides its flexibility, it also has the potential of reducing the cost of introducing new technologies supporting sophisticated future signal processing functions.

However, the main limitations of the current technologies employed in SDR are:

- A/D and D/A conversion (dynamic and sampling rate);
- power consumption and power dissipation;
- speed of programmable components; and
- cost.

The future progress in A/D conversion will have an important impact on the further development of SDR architectures. A high A/D sampling rate and resolution, i.e. high signal dynamic, may allow a direct down-conversion to be used with a very wideband RF stage [26] i.e. the sampling is performed at the RF stage without any analogue IF unit, 'zero IF' stage. The amount of power consumption and dissipation of today's components (e.g. processors, FPGAs) may prevent its use in the mobile terminal station due to low battery lifetimes. However, its use in base stations is currently state-of-the art, for instance, in the UMTS infrastructure (UMTS BS/Node-B) [1].

References

[1] 3GPP (TS 25.401), "UTRAN overall description," *Technical Specification*, Sophia Antipolis, France, 2002.

[2] Alamouti S. M., "A simple transmit diversity technique for wireless communications," *IEEE Journal on Selected Areas in Communications*, vol. 16, pp. 1451–1458, Oct. 1998.

[3] Bauch G., *"Turbo-Entzerrung" und Sendeantennen-Diversity mit "Space–Time-Codes" im Mobilfunk*, Düsseldorf: Fortschritt-Berichte VDI, series 10, no. 660, 2000, PhD thesis.

[4] Bauch G. and Hagenauer J., "Multiple antenna systems: capacity, transmit diversity and turbo processing," in *Proc. ITG Conference on Source and Channel Coding*, Berlin, Germany, pp. 387–398, Jan. 2002.

[5] Chuang J. and Sollenberger N., "Beyond 3G: wideband wireless data access based on OFDM and dynamic packet assignment," *IEEE Communications Magazine*, vol. 38, pp. 78–87, July 2000.

[6] Cimini L., Daneshrad B., and Sollenberger N. R., "Clustered OFDM with transmitter diversity and coding," in *Proc. IEEE Global Telecommunications Conference (GLOBECOM '96)*, London, UK, pp. 703–707, Nov. 1996.

[7] Dammann A. and Kaiser S., "Standard conformable diversity techniques for OFDM and its application to the DVB-T system," in *Proc. Global Telecommunications Conference (GLOBECOM 2001)*, San Antonio, USA, pp. 3100–3105, Nov. 2001.

[8] Dammann A. and Kaiser S., "Transmit/receive antenna diversity techniques for OFDM systems," *European Transactions on Telecommunications (ETT)*, vol. 13, pp. 531–538, Sept./Oct. 2002.

[9] Damman A. and Raulefs R., "Increasing time domain diversity in OFDM systems," In *Proc. IEEE Global Telecommunications Conference (GLOBECOM 2004)*, Dallas, USA, pp. 809–812, Nov. 2004.

[10] Damman A., Raulefs R., and Kaiser S., "Beamforming in combination with space–time diversity for broadband OFDM systems," in *Proc. IEEE International Conference on Communications (ICC 2002)*, New York, USA, pp. 165–172, May 2002.

[11] ETSI HIPERMAN (TS 102 177), "High performance metropolitan area network, Part 1: physical layer," Sophia Antipolis, France, 2004.

[12] Foschini G. J., "Layered space–time architecture for wireless communication in a fading environment when using multi-element antennas," *Bell Labs Technical Journal*, vol. 1, pp. 41–59, 1996.

[13] Heath R. W. and Paulraj A., "A simple scheme for transmit diversity using partial channel feedback," in *Proc. 32nd Asilomar Conference on Signals, Systems, Computers*, Pacific Grove, USA, pp. 1073–1078, Nov. 1998.

[14] Kaiser S., "OFDM with code division multiplexing and transmit antenna diversity for mobile communications," in *Proc. IEEE International Symposium on Personal, Indoor and Mobile Radio Communications (PIMRC 2000)*, London, UK, pp. 804–808, Sept. 2000.

[15] Kaiser S., "Spatial transmit diversity techniques for broadband OFDM systems," in *Proc. IEEE Global Telecommunications Conference (GLOBECOM 2000)*, San Francisco, USA, pp. 1824–1828, Nov./Dec. 2000.

[16] Kaiser S., "Spatial pre-coding with phase flipping for wireless communications," in *Proc. IEEE Wireless Communications and Networking Conference (WCNC 2007)*, Hong Kong, China, March 2007.

[17] Kaiser S., "Effects of channel estimation errors on spatial pre-coding schemes with phase flipping," in *Proc. IEEE Vehicular Technology Conference (VTC 2007 Spring)*, Dublin, Ireland, April 2007.

[18] Kaiser S., "Performance of spatial phase coding (SPC) in broadband OFDM systems," in Proc. *IEEE International Conference on Communications (ICC 2007)*, Glasgow, UK, June 2007.

[19] Kaiser S., "Potential of code division multiplexing (CDM) and spatial diversity in future broadband OFDM systems," in *Proc. IEEE Wireless Communications and Networking Conference (WCNC 2008)*, Las Vegas, USA, March/April 2008.

[20] Li Y., Chuang J. C., and Sollenberger N. R., "Transmit diversity for OFDM systems and its impact on high-rate data wireless networks," *IEEE Journal on Selected Areas in Communications*, vol. 17, pp. 1233–1243, July 1999.

[21] Lindner J. and Pietsch C., "The spatial dimension in the case of MC-CDMA," *European Transactions on Telecommunications (ETT)*, vol. 13, pp. 431–438, Sept./Oct. 2002.

[22] Lo T. K. Y., "Maximum ratio transmission," *IEEE Transactions on Communications*, vol. 47, pp. 1458–1461, Oct. 1999.

[23] Seshadri N. and Winters J. H., "Two signaling schemes for improving the error performance of frequency division duplex transmission system using transmitter antenna diversity," *International Journal of Wireless Information Network*, vol. 1, pp. 49–59, 1994.

[24] Tarokh V., Jafarkhani H., and Calderbank A. R., "Space–time block codes from orthogonal designs," *IEEE Transactions on Information Theory*, vol. 45, pp. 1456–1467, June 1999.

[25] Tarokh V., Seshadri N., and Calderbank A. R., "Space–time codes for high data rate wireless communications," *IEEE Transactions on Information Theory*, vol. 44, pp. 744–765, March 1998.

[26] Tsurumi H. and Suzuki Y., "Broadband RF stage architecture for software-defined radio in handheld terminal applications," *IEEE Communications Magazine*, vol. 37, pp. 90–95, Feb. 1999.

[27] Wiesler A. and Jondral F. K., "A software radio for second- and third-generation mobile systems," *IEEE Transactions on Vehicular Technology*, vol. 51, pp. 738–748, July 2002.

[28] Wolniansky P. W., Foschini G. J., Gloden G. D., and Valenzuela R. A., "V-BLAST: an architecture for realizing very high data rates over the rich-scattering wireless channel," in *Proc. International Symposium on Advanced Radio Technologies*, Boulder, USA, Sept. 1998.

Definitions, Abbreviations, and Symbols

Definitions

Adaptive antenna system (AAS): a system adaptively exploiting more than one antenna to improve the coverage and the system capacity.

Adaptive coding and modulation (ACM): a transmitter switching automatically the modulation constellation and the FEC coding rate in order to adapt them to the instantaneous channel conditions (e.g. weather or fading) for improving the system capacity.

Adjacent channel interference (ACI): interference emanating from the use of adjacent channels in a given coverage area, e.g. dense cellular system.

Asynchronous: users transmitting their signals without time constraints.

Base station (BS): equipment consisting of a base station controller (BSC) and several base station transceivers (BST).

Burst: transmission event consisting of a symbol sequence (preamble and the data symbols).

Cell: geographical area controlled by a base station. A cell can be split into sectors.

Co-channel interference (CCI): interference emanating from the re-use of the same frequency band in a given coverage area, e.g. dense cellular system.

DC subcarrier: in an OFDM or OFDMA signal, the subcarrier whose frequency would be equal to the RF center frequency of the station.

Detection: operation for signal detection in the receiver. In a multi-user environment a *single-user* (SD) or *multi-user* (MD) detection can be used. The multi-user detection requires the knowledge of the signal characteristics of all active users.

    Single-user detection techniques for MC-CDMA: MRC, EGC, ZF, MMSE.

    Multi-user detection techniques for MC-CDMA: MLSE, MLSSE, IC, JD.

Multi-Carrier and Spread Spectrum Systems   Second Edition   K. Fazel and S. Kaiser
© 2008 John Wiley & Sons, Ltd

Doppler spread: changes in the phases of the arriving waves that lead to time-variant multi-path propagation.

Downlink (DL): direction from the BS to the TS.

Downlink channel: channel transmitting data from the BS to the TS.

FEC block: block resulting from the channel encoding.

Fixed wireless access (FWA): wireless access application in which the location of the terminal station (TS) is fixed during operation.

Frame: ensemble of data and pilot/reference symbols sent periodically in a given time-interval, e.g. OFDM frame, MAC frame.

Frequency division duplex (FDD): the transmission of uplink (UL) and downlink (DL) signals are performed at different carrier frequencies. The distance between the UL and DL carrier frequencies is called duplex distance.

Full-duplex: equipment (e.g. TS) which is capable of transmitting and receiving data at the same time.

Full load: simultaneous transmission of all users in a multi-user environment.

Guard time: cyclic extension of an OFDM symbol to limit the ISI.

Handover: mechanism that a mobile station changes connection from one base station to the other (e.g. due to bad reception conditions).

Inter-channel interference (ICI): interference between neighbor sub-channels (e.g. OFDM sub-channels) in frequency domain, e.g. due to Doppler effects.

Interference cancellation (IC): operation of estimating and subtracting the interference in the case of *multi-user* signal detection.

Inter-symbol interference (ISI): interference between neighbor symbols (e.g. OFDM symbols) in the time domain, e.g. due to multi-path propagation.

Multi-path propagation: consequence of reflections, scattering, and diffraction of the transmitted electromagnetic wave at natural and man-made objects.

Multiple access interferences (MAI): interference resulting from other users in a given multiple access scheme (e.g. with CDMA).

OFDM frame synchronization: generation of a signal indicating the start of an OFDM frame made of several OFDM symbols. Closely linked to OFDM symbol synchronization.

OFDM symbol synchronization: FFT window positioning, i.e. start time of the FFT operation.

Path loss: mean signal power attenuation between the transmitter and receiver.

PHY mode: combination of a signal constellation (modulation alphabet) and FEC parameters.

Point to multi-point (PMP): a topological cellular configuration with a base station (BS) and several terminal stations (TSs). The transmission from the BS towards the TS is called downlink and the transmission from the TS towards the BS is called uplink.

Preamble: sequence of channel symbols with a given auto-correlation property assisting modem synchronization and channel estimation.

Profile: set of parameter combinations identifying the transmitter or receiver implemented features and behavior (e.g. WiMAX PHY or MAC Profiles).

Puncturing: operation for increasing the code rate by not transmitting (i.e. by deleting) some coded bits.

Rake: bank of correlators, e.g. matched filters, to resolve and combine multi-path propagation in a CDMA system.

Ranging: operation of periodic timing advance (or power) adjustment to guarantee the required radio link quality.

Sampling rate control: control of the sampling rate of the A/D converter.

Sector: geometrical area resulting from the cell splitting by the use of a sector antenna.

Shadowing: obstruction of the transmitted waves by, for example, hills, buildings, walls, and trees, which results in more or less strong attenuation of the signal strength, modeled by log-normal distribution.

Shortening: operation for decreasing the length of a systematic block code that allows an adaptation to different information bit/byte sequence lengths.

Tail bits: zero bits inserted for trellis termination of a convolutional code in order to force the trellis to go to the zero state.

Time division duplex (TDD): the transmission of uplink (UL) and downlink (DL) signals is carried out in the same carrier frequency bandwidth. The UL and the DL signals are separated in the time domain.

Spectral efficiency: efficiency of a transmission scheme given by the maximum possible data rate (in bit/s) in a given bandwidth (in Hz). It is expressed in bit/s/Hz.

Area spectrum efficiency: gives the spectral efficiency per geographical coverage area, e.g. cell or sector. It is expressed in bit/s/Hz/cell or sector.

Spreading: operation of enlarging/spreading the spectrum. Several spreading codes can be used for spectrum spreading.

Synchronous: users transmitting following a given time pattern.

Uplink (UL): direction from TS to BS.

Uplink channel: channel transmitting data from TS to BS.

Abbreviations

AAS	Adaptive Antenna System
ACF	Auto-Correlation Function
ACI	Adjacent Channel Interference
A/D	Analogue/Digital (converter)
AGC	Automatic Gain Control
AK	Authorization Key
ARIB	Association of Radio Industries and Businesses (Japanese association)
ARQ	Automatic Repeat re-Quest
ASIC	Application Specific Integrated Circuit
ASN	Access Service Network
A-TDMA	Advanced TDMA (EU RACE project)
ATIS	Alliance for Telecommunications Industry Solutions
ATM	Asynchronous Transfer Mode
ATPC	Automatic Transmit Power Control
ATTC	Automatic Transmit Time Control
AWGN	Additive White Gaussian Noise
B3G	Beyond 3G
BCH	Bose Chaudhuri-Hocqenghem (FEC code)
BER	Bit Error Rate
BLAST	Bell Labs Layered Space Time
BPSK	Binary Phase Shift Keying
BRAN	Broadband Radio Access Network
BS	Base Station (= Access Point, AP)
BSC	BS Controller
BST	BS Transceiver
BTC	Block Turbo Code
BU	Bad Urban (a radio channel model)
BWA	Broadband Wireless Access
CA	Certification Authority
CAZAC	Constant Amplitude Zero Auto Correlation
CC	Convolutional Code
CCSA	China Communications Standards Association
CCF	Cross-Correlation Function
CCI	Co-Channel Interference
CDD	Cyclic Delay Diversity
CDM	Code Division Multiplexing
CDMA	Code Division Multiple Access
CDMA-2000	Code Division Multiple Access standard 2000 (American 3G standard)
CF	Crest Factor (square root of PAPR)
C/I	Carrier-to-Interference power ratio
C/N	Carrier-to-Noise power ratio

C/(N+I)	Carrier-to-Noise and -Interference power ratio
CODIT	COde DIvision Testbed (RACE research project)
COST	European Cooperation in the Field of Scientific and Technical Research
CPE	Common Phase Error
CRC	Cyclic Redundancy Check
CS	Convergence Sub-layer
CSI	Channel State Information
CSN	Connectivity Service Network
CTC	Convolutional Turbo Code
D/A	Digital/Analogue (converter)
DAB	Digital Audio Broadcasting
D-AMPS	Digital-Advanced Mobile Phone Service
DC	Direct Current
DD	Delay Diversity
DDD	Discontinuous Doppler Diversity
DECT	Digital Enhanced Cordless Telecommunications
DES	Data Encryption Standard (e.g. 3-DES)
DFS	Dynamic Frequency Selection
DFT	Discrete Fourier Transform
DFTS-OFDM	DFT-Spread OFDM
DL	Downlink
DLC	Data Link Control
D-QPSK	Differential-QPSK
DS	Direct Sequence (DS-CDMA)
DSP	Digital Signal Processor
DVB	Digital Video Broadcasting
DVB-H	DVB standard for Handheld devices
DVB-RCT	DVB-Return Channel-Terrestrial
DVB-S	DVB standard for Satellite broadcasting
DVB-T	DVB standard for Terrestrial broadcasting
EAP	Extensible Authentication Protocol
ECB	Electronic Code Book
EDE	Encrypt-Decrypt-Encrypt
EDGE	Enhanced Data for Global Evolution
EGC	Equal Gain Combining
EGT	Equal Gain Transmission
EIRP	Effective Isotopic Radiated Power
EKS	Encryption Key Sequence
eNode-B	LTE Base Station
ETSI	European Telecommunication Standard Institute
EU	European Union
FCH	Frame Control Header
FDD	Frequency Division Duplex

FDMA	Frequency Division Multiple Access
FEC	Forward Error Correction
FFH	Fast-FH
FFT	Fast Fourier Transform
FH	Frequency Hopping (FH-CDMA)
FIR	Finite Impulse Response
FPGA	Field Programmable Gate Array
FRAMES	Future Radio Wideband Multiple Access System
FTP	File Transfer Protocol
FWA	Fixed Wireless Access
GMSK	Gaussian Minimum Shift Keying
3GPP	Third Generation Partnership Project
GPRS	General Packet Radio Services
GSM	Global System for Mobile communications
H-ARQ	Hybrid ARQ
H-FDD	Half-duplex Frequency Division Duplex
HHO	Hard Hand-Over
HIPERLAN	HIgh PERformance Local Area Network
HIPERMAN	HIgh PERformance Metropolitan Area Network
HM	HIPERMAN
HMAC	Hashed Message Authentication Code
HPA	High Power Amplifier
HSDPA	High Speed Downlink Packet Access (UMTS$^+$)
HSUPA	High Speed Uplink Packet Access (UMTS$^+$)
HSPA	High Speed Packet Access (UMTS$^+$)
HT	Hadamard Transform/Hilly Terrain (a radio channel model)
IBO	Input Back-Off
IC	Interference Cancellation
ICI	Inter-Carrier Interference
IDFT	Inverse Discrete Fourier Transform
IEEE	Institute of Electrical and Electronics Engineers
IF	Intermediate Frequency
IFDMA	Interleaved FDMA
IFFT	Inverse Fast Fourier Transform
IHT	Inverse Hadamard Transform
IMT-2000	International Mobile Telecommunications-2000
IMT-Advanced	International Mobile Telecommunications-Advanced
IP	Internet Protocol
I/Q	In-phase/Quadrature
IR	Infra-Red
IS	Interim Standard (e.g. American Standard IS-95)
ISDN	Integrated Service Digital Network

ISI	Inter-Symbol Interference
ISM	Industrial Scientific and Medical (ISM license-free band)
ISO	International Standards Organization
JD	Joint Detection
JTC	Joint Technical Committee
KEK	Key Encryption Key
LAN	Local Area Network
LDPC	Low Density Parity Check (FEC code)
LLF	Log Likelihood Function
LLR	Log-Likelihood Ratio
LMDS	Local Multi-point Distribution System
LO	Local Oscillator
LOS	Line Of Sight
LTE	Long Term Evolution (evolved UTRA)
MA	Multiple Access
MAC	Medium Access Control
MAI	Multiple Access Interference
MAP	Maximum *A Posteriori*
MBS	Mobile Broadband System
MC	Multi-Carrier
MC-CDMA	Multi-Carrier CDMA
MC-DS-CDMA	Multi-Carrier DS-CDMA
MCM	Multi-Carrier Modulation
MC-SS	Multi-Carrier Spread Spectrum
MC-TDMA	Multi-Carrier TDMA (OFDM and TDMA)
MF	Match Filter
MIMO	Multiple Input Multiple Output
MISO	Multiple Input Single Output
ML	Maximum Likelihood
MLD	ML Decoder (or Detector)
MLSE	ML Sequence Estimator
MLSSE	ML Symbol-by-Symbol Estimator
MIB	Management Information Base
MMAC	Multimedia Mobile Access Communication
MMDS	Microwave Multi-point Distribution System
MMSE	Minimum Mean Square Error
MPEG	Moving Picture Expert Group
M-PSK	Phase Shift Keying constellation with M points, e.g. 16-PSK
M-QAM	QAM constellation with M points, e.g. 16-QAM
MRC	Maximum Ratio Combining
MRT	Maximum Ratio Transmission
MSB	Most Significant Bit

MSK	Minimum Shift Keying
MT-CDMA	Multi-Tone CDMA
MU-MIMO	Multi-User MIMO
NLOS	Non Line Of Sight
Node-B	UMTS base station
NT	Network Termination
OBO	Output Back-Off
OFCDM	Orthogonal Frequency and Code Division Multiplexing
OFDM	Orthogonal Frequency Division Multiplexing
OFDMA	Orthogonal Frequency Division Multiple Access
OSI	Open System Interconnection
PAPR	Peak-to-Average Power Ratio
PD	Phase Diversity
PDC	Personal Digital Cellular (Japanese mobile standard)
PDCP	Packet Data Convergence Protocol
PDU	Protocol Data Unit
PER	Packet Error Rate
PHY	PHYsical (layer)
PKM	Privacy Key Management protocol
PLL	Phase Lock Loop
PMP	Point to Multi-Point
PN	Pseudo Noise
POTS	Plain Old Telephone Services
PPM	Pulse Position Modulation
PRBS	Pseudo Random Binary Sequence
P/S	Parallel-to-Serial (converter)
PSD	Power Spectral Density
PSTN	Public Switched Telephone Network
QAM	Quadrature Amplitude Modulation with square constellation
QEF	Quasi Error Free
QEGT	Quantized EGT
QoS	Quality of Service
QPSK	Quaternary Phase Shift Keying
RA	Rural Area (a radio channel model)
RACE	Research in Advanced Communications in Europe (EU research projects)
RAN	Radio Access Network
RF	Radio Frequency
RLC	Radio Link Control
RMS	Root Mean Square

RNC	Radio Network Controller
RRC	Radio Resource Control
RS	Reed-Solomon (FEC Code)
Rx	Receiver
SAID	Security Association IDentifier
SAP	Service Access Point
SAR	Segmentation And Reassembling
SC	Single Carrier
SCD	Sub-Carrier Diversity
SD	Selection Diversity
SDR	Software Defined Radio
SF	Spreading Factor
SFBC	Space Frequency Block Codes
SFC	Space Frequency Coding
SFH	Slow Frequency Hopping
SHF	Super High Frequency
SI	Self Interference
SIMO	Single Input Multiple Output
SIR	Signal-to-Interference power Ratio
SISO	Soft-In/Soft-Out (or Single Input Single Output)
SME	Small and Medium size Enterprise
SNI	Service Node Interface
SNR	Signal-to-Noise Ratio
S-OFDMA	Scalable OFDMA
SOHO	Small Office/Home Office
SOVA	Soft Output Viterbi Algorithm
SPC	Spatial Pre-Coding
S/P	Serial-to-Parallel (converter)
SS	Spread Spectrum
SS-MC-MA	Spread Spectrum Multi-Carrier Multiple Access
SSPA	Solid State Power Amplifier
STC	Space Time Coding
STBC	Space Time Block Coding
STTC	Space Time Trellis Coding
SU-MIMO	Single-User MIMO
TC	Turbo Code
TD	Total Degradation
TDD	Time Division Duplex
TDM	Time Division Multiplex
TDMA	Time Division Multiple Access
TEK	Traffic Encryption Key
TF	Transmission Frame
TIA	Telecommunication Industry Association (American Association)

TPC	Turbo Product Code
TPD	Time-variant Phase Diversity
TS	Terminal Station
TTA	Telecommunications Technologies Association (Korea)
TTC	Telecommunication Technology Committee (Japan)
TU	Typical Urban (a radio channel model)
TWTA	Travelling Wave Tube Amplifier
Tx	Transmitter
UE	User Equipment (LTE terminal)
UHF	Ultra High Frequency
UL	Uplink
UMTS	Universal Mobile Telecommunication System
UNI	User–Network Interface
UTRA	UMTS-Terrestrial Radio Access
UWB	Ultra Wide Band
VCO	Voltage Controlled Oscillator
VHF	Very High Frequency
VoIP	Voice over IP
VSF	Variable Spreading Factor
WARC	World Administration Radio Conference
WCDMA	Wideband CDMA
WiMAX	Wireless Interoperability for Microwave Access
WINNER	Wireless World Initiative New Radio
WH	Walsh–Hadamard
WLAN	Wireless Local Area Network
WLL	Wireless Local Loop
WMAN	Wireless Metropolitan Area Network
xDSL	Digital Subscriber Line (e.g. x: A = asymmetry)
ZF	Zero Forcing (equalization)

Symbols

$a^{(k)}$	source bit of user k
$\mathbf{a}^{(k)}$	source bit vector of user k
a_p	amplitude of path p
$b^{(k)}$	code bit of user k
$\mathbf{b}^{(k)}$	code bit vector of user k
B	bandwidth
B_s	signal bandwidth
c	speed of light
$c_l^{(k)}$	chip l of the spreading code vector $\mathbf{c}^{(k)}$
$\mathbf{c}^{(k)}$	spreading code vector of user k
$\mathbf{c}_n$	spatial pre-coding vector
C	capacity
$\mathbf{C}$	spreading code matrix
$d^{(k)}$	data symbol of user k
$\mathbf{d}^{(k)}$	data symbol vector of user k
D_O	diversity
D_f	frequency diversity
D_t	time diversity
dB	decibel
dBm	decibel relative to 1 mW
E{.}	expectation
E_b	energy per bit
E_c	energy per chip
E_s	energy per symbol
f	frequency
f_c	carrier frequency
$\mathbf{H}$	channel matrix
f_D	Doppler frequency
$f_{D,filter}$	maximum Doppler frequency permitted in the filter design
f_{Dmax}	maximum Doppler frequency
$f_{D,p}$	Doppler frequency of path p
f_n	nth sub-carrier frequency
F	noise figure in dB/feedback information
F_s	sub-carrier spacing
$G_{Antenna}$	antenna gain

$G_{l,l}$	lth diagonal element of the equalizer matrix $\mathbf{G}$
$\mathbf{G}$	equalizer matrix
$G^{[j]}$	equalizer matrix used for IC in the jth iteration
$h(t)$	impulse response of the receive filter or channel impulse response
$h(\tau,t)$	time-variant channel impulse response
$H(f,t)$	time-variant channel transfer function
$H_{n,i}$	discrete-time/frequency channel transfer function
$H_{l,l}$	lth diagonal element of the channel matrix $\mathbf{H}$
I_c	size of the bit interleaver
I_{TC}	size of the Turbo code interleaver
j	$\sqrt{-1}$
J_{it}	number of iterations in the multi-stage detector
K	number of active users
K_{Rice}	Rice factor
L	spreading code length
L_a	length of the source bit vector $\mathbf{a}^{(k)}$
L_b	length of the code bit vector $\mathbf{b}^{(k)}$
L_d	length of the data symbol vector $\mathbf{d}^{(k)}$
m	number of bits transmitted per modulated symbol
M	number of data symbols transmitted per user and OFDM symbol
$n(t)$	additive noise signal
$\mathbf{n}$	noise vector
N_c	number of sub-carriers
N_l	lth element of the noise vector $\mathbf{n}$
N_f	pilot symbol distance in frequency direction
N_{grid}	number of pilot symbols per OFDM frame
N_{ISI}	number of interfering symbols
N_p	number of paths
N_s	number of OFDM symbols per OFDM frame
N_t	pilot symbol distance in time direction
N_{tap}	number of filter taps
p(.)	probability density function
P{.}	probability
P_b	BER
P_G	processing gain
Q	number of user groups
$\mathbf{r}$	received vector after inverse OFDM
$\mathbf{r}^{(k)}$	received vector of the kth user after inverse OFDM
R	code rate
R_b	bit rate
rect(x)	rectangular function
R_l	lth element of the received vector $\mathbf{r}$
R_s	symbol rate

$\mathbf{s}$	symbol vector before OFDM
$\mathbf{s}^{(k)}$	symbol vector of user k before OFDM
S_l	lth element of the vector $\mathbf{s}$
$\text{sinc}(x)$	$\sin(x)/x$ function
t	time/number of error correction capability of an RS code
T	source symbol duration
T_c	chip duration
T_d	data symbol duration
T_{fr}	OFDM frame duration
T_g	duration of guard interval
T_s	OFDM symbol duration without guard interval
T_s'	OFDM total symbol duration with guard interval
T_{samp}	sampling rate
$\mathbf{u}$	data symbol vector at the output of the equalizer
U_l	lth element of the equalized vector $\mathbf{u}$
v	velocity
$v^{(k)}$	soft decided value of the data symbol $d^{(k)}$
$\mathbf{v}^{(k)}$	soft decided value of the data symbol vector $\mathbf{d}^{(k)}$
V_{guard}	loss in SNR due to the guard interval
V_{pilot}	loss in SNR due to the pilot symbols
$w^{(k)}$	soft decided value of the code bit $b^{(k)}$
$\boldsymbol{w}^{(k)}$	soft decided value of the code bit vector $\mathbf{b}^{(k)}$
w_n	power normalization factor on subcarrier n
$x(t)$	transmitted signal
$X(f)$	frequency spectrum of the transmitted signal
$y(t)$	received signal
α	roll-off factor
β	primitive element of the Galois field
Γ	log-likelihood ratio
δ	delay
$\Delta^2(.\,,.)$	squared Euclidean distance
θ	cross-correlation function
Λ	SPC threshold
σ^2	variance of the noise
τ	delay
τ_p	delay of path p
ϕ	auto-correlation function
φ_p	phase offset of path p
Ω	average power
$(.)^H$	Hermitian transposition of a vector or a matrix
$(.)^T$	transposition of a vector or a matrix
$(.)^{-1}$	inversion
$(.)^*$	complex conjugation
$\lvert.\rvert$	absolute value
$\lVert.\rVert$	norm of a vector
$\otimes$	convolution operation/Kronecker product

Index

3GPP 218

Adaptive techniques 190–192
  Adaptive channel coding and modulation 191
  Adaptive power control 192
  Nulling of weak sub-carriers 191
Advantages and drawbacks of MC-CDMA 49
Advantages and drawbacks of OFDM 34–35
Alamouti space-time block code (STBC) 307
AM/AM (AM/PM) conversion (HPA) 199–200
Analog-to-digital (A/D) conversion 135–136
Antenna diversity 263, 302–330
Antenna gain 209
Automatic gain control (AGC) 154

Bad urban (BA) channel model 20
Beamforming 227
Beyond 3G (B3G) 6, 276
Bit nodes 183–184
BLAST architecture 303–304
  Diagonal (D-BLAST) 304
  Vertical (V-BLAST) 304
Blind and semi-blind channel estimation 169–170
Block codes 175–178, 180–186
Block linear equalizer 68

Bluetooth 5
Boltzmann constant 209
Broadband (fixed) wireless access (BWA) 50, 238, 242
  Channel characteristics 23
  Network topology 244

CDMA-2000 3, 42–43
Cellular systems beyond 3G 6, 276
Certification 271
Channel coding and decoding 174–187
Channel estimation 154–174
  Adaptive design 159
  Autocorrelation function 157
  Boosted pilot symbols 161
  Cross-correlation function 159
  Downlink 170
  MC-SS systems 170–174
  One-dimensional 159
  Overhead due to pilot symbols 161
  Performance analysis 162–167, 170–173
  Pilot distance 160–161
  Robust design 160
  Two-dimensional 155–159
  Uplink 170
Channel fade statistics 18–19
Channel impulse response 16–17
Channel matrix 34, 57
Channel modeling 16–18
Channel selection in digital domain 334–335

Multi-Carrier and Spread Spectrum Systems Second Edition K. Fazel and S. Kaiser
© 2008 John Wiley & Sons, Ltd